CLINICAL ATLAS OF HUMAN CHROMOSOMES

CLINICAL ATLAS OF HUMAN CHROMOSOMES

SECOND EDITION

Jean de Grouchy, M.D.

Director of Research
Head, Laboratoire de Cytogénétique
   de Humaine Comparée
Hôpital Necker-Enfants-Malades
Paris, France

Catherine Turleau, M.D.

Research Assistant
Laboratoire de Cytogénétique
   de Humaine Comparée
Hôpital Necker-Enfants-Malades
Paris, France

Foreword by **Victor A. McKusick, M.D.**

William Osler Professor and Director
Department of Medicine
The Johns Hopkins University School of Medicine
Physician-in-Chief
Johns Hopkins Hospital
Baltimore, Maryland

A WILEY MEDICAL PUBLICATION
JOHN WILEY & SONS
New York · Chichester · Brisbane · Toronto · Singapore

Cover and interior design: Wanda Lubelska
Production Supervisor: Audrey Pavey

An edition in the French language has been published by
Expansion Scientifique under the title *Atlas des maladies
chromosomiques, deuxième edition.*

Library of Congress Cataloging in Publication Data:

Grouchy, Jean de.
  Clinical atlas of human chromosomes.

  (A Wiley medical publication)
  Translation of: Atlas des maladies chromosomiques.
Deuxième éd.
  Includes bibliographies and index.
  1. Human chromosome abnormalities—Atlases.
2. Cytogenetics—Atlases. 3. Medical genetics—Atlases.
4. Abnormalities, Human—Atlases. I. Turleau,
Catherine. II. Title. [DNLM: 1. Chromosome aberrations
—Atlases. 2. Chromosome abnormalities—Atlases.
3. Chromosome mapping—Atlases. 4. Cytogenetics—
Atlases. 5. Karotyping—Atlases. QS 17 G882a]

RB44.G7613 1984   616'.042   83-16839
ISBN 0-471-89205-X

Printed in the United States of America

10 9 8 7 6 5 4 3 2 1

FOREWORD

The organization that Drs. de Grouchy and Turleau have devised for their *Atlas* is innovative and highly useful. As a catalog of chromosomal anomalies, it is comparable to my catalog of Mendelian traits (*Mendelian Inheritance in Man: Catalogs of Autosomal Dominant, Autosomal Recessive and X-linked Phenotypes*, sixth edition, 1983). The de Grouchy–Turleau *Atlas*, however, is much more than a catalog of chromosomal anomalies; mapping information is also covered.

Within the last 15 years since the advent of "banding methods" for chromosome study, a remarkable development has been the description of many "new" chromosomal syndromes. Chromosomal aberrations that previously could not be precisely defined yielded to analysis by the banding techniques. And, when cases thus shown to be chromosomally identical were analyzed phenotypically, a clinical syndrome often emerged. It is these rapid advances that are embraced by the de Grouchy–Turleau *Atlas*. It is also useful to have the information on genes assigned to specific chromosomes, a field to which the authors have contributed much and also of phenomenal advance in the last few years.

The abundant illustrations of the clinical features, as well as of the karyologic features, and the succinct listing of the characteristics of each of the chromosomal syndromes should serve to fix them in mind. Many of the conditions are indeed physiognomonic diagnoses.

High-resolution cytogenetics, available for the last seven years or so, has further refined the process of description of "new" chromosomal syndromes and has permitted demonstration of the chromosomal basis of some disorders previously etiologically mysterious, for example, the Prader-Willi or Langer-Giedion syndrome. The borderline between mendelian disorders and chromosomal aberrations has become blurred; mutation is a continuum from microscopically evident change to nucleotide substitution. Also with high-resolution cytogenetics, chromosome changes specific to particular forms of neoplasia have been established in increasing numbers in the last few years. The detailed information on chromosome changes in cancer fills a large catalog of its own (Felix Mitelmann, *Cytogenetics and Cell Genetics* 36:5–516, 1983). At the same time oncogenes in humans have been discovered by the methods of molecular genetics and localized to regions that show microscopic abnormality in particular neoplasms.

The detailed study of chromosome aberrations has contributed to chromosomal mapping by providing information through the principle of deletion mapping (de Grouchy and Turleau contributed to this area by mapping, for example, the adenylate kinase gene to the end of the long arm of chromosome 9) and by dosage observations, for example, mapping of the glutathione reductase gene to chromosome 8 by the study of trisomy 8 cases. Balanced translocations and heteromorphisms have been useful for chromosome mapping by conventional family linkage studies—Duffy blood group was first assigned to chromosome 1 and haptoglobin to chromosome 16 by such an approach—but, in addition, balanced translocations have in recent times become unexpectedly informative in the mapping field, particularly in mapping the X chromosome, where assignment of the locus for Duchenne muscular dystrophy to Xp21 is a triumph. Application of the same principle to autosomal translocation is likely to be increasingly observed. Cytogeneticists have the opportunity to make observations important to chromosome mapping through study of these experiments of nature. Conversely the mapping information now available can be very useful to the cytogeneticist in analyzing the nature of chromosomal aberrations by dosage affect.

We welcome this new edition of a standard source book of human genetics.

Victor A. McKusick, M.D.

FOREWORD

PREFACE

A comparison between the maps of autosomal syndromes from the first edition and those that we have retained for this second edition provides an idea of what has been happening in human clinical cytogenetics in the last 5 years. More than 30 new syndromes have been characterized. At the same time, the number of known autosomal gene assignments has doubled. Many involve enzymatic markers showing gene dosage effect, which are invaluable in certain cytogenetic diagnoses. In our view, these two considerations fully justify the updating of this book.

We have preserved the underlying principles, as well as the general format, of the first edition, but we have introduced certain modifications. All references to chromosomal evolution in primates have been deleted. Knowledge in this field has made such enormous progress that references only to the higher primates has lost all significance.

The nomenclature of the syndromes used in the first edition had become too imprecise. The choice of a new nomenclature, however, has proven to be a very delicate matter, and it is likely that no perfect system can really be found. To as great an extent as possible, we have attempted to describe precisely the region in question for partial monosomies and trisomies. Thus, what was formerly called *partial 7q trisomy* now includes *7q2 trisomy* and *7q3 trisomy*. The word "partial" has been eliminated, with a few exceptions.

Ring chromosomes are responsible for phenotypes for which it is often difficult to find a characteristic syndrome. Thus, in a number of cases, we have simply indicated the most frequently observed signs.

The descriptions of the very important syndromes, such as trisomies 21, 13, and 18, have been only slightly modified. Chapters concerning syndromes that had previously been based only on a small number of cases, but for which we now know an appreciable number, have been completely rewritten. New syndromes, of course, are the subject of new chapters. We have also devoted a chapter to polyploidy.

As we stated previously, the gene map of man has become extremely "rich." The following table, which is limited to autosomal assignments, will provide an idea of this richness. The names of cities correspond to places where human gene mapping (HGM) conferences have been held.

Number of Autosomic Assignments

Conference	Confirmed	Tentative	Contradictory	Total
New Haven (1973)	31	28	5	64
Rotterdam (1974)	48	32	6	86
Baltimore (1975)	72	46	7	125
Winnipeg (1977)	83	82	11	176
Edinburgh (1979)	123	87	20	230
Oslo (1981)	179	115	43	337
Los Angeles (1983)	247	161	80	488

Among the major contributions of recent years and the most rapidly expanding fields of research, the discovery and assignment of oncogenes and DNA polymorphisms must be emphasized.

In light of this, we decided to attribute an overriding importance to gene assignments. For each chromosome, we have deleted the diagram indicating the assignments and have replaced it by an enumeration of genetic markers, their abbreviations, and their assignments, as is current practice. In the appendix, just as in the earlier edition, we have listed the markers and their symbols. The significance of the symbols used, detailing the mode of determination of each assignment, is also given.

The basic essential document that enabled us to carry out this work was Victor McKusick's letter of March 1, 1983 on the human gene map, which was a summary of all assignments known up to that time. We are deeply grateful to him for allowing us the use of this document.

We have completed the gene map by introducing the map of gene illnesses for which assignments are now known. This is reproduced from Victor McKusick's "The Morbid Anatomy of the Human Genome," and again we wish to thank him for the use of this original document.

We have added a new appendix entitled "Types and Mechanisms of Formation of Chromosomal Aberrations." This section was often requested by readers of the first edition. Needless to say, it is not intended for cytogeneticists, but rather for the non-specialist who wishes to become familiar with those mechanisms that cause the various chromosomal aberrations that are themselves responsible for the chromosomal illnesses described in the atlas.

Cytogenetic techniques, and especially the perfecting of high resolution techniques, have made considerable progress over the last 5 years. The appendix dedicated to these techniques has thus been completed by a description of these and other techniques, such as NORs banding, which displays nucleolar organizers of acrocentric chromosomes. In the same way, we have included ideograms and karyotypes in R and G bandings that correspond to these high resolution techniques. These ideograms serve as a frontispiece for each chromosome chapter. The karyotypes are regrouped in the appendix.

Will this clinical atlas of human chromosomes one day appear in a third edition? It seems unlikely that chromosome pathology as described herein will undergo profound changes in the next few years. The limits of chromosome segments responsible for the various syndromes might, of course, be further refined, but it is probable that the basic syndromes have now been characterized. We are thus hopeful that this second edition will enjoy a long and happy life!

Jean de Grouchy
Catherine Turleau

PREFACE TO THE FIRST EDITION

Until the end of the fifties, not so long ago, human chromosomes were mysterious hieroglyphics, kinds of small inverted V's, lined up at the bottom of a page in genetic textbooks, with a haughty caption: "The 48 Chromosomes of Man."

Times have changed. To begin with, two chromosomes have been lost. But most important, chromosomes have become entities, even very important individuals. They are photographed, cut apart, and reassembled on the scientist's whim. They are dressed in bright and gaudy colors, made to fluoresce, and even costumed as harlequins.

Each of the 46 human chromosomes now has a personality, preventing confusion with any other. This personality has several aspects. First and foremost is banding in the different systems (Q-, G-, R-banding, etc.), which is always specific for a given chromosome.

Next is the way in which the chromosome has evolved since the ancestral species common to man and the other hominoid primates, namely our closest cousins, the chimpanzee, the gorilla, and the orangutan. Certain chromosomes have remained extraordinarily stable and can be considered true "paleochromosomes." Chromosome 6 is a good example. Others, such as chromosome 9, have undergone numerous rearrangements and today are different in the four species.

Another aspect of the chromosome's personality derives from the number and nature of genes it carries. At the Congress of Human Genetics held in Paris in 1971, F. Ruddle reviewed the gene map of man. At that time two autosomal localizations were known, those of thymidine kinase on chromosome 17 and of lactico dehydrogenase A on chromosome 11. Today, only 5 years later, more than 100 genes are localized. Among these, some are dear to the heart of every geneticist: the Rh factor on chromosome 1, the HLA locus on chromosome 6, and—the latest localization—the ABO blood group on chromosome 9.

Without doubt, a leading element of the personality of a chromosome is the pathology for which it is responsible. In this respect, the most notorious chromosome is also the smallest, chromosome 21. Many, such as chromosomes 4, 5, 8, 9, 13, and 18, have a very rich pathology. It includes well-defined syndromes characterized by the craniofacial dysmorphia, by the degree and nature of the encephalopathy, by the presence or absence of growth retardation, by the age at which the patients are first examined, and by many other features. The symptoms of chromosome 8 trisomy resemble neither those of the "cri du chat" syndrome nor those of 9p monosomy syndrome. By contrast, other chromosomes, such as the largest, chromosome 1, have a very scant pathology, even nonexistent. They may have functions of such importance that any unbalance due to monosomy or trisomy is incompatible with survival.

These facets of the personality of chromosomes are the basis of this book, which we envisaged titling: "From 1 to 22, and X and Y"—but this would not have been serious enough.

In brief, a chapter will be devoted to each chromosome, according to its order in the human karyotype. All chapters are designed in the same manner. Each begins with a short résumé of the chromosome's personality. Reference to its evolution in the hominoid primates is based on our own work and that of B. Dutrillaux. Following this is a "physical" description, including the chromosome in its standard staining; the schematic of regions and bands as specified by the nomenclature of Paris, 1971; and several examples of banding obtained by different techniques. When justifiable, this morphological identification is completed by a description of the polymorphism of the chromosome.

This "mark of distinction" of the chromosome is followed by its gene map. The latter is in line with the information bulletin of February 1, 1976, "The Human Gene Map," issued by V. A. McKusick, as well as with the reports of the three conferences dedicated to "Human Gene Mapping"—at New Haven (1973), Rotterdam (1974), and Baltimore (1975). Only more recent papers, not covered in these proceedings, have been listed.

Among gene localizations, certain are solidly established, having been confirmed by several laboratories. Others are less reliable: they have been suggested by a single group of investigators and are not yet confirmed, some having even been contradicted. It is sometimes difficult to draw the dividing line between the two groups. In cases of uncertainty we have exercised prudence in considering the localization as "provisional." Well-established localizations are given on the right-hand side of the chromosomal schematic and the others on the left. In some cases, localizations are given on both sides, indicating that they are definite for the given chromosome, but that the more precise localization has not yet been confirmed.

The names and abbreviations of genetic markers, principally enzymes, are still highly inconsistent; they vary between publications as well as within the same publication. As far as possible, we have tried to retain the names and abbreviations most often employed at the time of writing.

In the majority of the chapters the most important part centers on pathology. The concept of syndrome is difficult to define. The best definition is perhaps "a pattern of symptoms common to a set of individuals." It is evident that the description of a new syndrome necessitates compiling a sufficient number both of clinical features and of observed individuals. Obviously, the description of chromosome 21 trisomy, the "cri du chat" disease, and the like presents no difficulty. By contrast, there are isolated observations whose correlation with a well-defined syndrome is much more open to question. We have endeavored to ascertain a minimum of three observations of patients with clearly comparable phenotypes. The figure on page xvii indicates the syndromes we have retained as valid.

Wherever it was possible to confirm the concept of "type and countertype" defined by J. Lejeune, comparing the syndromes due to trisomy and to monosomy for a given chromosomal segment, we have summarized in a table the main contrasting features.

Regarding the bibliography, we have not attempted to provide all known references, but only the principal ones. For "classical" syndromes we have listed the references, particularly those of recent reviews, that enable the reader to locate the entire bibliography. Where syndromes are based on a small number of observations, all known references are given.

Each syndrome is illustrated by photographs of patients. Figure numbers with letters (e.g., 1a, 1b) indicate photographs of the same patient. The sources of these photographs—publications, the collection of the Institut de Progénèse (Prof. J. Lejeune), and our personal collection—are listed at the end of the book.

The reader may be surprised by the lack of examples of chromosomal rearrangements in the illustrations. We believe that such illustrations would add little to the book's informative value. In complete trisomies, the additional chromosome is not different from the chromosomes of the normal pair. In translocations, resulting in partial trisomies or monosomies, each observation constitutes a special case. When a preferential breakage point seems to exist on a chromosome, we have indicated this.

Considerable importance has been given to the appendixes. They include:

- A description of the current cytological techniques. This is sufficiently complete and detailed to allow their practical application. In each case we have described in detail a proven technique and have mentioned the essential points of several variants.

- A description of the basic concepts necessary for the interpretation of palmar and digital dermatoglyphics.
- The chromosomal nomenclature as defined by the conventions of Denver, London, Chicago, and Paris.
- A list of localized genetic markers and the identification of the corresponding chromosome.
- A "syndrome finder," which classifies the most important symptoms and the syndromes they suggest.

This book was not designed as a treatise of classical cytogenetics. Many such works are indeed available, from the classic *Les chromosomes humains* by R. Turpin and J. Lejeune (1965) to the more recent *Human Cytogenetics* by J. Hamerton (1971). The fundamental principles they expound have evolved very little since then. By contrast, genetic localizations and chromosomal pathology have evolved considerably; they justify the writing of this new book, which has been designed as an atlas.

The reader will no doubt see the resemblance to geographical atlases—the various geographic, political, economic, and historic maps that illustrate the advances and retreats of national boundaries under the influence of the centuries. In the same way, the reader will find a morphological map of each chromosome with its topographical bands, as revealed by the different banding techniques; a map of pathology, indicating the different syndromes for which a given chromosome may be responsible; and a gene map, describing resources—i.e., a given chromosome's wealth in genes.

Everyone consults an atlas at one time or another. The present atlas is intended for a vast audience of biologists, doctors, and above all for pediatricians. By vocation the pediatrician is an amateur of rare diseases. Today, he is invited to recognize the abundance of new cytogenetic syndromes, challenging his perspicacity.

Destined for the pediatrician's consulting room, this atlas is also intended for the cytogeneticist, as much for his office as for his laboratory bench, where it can provide the necessary formulas for obtaining chromosomal preparations of good quality.

It is also intended to provide the physician and cytogeneticist with the means of a common language, comprehensible to both. The doctor must know when to ask for a karyotypic examination and also how to interpret the results provided by the cytogeneticist. This is the aim of the nomenclature presented in detail in the appendix.

Other important specialists should also be concerned—for example, the biochemist, the hematologist, or the immunologist, of whom the pediatrician or cytogeneticist will solicit an assay of a given enzymatic activity, which may be increased or decreased. They may also be asked to determine a given blood group, whose locus is situated on the chromosome or chromosomal segment present singly or in triplicate, or again to assay a given gammaglobulin. Such is the aim of the gene map of each chromosome, indicating which abnormal biological activity should be most profoundly studied.

We hope that the reader will approve our ambitions and at the same time perhaps be further impressed by the remarkable organization of man's hereditary material—an organization that certainly remains full of surprises.

Jean de Grouchy
Catherine Turleau

ACKNOWLEDGMENTS

We would like to express our gratitude to the numerous authors who have sent us the original photographs of their patients and thus allowed us to assemble an exceptionally rich iconography. Professor J. Lejeune and Dr. Marie-Odile Rethoré have lent us many photos of their patients from the Institut de Progénèse, Paris. Dr. B. Dutrillaux has allowed us to use unpublished documents for our chromosome morphology charts. Our appreciation and thanks are due to all of these contributors.

Our thanks also go to our photography expert, Mr. J. Berghege, who had the heavy task of reproducing all the necessary documents.

The idiograms at the beginning of each chapter are from "An International System for Human Cytogenetic Nomenclature High-Resolution Banding, *Cytogenetics and Cell Genetics,* 1981, Vol. 31, no. 1, Karger, pp. 1–32. They are reproduced with the grateful authorization of the authors and the editor.

Although she had many occasions, Mrs. Michelle Muesser has never, not even for one second, lost her good humor, nor her patience, nor her cold blood, during the preparation of the manuscript. We owe her very particular thanks.

Lastly, we are indebted to Mrs. Jerry Bram, who helped us greatly in preparing the English manuscript.

Jean de Grouchy
Catherine Turleau

CONTENTS

POLYPLOIDY $\longrightarrow$ 412

APPENDIXES $\longrightarrow$ 417

CLINICAL ATLAS OF
HUMAN CHROMOSOMES

CHROMOSOME 1

REFERENCES

GARDNER R.J.M., McCREANOR H.R., PARSLOW M.I., VEALE A.M.O.: Are 1q+ chromosomes harmless? *Clin. Genet.*, **6**:383–393, 1974.

NIELSEN J., FRIEDRICH U., HREIDARSSON A.B.: Frequency and genetic effect of 1qh+. *Humangenetik*, **21**:193–196, 1974.

TUUR S., KAOSSAR M., MIKELSAAR A.V.: 1q+ variants in a normal adult population (one with a pericentric inversion). *Humangenetik*, **24**:217–220, 1974.

Chromosome 1 is, by definition, the largest chromosome in the karyotype. It is also the chromosome for which the richest gene map is known. Nearly 35 genes have been localized with a fair amount of certainty. Among them are the Rh blood group and such diseases as Gaucher's disease, a glycogenosis, Charcot-Marie-Tooth disease.

Structural and numerical anomalies are probably lethal in most cases. Complete trisomy is observed only in spontaneous abortions.

Very few carriers of structural rearrangements are known. Terminal or interstitial trisomies of the long arm, as well as monosomy for the distal extremity of 1q may be described. Several observations of interstitial deletions of 1q have been reported, but have not been included here because of their heterogeneity. No anomaly of the short arm has been reported, but r(1) has occasionally been observed.

a   b

As for all chromosomes with a secondary constriction, chromosome 1 has a polymorphism. The constriction may vary in length. The figure shows (1) the appearance usually observed and (2) a greatly exaggerated constriction (*uncoiler*) (C-banding).

This marker chromosome is generally not considered harmful (Gardner et al., 1974; Nielsen et al., 1974; Tüür et al., 1974). It is transmitted as a dominant trait and allows linkage studies. In the case of chromosome 1, such investigations have demonstrated close linkage between the Duffy blood-group locus and the constriction.

A nomenclature is available to describe such polymorphism (see appendix).

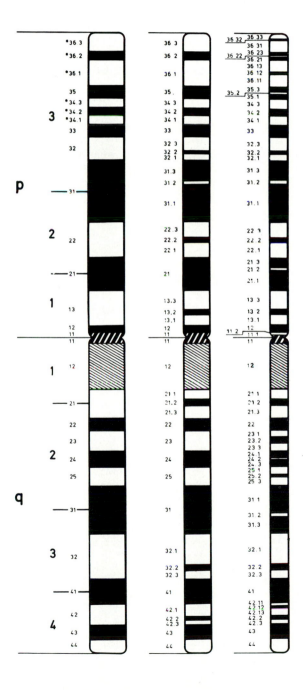

GENE MAP

1p36	A12M2	(P)	Adenovirus-12 chromosome modification sites 1A (10292) V
1p36 (at end of 1p)	GDH		Glucose dehydrogenase (13809) S, F
1p36-1pter	ENO1		Enolase-1 (17243) S, F, R
1p34-1pter	PGD		6-Phosphogluconate dehydrogenase (17220) F, S
1p34	NB	(P)	Neuroblastoma (25670) Ch
(1p32-1pter)	EL1		Elliptocytosis-1 (13050) F
1p32-1pter	RH		Rhesus blood group (11170) F-S, D
1p32-1pter	GALE		UDP galactose-4-epimerase (23035) S, LD
1p34	FUCA		Alpha-L fucosidase (23000) S, F, R
(1p32-1p34)	SC		Scianna blood group (11175) F
(1p32-1p34)	RD	(P)	Radin blood group (11162) F
1p32 (distal to PGM1)	UMPK		Uridine monophosphate kinase (19171) S, R
1p34	AK2		Adenylate kinase-2, mitochondrial (10302) S, F, R
1p22.1	PGM1		Phosphoglucomutase-1 (17190) F, S, R
1p11-1qter	GBA	(P)	Acid beta-glucosidase (23080) S
1p13 or 1q25	ASG	(L)	Aspermiogenesis factor (10845) Ch
1p	DO	(I)	Dombrock blood group (11060) F (see chr. 4)
(1p1)	AMY2		Amylase, pancreatic (10465) F, A
(1p1)	AMY1		Amylase, salivary (10470) F, A
1p	RP1	(L)	Retinitis pigmentosa-1 (18010) F
1qh			Satellite DNA III (D121) (12634) A
(1q2)	EL2	(L)	Elliptocytosis-2 (13060) F
1q (close to Fy)	CAE		Cataract, zonular pulverulent (11620) F
1q2 (distal to 1qh)	FY		Duffy blood group (11070) F, Fc
(1q2)	CMT1		Charcot-Marie-Tooth disease, slow nerve conduction type (11820) F
1q21-1q23	UGP1		Uridyl diphosphate glucose pyrophosphorylase-1 (19175) S, R

1q22	A12M3	(P)	Adenovirus-12 chromosome modification site −1B (10294)
(I)1q25;1q42	PEPC		Peptidase C (17000) S, R
1q42-1q43	RN5		5S ribosomal RNA gene (s) (18042) A
1q42-1qter	FH		Fumarate hydratase (13685) S, R
1q31-1q42	GUK1		Guanylate kinase-1 (13927) S, D
	GUK2		Guanylate kinase-2 (13928) S, D
1q42 and 1q22	A12M1	(P)	Adenovirus-12 chromosome modification site-1B and 1C (10293) V
1q23-1q25	AT3		Antithrombin III (10730) F, D, A
	MTR	(P)	5-Methyltetrahydrofolate: L-homocysteine s-methyl-transferase (tetrahydro-pteroylglutamate methyl-transferase) (15657) S
	SDH	(P)	Succinate dehydrogenase (18547) S
1cen-1q32	PFKM	(P)	Phosphofructokinase, muscle type (23280) S
	XPA	(P)	Xeroderma pigmentosum, complementation group A (27870) S

MORPHOLOGY AND BANDING

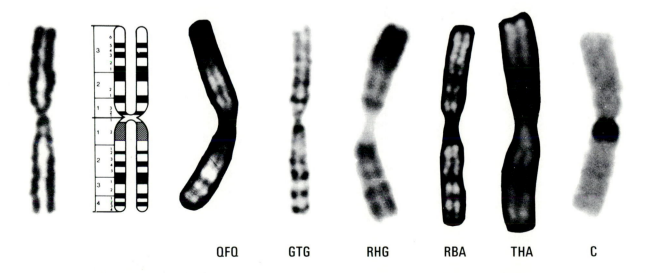

QFQ   GTG   RHG   RBA   THA   C

1q TRISOMIES

1q32→qter Trisomy

Abnormally wide fontanels
Triangular face
Severe micrognathia
Short life expectancy

REFERENCES

BONFANTE A., STELLA M., ROSSI G.: Partial trisomy of the long arm of chromosome 1 due to a familial translocation t(1;10) (q32; q26). *Hum. Genet.*, **45**:339–343, 1978.

BOURROUILLOU G., COLOMBIES P., BLANC P.: Trisomie 1q secondaire à une translocation réciproque maternelle. *C.R. Séances Soc. Biol.*, **172**:359–362, 1978.

FRYNS J.P., MUELENAERE A. de PEDERSEN J., VAN DEN BERGHE H.: Partial distal 1q trisomy. A distinct clinical dysmorphic syndrome in adulthood. *Ann. Génét.*, **23**:181–182, 1980.

LEISTI J., AULA P.: Partial trisomy 1 (q42→ter). *Clin. Genet.*, **18**:371–378, 1980.

LUNGAROTTI M.S., FALORNI A., CALABRO A., PASSALACQUA F., DALLAPICCOLA B.: De novo duplication 1q32-q42: variability of phenotypic features in partial 1q trisomics. *J. med. Genet.*, **17**:398–401, 1980.

NEU R.L., GARDNER L.I.: A partial trisomy of chromosome 1 in a family with a t(1q−; 4q+) translocation. *Clin. Genet.*, **4**:474–479, 1973.

REHDER H., FRIEDRICH U.: Partial trisomy 1q syndrome. *Clin. Genet.*, **15**:534–540, 1979.

STEFFENSEN D.M., CHU E.H.Y., SPEERT D.P., WALL P.M., MEILINGER K., KELCH R.P.: Partial trisomy of the long arm of human chromosome 1 as demonstrated by in situ hybridization with 5S ribosomal RNA. *Hum. Genet.*, **36**:25–33, 1977.

TAYSI K., SEKHON G.S.: Partial trisomy of chromosome 1 in two adult brothers due to maternal translocation (1q−;6p+). *Hum. Genet.*, **4**:277–285, 1978.

VAN DEN BERGHE H., EYGEN H. van, FRYNS J.P., TANGHE W., VERRESEN H.: Partial trisomy 1, karyotype 46,XY,12−,t(1q, 12p)+. *Humangenetik* **18**:225–230, 1973.

YUNIS E., EGEL H., ZUNIGA R., RAMIREZ E.: «de novo» trisomy 1q32→1qter and monosomy 3p25→3pter. *Hum. Genet.*, **36**:113–116, 1977.

The first two observations of trisomy 1qter were by Van den Berghe et al. (1973) and Neu and Gardner (1973). Later observations are somewhat disparate due to the variability of the breakpoint on 1q and the extent of the associated monosomy. The syndrome we will describe corresponds to cases of 1q32→qter trisomy associated with slight monosomy (Stephensen et al., 1977; Yunis et al., 1977; Bonfante et al., 1978; Taysi and Sekkon, 1978; Rehder and Friedrich, 1979; Fryns et al., 1980; Lungarotti et al., 1980). We will then briefly indicate the corresponding clinical symptoms when the breakpoint is more proximal, in 1q23 or 5. We have not considered those observations wherein the phenotype more closely resembles that of associated monosomy, as, for example, 4p monosomy (Bourrouillou et al., 1978; Leisti and Aula, 1980).

GENERALIZATIONS

Mean birth weight: 2,500 g

PHENOTYPE

Trisomy 1q32→qter is most often characterized by severe malformations leading to death in the first few days of life. Newborns draw attention immediately by their low birth weight, facial dysmorphism, and the gravity of their clinical condition.

Craniofacial Dysmorphism

The skull appears voluminous, with a wide forehead. The anterior fontanelle, as well as the metopic suture are gaping. Brachycephaly is present. The face is triangular, with midface hypoplasia and a pointed chin.

Palpebral fissures slant downward and outward. The eyes are sometimes small, and hypertelorism is often present.

The nose is small and beaked, with a wide flattened bridge.

The philtrum is long.

Micrognathia is very distinct.

The ears are low-set, small, poorly folded, and "simple."

Neck, Thorax, and Abdomen

An excess of skin is visible on the neck.

The thorax is sometimes wide, with a deviated sternum or nipples abnormally widespaced. Hernias are frequent.

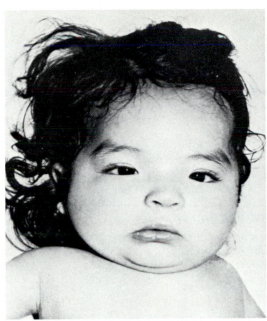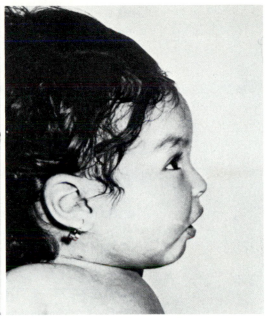

1.1

1.2

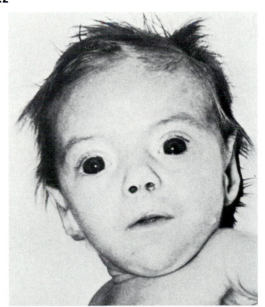

Limbs

Fingers and toes are long and tapering and may overlap.

Genitalia

Cryptorchidism is frequent.

Malformations

In four cases out of six, cardiac malformation was present, usually a truncus arteriosus.

Other malformations are mainly digestive: intestinal stenosis and atresia, bile duct stenosis, pyloric stenosis.

Mental Retardation and Prognosis

The infants usually died early after birth. Two of the patients, half-brothers, are now adults aged 30 and 39. One of them has an IQ of 45; the IQ of the other is lower. Neither has had cardiac malformation.

CYTOGENETIC STUDY

Three cases have arisen *de novo* (two duplications and one t(1;3)). The four others are due to parental translocations (with 13q, 10q and 6p).

When trisomy results from a parental rearrangement, a family history of early abortions is very frequent.

DERMATOGLYPHICS

Dermatoglyphics are not characteristic. A single transverse palmar crease is often noted.

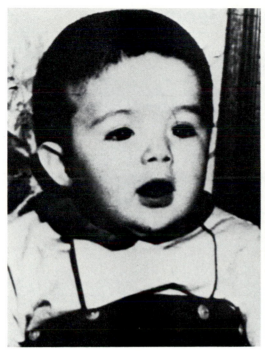

1.3

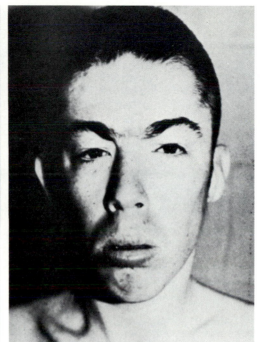

1.4

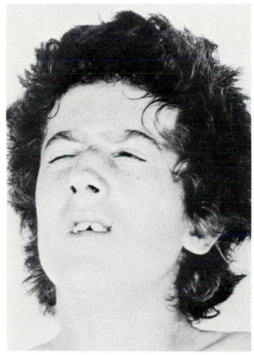

1q23 or 5→qter Trisomy

REFERENCES

CHEN H., GERSHANIK J.J., MAILHES J.B., SANUSI I.D.: Omphalocele and partial trisomy 1q syndrome. *Hum. Genet.*, **53**:1–4, 1979.

GARRETT J.H., FINLEY S.C., FINLEY W.H.: Fetal loss and familial chromosome 1 translocations. *Clin. Genet.*, **8**:341–348, 1975.

NEU R.L., GARDNER L.I.: A partial trisomy of chromosome 1 in a family with a t(1q−; 4q+) translocation. *Clin. Genet.*, **4**:474–479, 1973.

NORWOOD T.H., HOEHN H.: Trisomy of the long arm of human chromosome 1. *Humangenetik*, **25**:79–82, 1974.

VAN DEN BERGHE H., EYGEN M. van, FRYNS J.P., TANGHE W., VERRESEN H.: Partial trisomy 1, karyotype 46, XY,12−, t(1q,12p)+. *Humangenetik*, **18**:225–230, 1973.

In five cases, the trisomic segment was longer with a break in q25 (Neu and Gardner, 1973; Garret et al., 1975; Chen et al., 1979), or in q23 (Van den Berghe et al., 1973; Norwood and Hoehn, 1974). Inner organ malformations were much more serious and rapidly resulted in death.

Mean birth weight: 2,350 g for four children born between 37 and 40 weeks.

Craniofacial dysmorphism is similar to that of 1q32→qter trisomy.

Cardiac malformations, of the truncus arteriosus type, are especially serious. Other malformations include a bilateral harelip; omphalocele; constant hypoplasia of the thymus; absence of gall bladder in the two cases where breakpoints are in q23; adrenal hypoplasia; renal anomalies; anomalies of the spleen and pancreas; and aplasia of the cerebellum.

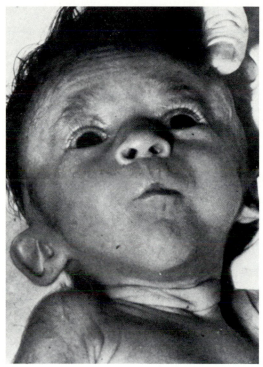
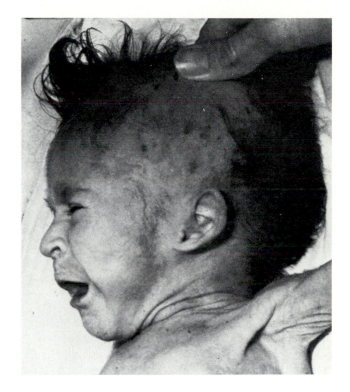

1.5

1q25→q32 Trisomy

REFERENCES

GARVER K.L., CIOCCO M., TURACK N.A.: Partial monosomy or trisomy resulting from crossing over within a rearranged chromosome 1. *Clin. Genet.*, **10**:319–324, 1976.

PALMER C.G., CHRISTIAN J.C., MERRITT A.D.: Partial trisomy 1 due to a «shift» and probable location of the Duffy (Fy) locus. *Am. J. hum. Genet.*, **29**:371–377, 1977.

PAN S.F., FATORA S.R., SORG R., GARVER K.L., STEELE M.W.: Meiotic consequences of an intrachromosomal insertion of chromosome n° 1: a family pedigree. *Clin. Genet.*, **12**:303–313, 1977.

SCHINZEL A.: Possible trisomy 1q25→1q32 in a malformed girl with a de novo insertion in 1q. *Hum. Genet.*, **49**:167–173, 1979.

Six cases of intercalary 1q25→q32 trisomy have been reported (Garver et al., 1976; Palmer et al., 1977; Pan et al., 1977) and were the subject of a review by Schinzel (1979).

Main symptoms are deep set eyes; abnormally wide nose bridge; small mouth; very severe retrognathia; a high-arched palate; low set ears, simple, in posterior rotation with an abnormal fold of the upper part of the helix; high frequency of cardiopathies; flexion contractures of the fingers; mental retardation of varying degrees, sometimes moderate; delayed or accelerated growth rate; and apparently normal survival.

1.6

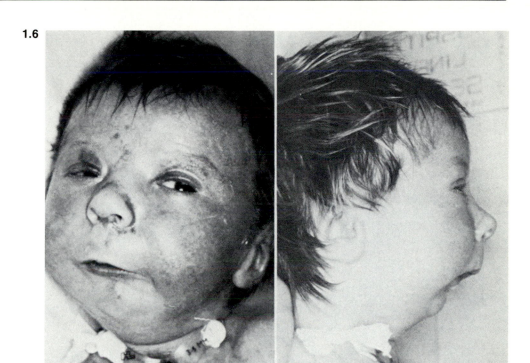

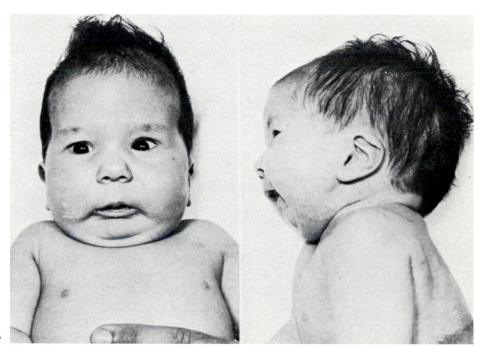

1.7

1q4 MONOSOMY

Microbrachycephaly
Epicanthus
Micrognathia

REFERENCES

ANDRLE M., ERLACH A., MAYR W.R., RETT A.: Terminal deletion of (1) (q42) and its phenotypical manifestations. *Hum. Genet.,* **41**:115–120, 1978.

DIGNAN P. St-J., SOUKUP S.: Terminal long-arm deletion of chromosome 1 in a male infant. *Hum. Genet.,* **48**:151–156, 1979.

KESSEL E., PFEIFFER R.A., BLANKE W., SCHWARZ J.: Terminal deletion of the long arm of chromosome 1 in a malformed newborn. *Hum. Genet.,* **42**:333–337, 1978.

MANKINEN C.B., SEARS S.W., ALVAREZ V.R.: Terminal (1) (q43) long arm deletion of chromosome n° 1 in a three year-old female. *Birth. Defects, Orig. Art. Ser., XII-5*:131–136, 1976.

Six cases of terminal deletion of 1q are known. The breakpoint, according to the authors, is located in either 1q42 or 1q43, but it may well be the same deletion (Mankinen et al., 1976; Kessel et al., 1978; Andrle et al., 1978; Dignan and Soukup, 1979; two personal observations).

GENERALIZATIONS

- Sex ratio: 3M/3F
- Mean birth weight: 2,800 g
- Mean crown-heel length at birth: 46.7 cm
- Mean head circumference at birth: 32.2 cm

PHENOTYPE

Microcephaly is evident from birth on. Two infants had generalized hypotonia and a shrill cry. Delay in growth rate persists, especially in height. Mental retardation is severe.

Craniofacial Dysmorphism

Microcephaly remains severe, with brachycephaly. Hair is scant and fine. The face is round with heavy cheeks. Palpebral fissures are slightly slanted upward and outward. A very pronounced bilateral epicanthus exists, leading to pseudohypertelorism. Strabismus has been noted in several cases.

The nose is short and bulbous with a flattened bridge. The philtrum is prominent and well defined.

Micrognathia is generally noted: the children appear to be chewing on their lower lip. In one case, prognathism was present. Ears are generally well folded and of normal implantation.

Neck, Thorax, and Abdomen

The neck may be short. Anomalies of the vertebrae have been noted, as well as supranumerary ribs and scoliosis in one case.

Limbs

Minor malformations have been described: joint deformities, tapering fingers, syndactyly, brachydactyly, club foot, and hypoplasia of the nails.

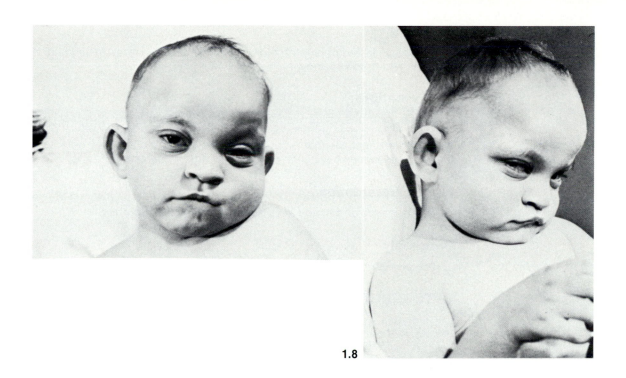

1.8

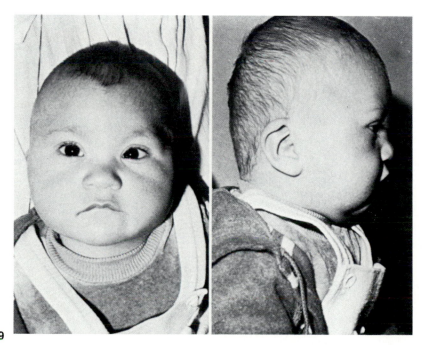

1.9

Genitalia

Hypospadias and/or cryptorchidism were reported in three male children. Vaginal stenosis was noted in one girl.

Malformations

In one case, the infant died after 17 days of malformations which curiously resembled those common to 1qter trisomy, involving a pseudotruncus arteriosus and the absence of gall bladder (Kessel et al., 1978). Ventricular septal defect, cleft palate, and a single kidney have been reported in other observations.

Mental Retardation

This is severe and nearly always accompanied by seizures.

Prognosis

The oldest child is now 5.

CYTOGENETIC STUDY

Deletion always arises *de novo* and involves only two bands, q43 and q44. For this reason, it should be carefully searched for.

DERMATOGLYPHICS

A certain amount of disorganization is noted in the flexion creases and dermal ridges. A single transversal palmar crease, or its equivalent, is usually present. The axial triradius may be in t″.

1.10

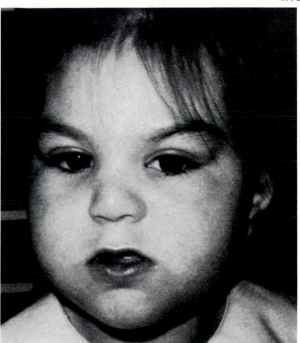

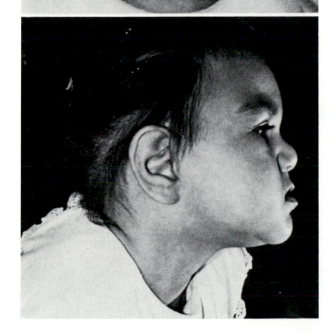

1.11

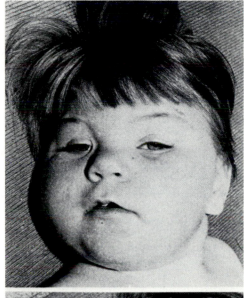

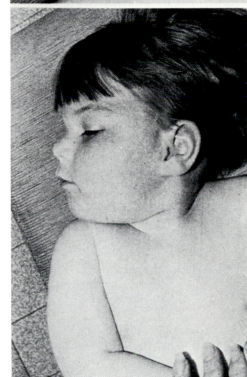

r(1)

REFERENCES

BOBROW M., EMERSON P.M., SPRIGGS A.I., ELLIS H.L.: Ring-1 chromosome, microcephalic dwarfism, and acute myeloid leukemia. *Amer. J. Dis. Child.*, **126**:257–260, 1973.

COOKE P., GORDON R.R.: Cytological studies on a human ring chromosome. *Ann. Hum. Genet.*, **29**:147–150, 1965.

GORDON R.R., COOKE P.: Ring 1 chromosome and microcephalic dwarfism. *Lancet* **2**:1212–1213, 1964.

KJESSLER B., GUSTAVSON K.H., WIGERTZ A.: Apparently non-deleted ring-1 chromosome and extreme growth failure in a mentally retarded girl. *Clin. Genet.*, **14**:8–15, 1978.

WOLF C.B., PETERSON J.A., LOGRIPPO G.A., WEISS L.: Ring 1 chromosome and dwarfism—a possible syndrome. *J. Pediat.*, **71**:719–722, 1967.

Several observations of ring 1 chromosome are known (Gordon and Cooke, 1964; Wolf et al., 1967; Bobrow et al., 1973). It is difficult to describe a well-defined clinical syndrome. Nonetheless, some features common to the patients can be isolated:

- Very low birth weight
- Considerably delayed development, affecting weight more than stature
- Moderate mental retardation with a lively and pleasant personality
- Absence of major congenital malformation

1.12

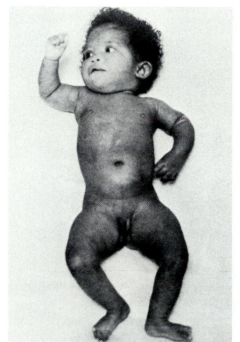

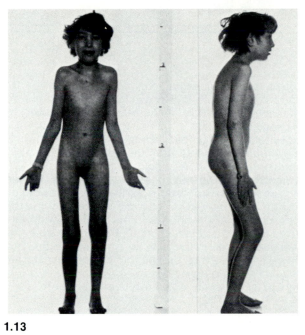

1.13

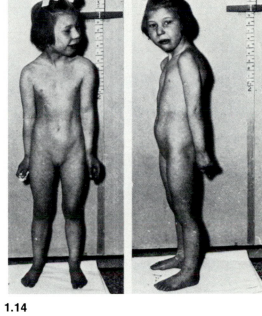

1.14

1.15

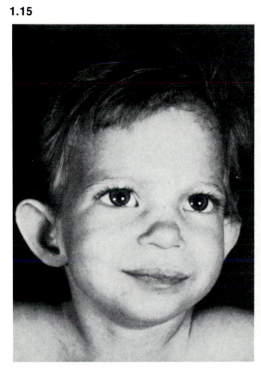

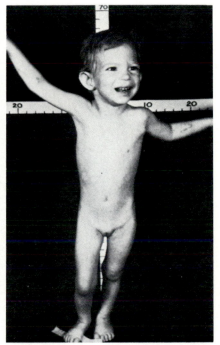

CHROMOSOME 2

REFERENCE

BACCHICHETTI C., LENZINI E., PESERICO A., TENCONI R.: Study on segregation and risk for abnormal offspring in carriers of pericentric inversion of the (p11→q13) segment of chromosome 2. *Clin. Genet.*, **18**: 402–407, 1980.

Fewer gene localizations are known for chromosome 2 than for chromosome 1. They include, however, the very important gene family that codes for the kappa light chain of immunoglobulins. The gene for aniridia, the dominant disease, has tentatively been localized on the short arm by linkage with ACP1.

From a morphological point of view, this chromosome includes a fragile site in 2q1, which probably represents the evolutionary stigma from the fusion that arose in the human phylum between two acrocentrics found in pongidae.

Pericentric inversions of chromosome 2 have been observed in couples consulting for reproductive failure, including spontaneous abortions, birth of deformed children, stillbirths, or death soon after birth. The breakpoints of these pericentric inversions would seem to be found fairly constantly in p11 and q13. This localization might be linked with events having occurred during evolution, and could be responsible for the increase in the risk of spontaneous abortion. The risk of "aneusomie de recombinaison" in living children, however, appears to be negligible (Bacchichetti et al., 1980).

Only two syndromes, trisomy 2p2 and trisomy 2q3, have been considered.

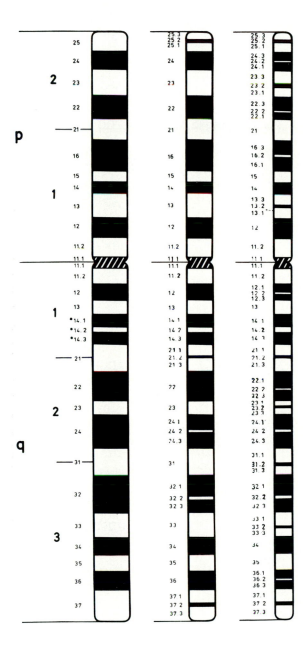

GENE MAP

(2p)	AKE	(L)	Acrokeratoelastoidosis (10185) F
2p	AHH	(P)	Aryl hydrocarbon hydroxylase (10833) S
(2p23)	AN1	(L)	Aniridia (10620) F
2p25	ACP1		Acid phosphatase-1 (17150) D, S
2p23? (Close to ACP1)	POC	(P)	Proopiomelanocortin (17683) REa
2p23	MDH1		Malate dehydrogenase, soluble (15420) S
2p11-2p22	GLAT	(P)	Galactose enzyme activator (13703) S
2p13-2cen	IGK		IMMUNOGLOBULIN KAPPA LIGHT CHAIN GENE FAMILY (Km; Inv) A, Rea
	IGKV		Variable region of kappa light chain (many genes) (14721)
	IGKJ		J region of kappa light chain (several genes) (14719)
	IGKC		Constant region of kappa light chain (14720)
(2p13-2cen)	JK		Kidd blood group (11100) F
Linked to JK	CO	(L)	Colton blood group (11045) D, F
2p23-2q32?	ADCP2		Adenosine deaminase complexing protein-2 (10272) S
2q11	FS		Fragile site 2q11 (13661)
2q21.3	GRS	(L)	Gardner syndrome (17530) Ch
2q32-2qter	IDH1		Isocitrate dehydrogenase, soluble (14770) S
2q32-2qter	RPE	(P)	Ribulose 5-phosphate 3-epimerase (18048) S
	IGAS	(P)	Immunoglobulin heavy chain attachment site (14710)S
	ACEE	(P)	Regulator of acetylcholinesterase (10065) D
	UGP2	(P)	Uridyl diphosphate glucose pyrophosphorylase-2 (19176) S
	FN	(P)	Fibronectin (13560) S

MORPHOLOGY AND BANDING

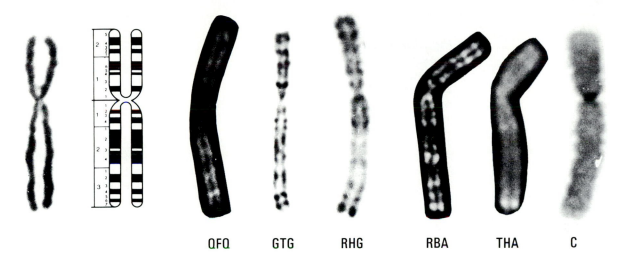

QFQ   GTG   RHG   RBA   THA   C

2p2 TRISOMY

High forehead
Hypertelorism
Micrognathia
Slender children
Arachnodactyly

REFERENCES

ARMENDARES S., SALAMANCA-GOMEZ F.: Partial 2p trisomy (p21→pter) in two siblings of a family with a 2p−;15q+ translocation. *Clin. Genet.*, **13**:17–24, 1978.

FRANCKE U.: Clinical syndromes associated with partial duplications of chromosomes 2 and 3: dup (2p), dup (2q), dup (3p), dup (3q). *Birth Defects, Orig. Art. Ser., XIV-6C*:191–217, 1978.

MAYER U., SCHWANITZ G., GROSSE K.P., ETZOLD R.: Trisomie partielle 2p par translocation familiale 2/6. *Ann. Génét.*, **21**:172–176, 1978.

NEU R.L., DENNIS N.R., FISHER J.E.: Partial 2p trisomy in a 46,XY,der(5),t(2;5)(p23;p15)pat infant; autopsy findings. *Ann. Génét.*, **22**:33–34, 1979.

SEKHON G.S., TAYSI K., RATH R.: Partial trisomy for the short arm of chromosome 2 due to familial balanced translocation. *Hum. Genet.*, **44**:99–103, 1978.

Fifteen cases of 2p2 trisomy are known. The first ten were the subject of a review by Francke (1978) and five others have been published since then (Mayer et al., 1978; Sekkon et al., 1978; Armendares and Salamanca-Gomez, 1978; Neu et al., 1979).

According to these authors, trisomy involves the p21 or p23→pter region. In all cases, it is the result of a familial rearrangement.

GENERALIZATIONS

Sex ratio: 2M/1F
Mean birth weight: 2,680 g
Crown-heel length at birth appears normal

PHENOTYPE

Craniofacial Dysmorphism

The forehead is high with frontal bossing and, occasionally, a prominent metopic suture.

Hypertelorism is always present and is sometimes severe.

The bridge of the nose is wide and flat. The nose is short with a prominent tip and anteverted nares.

The palpebral fissures are slightly slanted downward and outward. Strabismus is nearly always present; epicanthus and various anomalies such as myopia, cysts, lacrymal canal stenosis, and puffiness around the eyelids have been observed in the very young child.

The mouth is small with downturned corners.

The chin is small and pointed.

The ears are low-set and large, with a prominent anthelix and a poorly developed lobe.

Neck, Thorax, and Abdomen

Patients are long-limbed, with a long narrow thorax. Scoliosis is nearly always present. Dimples have been reported.

Limbs

Limbs and fingers are long (arachnodactyly). The fingers are sometimes spatulated with clinodactylies. Toes are long and overlap. Flat feet are sometimes observed.

The musculature of the limbs and of the trunk is weak.

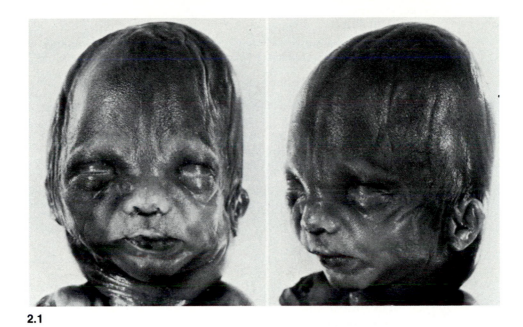

2.1

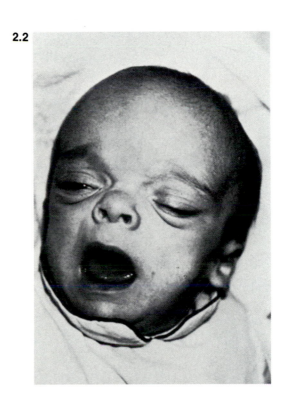

2.2

Genitalia

Genital anomalies are regularly found in boys: micropenis, hypoplasia of the scrotum, cryptorchidism. Hypoplasia of the external genitals has been observed in girls.

Malformations

Skeletal anomalies are always present; these include scoliosis (already mentioned), vertebral and rib anomalies, congenital dislocation of the hip, and abnormal segmentation of the sternum.

Cardiac malformations are frequently found, including dextrocardia and aortic stenosis. In a stillborn infant, the autopsy showed an extremely complex congenital heart disease, as well as cerebral abnormalities (Neu et al., 1979).

Other malformations include hemangiomas, detachment of retina, microphthalmy, and hydrocephaly.

Mental Retardation

This is very severe in all cases.

Prognosis

Staturo-ponderal retardation is always severe and is lower than the third percentile.

The children observed ranged in age from several months to 18 years. One child died at birth.

Remark

Trisomy 2pter in some ways resembles Aarskog's syndrome: common features are a short stature, hypertelorism, hypoplasia of the maxillary, a hollow chest, scoliosis, and cryptorchidism. However, Aarskog's syndrome is due to the mutation of a sex-linked gene, and is not accompanied by any pronounced mental debility; facial dysmorphism is different, and brachycephaly is present (Francke, 1978).

CYTOGENETIC STUDY

In all cases, trisomy results from malsegregation of a parental translocation. According to the author's interpretations, the breakpoint varies from p21 to p23. The other chromosome involved in the translocation is variable, and the breakpoint is generally terminal, causing negligible associated monosomy.

DERMATOGLYPHICS

Dermatoglyphics are not characteristic. A single transverse palmar crease is noted.

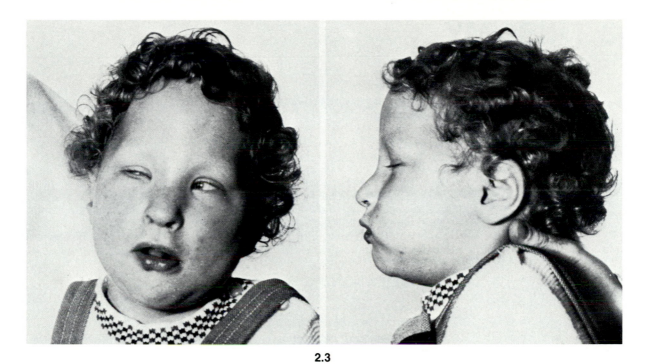

2.3

2.4

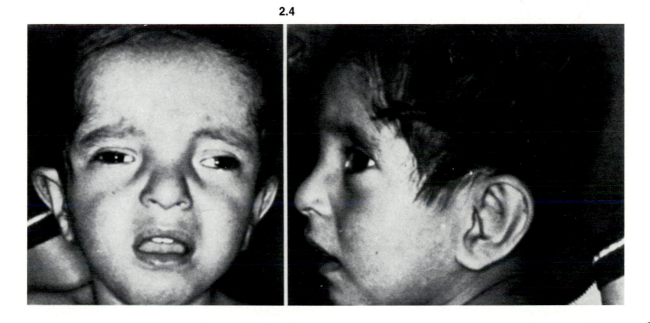

2q3 TRISOMY

Hypertelorism
Short nose, anteverted nares
Thin upper lip
Everted lower lip
Micrognathia

REFERENCES

DENNIS N.R., NEU R.L., BANNERMAN R.M.: Duplication 2q33→2q37 due to paternal ins(12;2) translocation. *Am. J. med. Genet.*, **1**:271–277, 1978.

FORABOSCO A., DUTRILLAUX B., TONI G., TAMBORINO G., CAVAZZUTI G.: Transloca- tion équilibrée t(2;13) (q32;q33) familiale et trisomie 2q partielle. *Ann. Génét.*, **16**:255– 258, 1973.

FRANCKE U.: Clinical syndromes associated with partial duplications of chromosomes 2 and 3: dup(2p), dup(2q), dup(3p), dup(3q). *Birth Defects, Orig. Art. Ser.*, *XIV-6C*:191– 217, 1978.

GILIBERTI P., CELONA A., DELLA PIETRA M., DE MASI R.V., FIORETTI G., PAGANO L., RENDA S., VETRELLA A., VENTRUTO V.: Familial translocation 2;17 with partial trisomy 2q32→2qter. *Ann. Génét.*, **23**:249– 250, 1980.

HOWARD-PEEBLES P.N., GOLDSMITH J.P.: Duplication of region 2q31→2qter in a family with 2/9 translocation. *Hum. Hered.*, **30**:84– 88, 1980.

LAURENT C., BIEMONT M.C., GUIBAUD P., GUILLOT J., NOEL B., QUACK B., GENEVIEVE M., CRESSENS M.L.: Sept cas de trisomie 2q34→2qter par transmission fa- miliale d'une translocation t(2;8) (q34;p23). *Ann. Génét.*, **21**:13–18, 1978.

RICCI N., DALLAPICCOLA B., COTTI G.: Translocation 2-D familiale. *Ann. Génét.*, **11**: 111–113, 1968.

WARREN R.J., PANIZALES E.G., CANT- WELL R.J.: Inherited partial trisomie 2: 46, XX, 1pt; t (1;2) (p36; q31). *Birth Defects, Orig. Art. Ser.*, *X1-5*:177–179, 1975.

WISNIEWSKI L., CHAN R., HIGGINS J.V.: Partial trisomy 2q and familial translocation t(2;18) (q31; p11). *Hum. Genet.*, **45**:225–228, 1978.

ZABEL B., HANSEN S., HARTMANN W.: Partial trisomy 2q and familial translocation t(2; 12) (q31; q24). *Hum. Genet.*, **32**:101–104, 1976.

Fifteen cases of 2q3 trisomy are known (Ricci et al., 1968; Forabosco et al., 1973; Warren et al., 1975; Zabel et al., 1976; Dennis et al., 1978; Laurent et al., 1978 [seven family members]; Wisniewski et al., 1978; Howard-Peebles and Goldsmith, 1980; Giliberti et al., 1980). (Also see Francke's review, 1978.)

Breakpoints vary according to the observations: q34 for the familial rearrangement observed by Laurent et al. (1978), and q31 in most other cases.

GENERALIZATIONS

- Sex ratio: 1M/1F
- Mean birth weight: 2,470 g (calculation based on eight cases)
- Crown-heel length at birth appears to be normal in the few cases for which it is known.

PHENOTYPE

The phenotype was well described by Laurent et al. (1978).

Craniofacial Dysmorphism

Microcephaly may be present, with frontal bossing and temporal depression.

Extreme hypertelorism is present. Palpebral fissures are small, narrow, and usually slant downward and outward.

The nose is short, with a flat wide nasal bridge. Nares are small and anteverted.

The philtrum is long, with undefined pillars.

The fleshy part of the upper lip is scarcely visible. The notch of Cupid's bow is deep. The lower lip is slightly everted.

The chin is small.

The ears are low-set and large, and an eversion is visible in the middle part of the helix.

In older children, this phenotype is modified, especially concerning the nose and mouth. The nose becomes beaked with an abnormally prominent bridge, and the mouth becomes markedly wide.

Neck, Thorax, and Abdomen

Kyphoscolioses, pectus excavatum, widely spaced nipples, and shortness of the neck have all been noted.

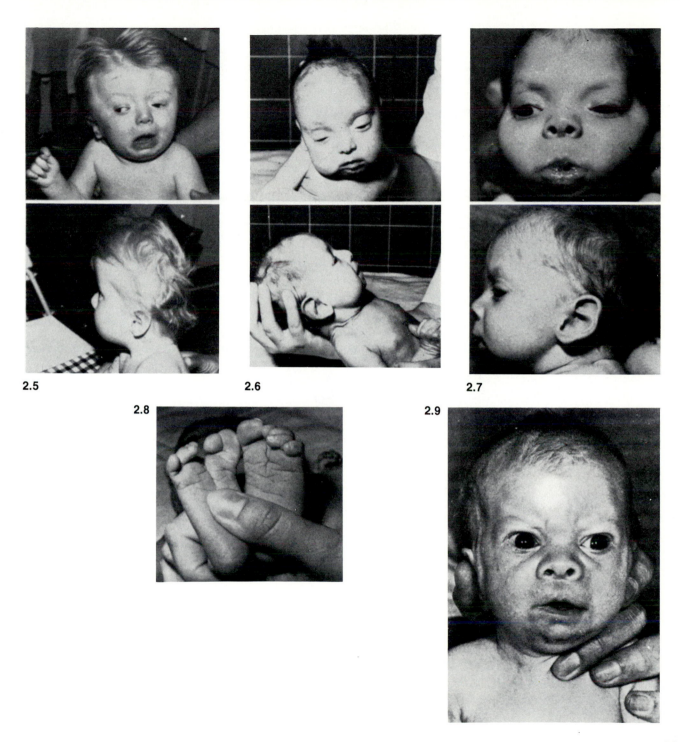

2.5

2.6

2.7

2.8

2.9

Limbs

The following minor anomalies have been reported: duplication of the thumb, or thickening of the last phalange, clinodactyly, hypoplasia of the second phalange of the Vth, and foot malformations.

Genitalia

Hypertrophy of the clitoris has been reported in several cases, as well as hypospadias and cryptorchidism.

Malformations

These are rare, and consist of congenital heart disease (aortic stenosis, ventricular septal defect, and patent ductus arteriosus) or of deforming uropathy.

Mental Retardation

The IQ is generally around 50.

Prognosis

The oldest patient was 28 years old. Somatic development appeared normal, apart from the appearance of a kyphoscoliosis.

Two children died of congenital heart disease, and two others of pneumonia.

CYTOGENETIC STUDY

All cases result from familial translocations or insertions. The second chromosome is variable. The associated monosomy is generally of little importance, since the break is terminal.

Breakpoints on chromosome 1 vary between q31 and q34, but the phenotype seems to be fairly similar.

DERMATOGLYPHICS

An excess of arches has been reported, but does not seem to be characteristic. A single transverse palmar crease is often noted.

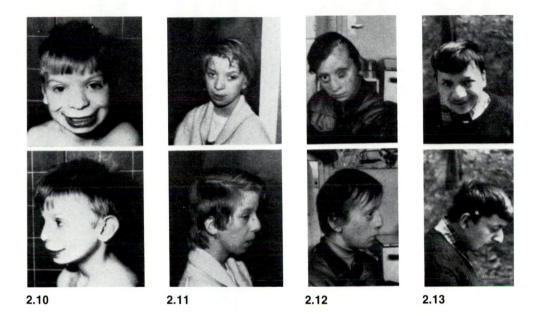

2.10   2.11   2.12   2.13

2.14

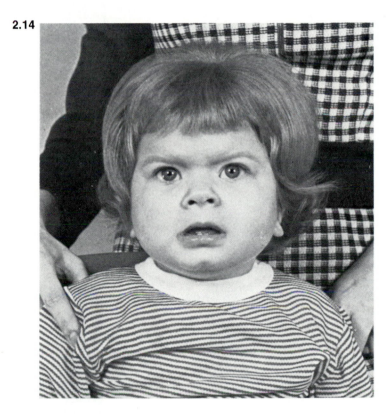

CHROMOSOME 3

Chromosome 3 is not yet rich in gene assignments. Among the diseases, generalized gangliosidosis, due to a deficiency of β-galactosidase, should be noted.

Trisomy 3q2 is one of the new syndromes to have been observed in a large number of cases. We have included here those cases resulting from "aneusomie de recombinaison" of a parental p25q21 to 5 pericentric inversion. The frequency of this type of rearrangement is striking, and appears to be highly characteristic of chromosome 3. Trisomy 3q2 also deserves mention because of its high degree of phenotypical similarity with Cornelia de Lange syndrome.

Other syndromes that have been considered involve the short arm: trisomy 3p2 and monosomy 3p25→pter.

A polymorphism of the centromere is present, and is visible especially in Q- and C-banding. A break in 3p14 is frequently observed in culture.

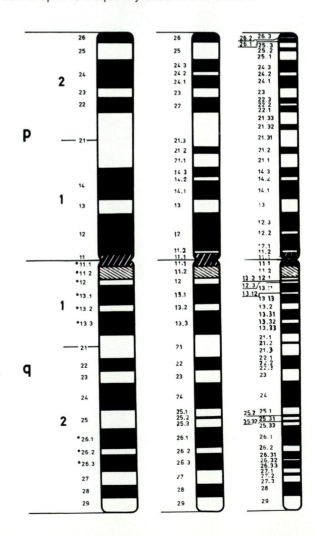

GENE MAP

3p21.1 (or 7p13)	GS	(L)	Greig craniopolysyndactyly syndrome (17570) Ch
3p14-3p23	SCCL	(P)	Small-cell cancer of lung (18228) Ch
3p12-3q13	GLB1		Beta-galactosidase-1 (23050) S
3p21-3q21	ACY1		Aminoacylase-1 (10462) S
3p13-3q12	GPX1		Glutathione peroxidase-1 (23170) S
	AF8T	(P)	Temperature sensitive (tsAF8) complement (11695) S
	HV1S	(P, I)	Herpes virus sensitivity (14245) S (see chr. 11)
	TFR	(P)	Transferrin receptor (19001) S
	TF	(P)	Transferrin (19000)S, H
	CHE1	(P)	Pseudocholinesterase-1 (17740) F
3q27-3q28	SST	(P)	Somatostatin (18245) REa

MORPHOLOGY AND BANDING

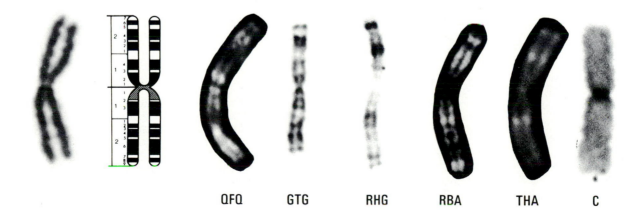

QFQ   GTG   RHG   RBA   THA   C

3q2 TRISOMY

REFERENCES

ALLDERDICE P.W., BROWNE N., MURPHY D.P.: Chromosome 3 duplication q21→qter deletion p25→pter syndrome in children of carriers of a pericentric inversion inv (3) (p25q21). *Am. J. Hum. Genet.*, **27**:699–718, 1975.

BOUE J., HIRSCHHORN K., LUCAS M., MOS-ZER M., BAC Ch.: Aneusomies de recombinaison. Conséquences d'une inversion péricentrique d'un chromosome 3 paternel. *Ann. Pédiatr.*, **21**:567–573, 1974.

CHIYO H.A., KUROKI Y., MATSUI I., NIITSU N., NAKAGOME Y.: A case of partial trisomy 3q. *J. med. Genet.*, **13**:525–527, 1976.

FALEK A., SCHMIDT R., JERVIS G.A.: Familial de Lange syndrome with chromosome abnormalities. *Pediatrics*, **37**:92–101, 1966.

FEAR C., BRIGGS A.: Familial partial trisomy of the long arm of chromosome 3 (3q). *Arch. Dis. Child.*, **54**:135–138, 1979.

FINEMAN R.M., HECHT F., ABLOW R.C. et al.: Chromosome 3 duplication q/duplication p syndrome. *Pediatrics*, **61**:611–618, 1978.

FRANCKE U.: Clinical syndromes associated with partial duplications of chromosomes 2 and 3: dup(2p), dup(2q), dup(3p), dup(3q). *Birth Defects, Orig. Art. Ser.*, *XIV-6C*:191–217, 1978.

FRYNS J.P., VAN EYGEN M., LOGGHE N., VAN DEN BERGHE H.: Partial trisomy for the long arm of chromosome 3 [3q(q21→qter)+] in a newborn with minor physical stigmata. *Hum. Genet.*, **40**:333–339, 1978.

KONDO I., HIRANO T., HAMAGUCHI H., OHTA Y., HAIBARA S., NAKAI H., TAKITA H.: A case of trisomy 3q21→qter syndrome. *Hum. Genet.*, **46**:141–147, 1979.

MULCAHY M.T., PEMBERTON P.J., SPRAGUE P.: Trisomy 3q: two clinically similar but cytogenetically different cases. *Ann. Génét.*, **22**:217–220, 1979.

NEU R.L., BARLOW M.J., GARDNER L.I.: A case of 46,XY,t(3q−;14q+)mat. *Clin. Genet.*, **4**:158, 1973.

SALAZAR D., ROSENFELD W., JHAVERI R.C., VERMA R.S., DOSIK H.: Partial trisomy of chromosome 3 (3q12→qter) owing to 3q/18p translocation. A trisomy 3q syndrome. *Am. J. Dis. Child.*, **133**:1006–1008, 1979.

SCHWANITZ G., SCHMID R.D., GROSSE G., GRAHN-LIEBE E.: Translocation familiale 3/22 mat avec trisomie partielle 3q. *J. Génét. hum.*, **25**:141–150, 1977.

SOD R., GIORGIUTTI E., MATAYOSHI T., GELMAN de KOHAN Z., MUNOZ E.: Familial transmission of a 3q22p translocation with partial trisomy of chromosome 3 in the propositus. *J. Génét. hum.*, **26**:173–176, 1978.

STENGEL-RUTKOWSKI S., MURKEN J.D., PILAR V., DUTRILLAUX B., RODEWALD A., GOEBEL R., BASSERMANN R.: New chromosomal dysmorphic syndromes. 3. Partial trisomy 3q. *Eur. J. Pediatr.*,**130**:111–125, 1979.

WILSON G.N., HIEBER V.C., SCHMICKEL R.D.: The association of chromosome 3 duplication and the Cornelia de Lange syndrome. *J. Ped.*, **93**:783–788, 1978.

YUNIS E., QUINTERO L., CASTENEDA A., RAMIREZ E., LEIBOVICI M.: Partial trisomy 3q. *Hum. Genet.*, **48**:315–320, 1979.

Resemblance to Cornelia de Lange syndrome
Hirsutism and synophrys

Trisomy for the 3q2 region is one of the most important syndromes to have been characterized over the last few years.

The following two remarks are pertinent:

1. Eight cases of trisomy 3q2 are associated with a 3pter monosomy resulting from an "aneusomie de recombinaison" of a parental pericentric inversion. Monosomy 3p is limited to the distal part of the chromosome and the phenotype is not very different from that of trisomy 3q2 (Boué et al., 1974; Allderdice et al., 1975; Fineman et al., 1978; Mulcahy et al., 1978). The observation of Allderdice et al. (1975) involved a family with 13 possible patients. We have considered the three patients for whom a cytogenetic diagnosis, as well as a clinical description, are available.

Fifteen observations of trisomy 3q2 resulting from another mechanism are known (Neu et al., 1973; Chiyo et al., 1976; Schwanitz et al., 1977; Fineman et al., 1978; Fryns et al., 1978; Sod et al., 1978; Wilson et al., 1978; Fear and Briggs, 1979; Kondo et al., 1979; Mulcahy et al., 1979; Salazar et al., 1979; Stengel-Rutkowski et al., 1979; Yunis et al., 1979).

2. Trisomy 3q2 has often resulted in a diagnosis of Cornelia de Lange syndrome. This phenotypical similarity was already pointed out by Falek et al. in 1966, before banding techniques were available. At the end of this chapter we will discuss their common symptomatology.

GENERALIZATIONS

- Sex ratio: 1M/1F
- Mean birth weight: 3,150 g
- Crown-heel length at birth appears to be normal
- Head circumference at birth also appears to be normal

PHENOTYPE

The most characteristic signs of trisomy 3q2 are hirsutism and thick, converging eyebrows (synophrys), which have led to its comparison with Cornelia de Lange syndrome.

Craniofacial Dysmorphism

The skull is abnormal in shape, brachycephalic, acrocephalic, or asymmetric. Microcephaly is sometimes present. The fontanels are occasionally wide.

Hair implantation is abnormal, and often very low set upon the neck, where it may be inverted; it may extend toward the face and to the temporal and zygomatic regions.

The face is square-shaped.

Eyebrows are thick and converged and this synophrys is very characteristic. Lashes are long and curved. The glabella may be prominent.

The palpebral fissures are slanted upward and outward. Epicanthus and hypertelorism are frequently found, as well as ocular malformations including unilateral

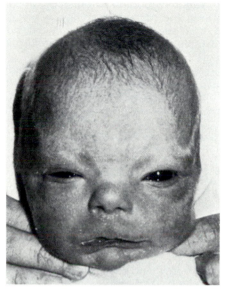

3.1

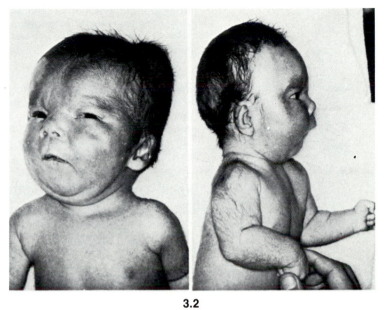

3.2

3.3

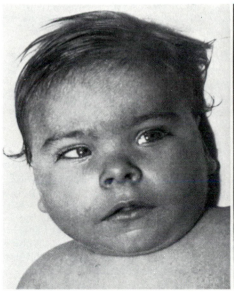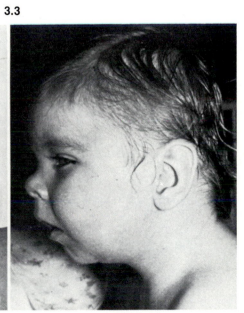

anophthalmia, glaucoma, cataracts, lens anomalies, coloboma, strabismus, obstruction of the tear ducts, nystagmus, megalocornea, and corneal opacities.

The bridge of the nose, already described, may be either prominent or flat, but is always wide. The nose is short with anteverted nares, also a highly characteristic sign.

The philtrum is prominent.

The corners of the mouth are downturned. The upper lip is thin, and these children often appear to be chewing on their lower lip.

Retromicrognathia is always present, and may contrast with the prominence of the maxillas. A high arched palate is nearly always found, and cleft palate may occasionally exist.

The ears are often low-set with no particular characteristics. Preauricular fistulae have been noted in two cases.

Neck, Thorax, and Abdomen

The neck is short, with an excess of skin, which has at times led to a diagnosis of the Bonnevie-Ulrich syndrome.

Nipples are hypoplastic and extremely wide-spaced.

Deformities of the spine, with absence of ribs and hemivertebrae visible on radiographies, are often observed. Shoulders tend to slope.

A persistence of the lanugo is observed in newborns, and hirsutism is seen in older patients. The latter is fairly generalized over the trunk and limbs.

A coccygian dimple has been reported in several cases, and a pilonidal sinus in two cases.

Omphalocele and umbilical hernias have been observed in several cases.

Limbs

Limbs are short, and fingers and toes are nearly always abnormal. Camptodactylies and syndactylies have been observed. Clinodactyly of the Vth is regularly found. A supernumerary rudimentary finger has been described in two cases. Fingernails may be hypoplastic or absent. X-ray films may show hypoplasias, or even absence of the phalanges. The thumb is often in too proximal a position.

Genitalia

In boys, hypogenitalism, with a hypoplastic scrotum and cryptorchidism, is constantly found; a micropenis is occasionally seen.

In girls, absence of the ovaries, along with a unicorn uterus, was noted in one case. In another case, bifidity of the vagina and uterus existed.

Malformations

Ocular malformations, which occur very frequently, have already been mentioned.

In one-third of the cases, a very complicated cardiac malformation is observed. This is nearly always present when the breakpoint is located in q21 or more proximal.

No congenital heart disease has been reported for the most distal trisomies whose breakpoints are in q25 or more distal.

Kidney malformations are fairly frequent and include polycystic, aplastic, or accessory kidneys. Digestive malformations have been reported in several cases, as well as one case of spina bifida. Various cerebral malformations have been described, and in particular, digitiform impressions.

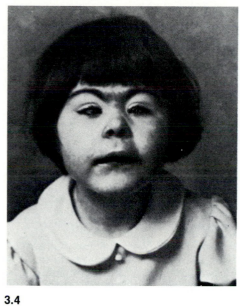

3.4

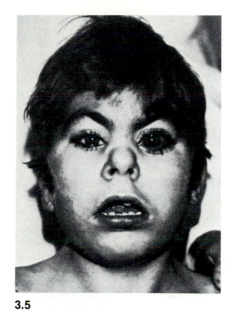

3.5

3.6

Prognosis

The prognosis depends on the size of the trisomic segment. When the trisomy involves the entire 3q2 region, cardiac malformations and early death are frequent. Birth weight is generally lower than the mean, and from birth on, signs of brain damage with respiratory distress and convulsions are present. Respiratory infections are also very frequent.

When the trisomy involves half of the 3q2 region, survival seems to be normal. Children have been observed between the ages of 1 and 11. Staturoponderal retardation is nearly always present, and the IQ is less than 50.

In two cases, the patients were trisomic only for the distal third of 3q2, and were seen at ages 13 and 19. Their somatic development was normal, with an IQ of around 50–60.

Trisomy 3q2 and the Cornelia de Lange Syndrome

The observation by Falek et al. (1966) was the first to be made of a familial chromosomal rearrangement associated with a symptomatology resembling that of Cornelia de Lange syndrome. The resemblance between trisomy 3q2 and the latter is now a classical notion. Among the signs that permit us to differentiate between these two symptoms, the following should be noted (Francke, 1978):

A low birth weight, which is always encountered in Cornelia de Lange syndrome, is not always found in trisomy 3q2

Minor differences exist in facial dysmorphism: acrocephaly and slanting of the palpebral fissures upward and outward would seem to be more characteristic of trisomy 3q2; the very striking convergence of the eyebrows and the median notch on the upper lip are more symptomatic of Cornelia de Lange syndrome

Ocular complications are especially frequent in trisomy 3q2

Congenital heart disease is also more frequent and more severe in trisomy 3q2 as is early death

Nonetheless, the similarities between these two entities is very marked, and the clinical diagnosis is not necessarily different. It is possible that cases of recurrence described in Cornelia de Lange syndrome were due to unidentified familial chromosomal rearrangements.

CYTOGENETIC STUDY

Trisomy can be due to an inversion or to a familial translocation, a de novo translocation, or a direct or inverted duplication.

The size of the trisomic segment varies as a function of the breakpoint on 3q. This can be situated in one of the following:

- q21 (or at a more proximal point). The trisomy would then concern the whole of the 3q2 region. This is the most frequently encountered situation.
- q25. Trisomy then involves half of the 3q2 region.
- q27. In one familial observation, trisomy only involved the distal third of 3q2.

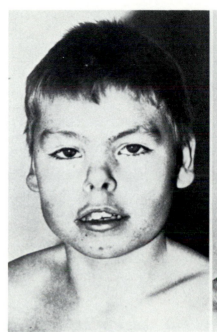 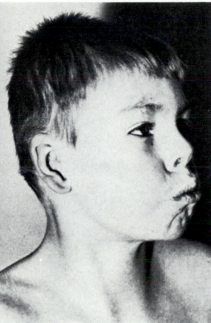 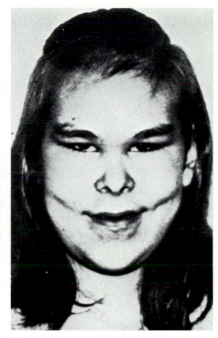

3.7

3.8

In the case of duplication/deficiency, the breakpoint on 3p is always in p25. For 3q, it is either at the proximal limit of the q2 region, in q13 or q21, or in the middle of this region, in q24 or q25.

DERMATOGLYPHICS

Dermatoglyphics do not appear to be characteristic. A high frequency of single uni- or bilateral transverse palmar creases is observed.

39

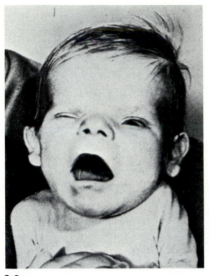

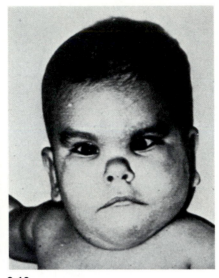

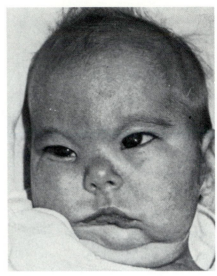

3.9

3.10

3.11

3.12

3.13

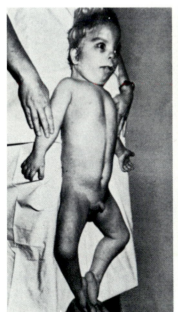

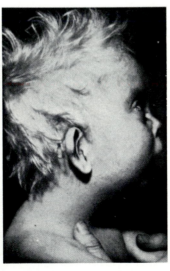

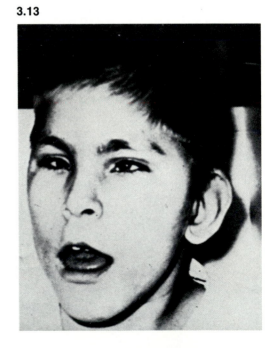

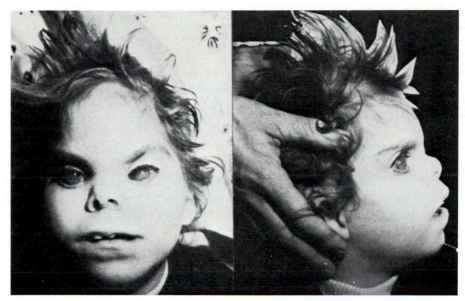

3.14

3.15

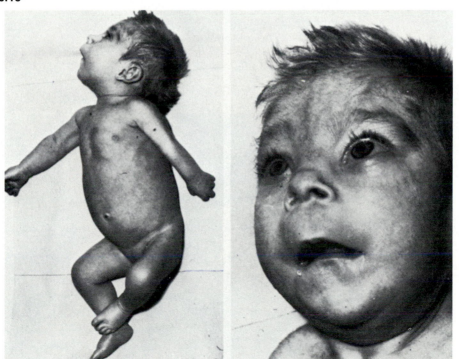

3p2 MONOSOMY

REFERENCES

FINEMAN R.M., HECHT F., ABLOW R.C. et al.: Chromosome 3 duplication q/deletion p syndrome. *Pediatrics,* **61**:611–618, 1978.

GONZALES J., LESOURD S., BRACONNIER A.: Délétion partielle du bras court du chromosome 3. À propos d'un cas. *Ann. Génét.,* **23**:119–122, 1980.

VERJAAL M., NEF J. de: A patient with a partial deletion of the short arm of chromosome 3. *Am. J. Dis. Child.* **132**:43–45, 1978.

Three cases of distal deletion, with the same breakpoint in p25, have been reported (Fineman et al., 1978; Verjaal and de Nef, 1978; Gonzales et al., 1980). In a very striking manner, the phenotype corresponding to this monosomy shows a number of similarities with that of trisomy 3q2. Though it is not possible to describe a syndrome based on three observations, the following should be noted to allow a comparison with the duplication/deficiency syndrome:

- Low birth weight
- Growth rate of less than the third percentile
- Anomalies in the shape of the skull, including asymmetry, frontal bossing, distended veins, and microcephaly
- A slightly triangular-shaped face, with a pointed chin and prominent jawbones
- Ptosis and hypertelorism
- Hypertrichosis with a low hair line, and thick, converging eyebrows
- A short, thick nose and deviated septum
- A mouth with downturned corners and thin lips, and a high arched palate
- Anomalies of the limbs and hands, including clinodactyly and a supernumerary finger and toe
- Thoracic anomalies, including pectus excavatum and scoliosis
- Hypogenitalism, vulvar anus, sacral dimple
- Preauricular sinuses
- Severe mental retardation
- No marked inner organ malformation has been noted. The three children were alive at ages 1, 18, and 22.

3.16

3.17

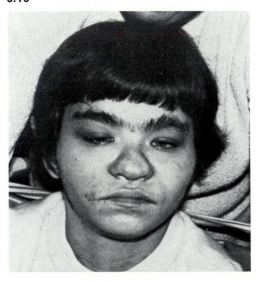
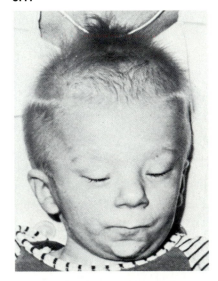

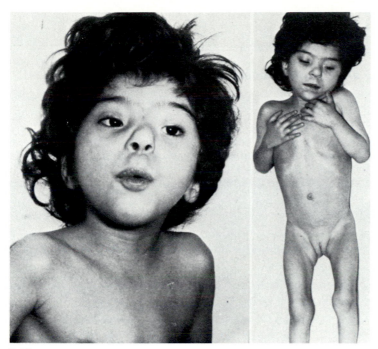 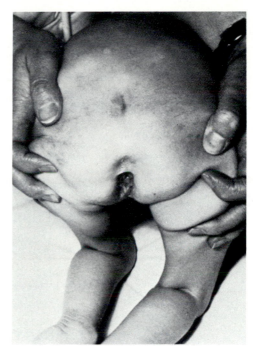

3.18

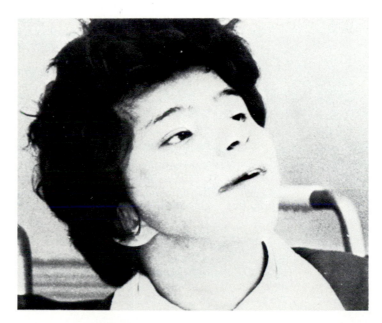

3p2 TRISOMY

Frontal bossing and temporal indentation
Large mouth
Small chin

REFERENCES

CHARROW J., COHEN M.M., MEEKER D.: Duplication 3p syndrome: report of a new case and review of the literature. *Am. J. med. Genet.*, **8**:431–436, 1981.

FRANCKE U.: Clinical syndromes associated with partial duplications of chromosomes 2 and 3: dup(2p), dup(2q), dup(3p), dup(3q). *Birth Defects: Orig. Art. Ser., XIV-6C*:191–217, 1978.

HERSH J.H., GREENSTEIN R.M., PERKINS J.C., REARDON P.C.: A case of 47,XY, +der(15),t(3;15) (p25;q11)pat presenting as partial 3p trisomy syndrome with multiple joint contractures. *J. med. Genet.*, **17**:396–398, 1980.

PARLOIR C., FRYNS J.P., VAN DEN BERGHE H.: Partial trisomy of the short arm of chromosome 3 (3p25→3pter). *Hum. Genet.*, **47**:239–244, 1979.

RETHORE M.O., LEJEUNE J., CARPENTIER S., PRIEUR M., DUTRILLAUX B., SERINGE Rh., ROSSIER A., JOB J.C.: Trisomie pour la partie distale du bras court du chromosome 3 chez trois germains. Premier exemple d'insertion chromosomique: ins(7;3) (q31;p21 p26). *Ann. Génét.*, **15**:159–165, 1972.

SCHINZEL A., HANSON J.W., PAGON R.A., HOEHN H., SMITH D.W.: Trisomy 3 (p23-pter) resulting from maternal translocation, t(3;4) (p23;q35). *Ann. Génét.* **21**:168–171, 1978.

Since the first publication by Rethoré et al. (1972), 14 children have been found to be trisomic for 3p2. The first 10 were the object of reviews by Francke (1978) and Schinzel et al. (1979). The four other observations were reported by Parloir et al. (1979, two patients), Mersch et al. (1980), and Charrow et al. (1981).

GENERALIZATIONS

The sex ratio would seem to weigh heavily in favor of boys. Out of 14 cases, there were only two girls and one case of sexual ambiguity with an XY sex complement.

- Mean birth weight: 3,235 g
- Mean crown-heel length at birth: 48.2 cm
- Mean head circumference at birth: 35.1 cm

PHENOTYPE

Craniofacial Dysmorphism

The skull is normal in volume. The forehead is high, with frontal and parietal bossing, giving the very characteristic appearance of temporal indentation. The face is square with heavy cheeks.

Palpebral fissures are normally oriented. The eyes are often small, and a fairly pronounced hypertelorism is present. The eyebrows are scant toward the interior.

The nose is short and rather thick with a fleshy tip.

The philtrum is short and prominent. In several cases, a uni- or bilateral harelip and a cleft palate are present.

The mouth is often large, with downturned corners.

The chin is frequently small and receding.

The ears are poorly folded, with no particular characteristics, and are sometimes low-set.

Neck, Thorax, and Abdomen

In infants, the neck is short with cutaneous folds.

Minor anomalies of the thorax have been reported including pectus excavatum, hypoplastic nipples which are abnormally wide-spaced, and minor radiologic anomalies of the vertebrae, ribs, and sternum.

Genitalia

Apart from the aforementioned case of sexual ambiguity, a small penis, hypoplastic scrotum, and testicular ectopy are nearly always observed.

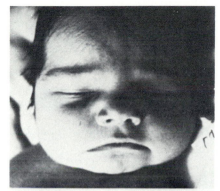

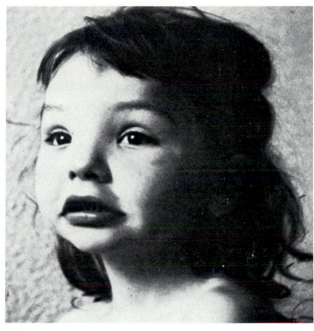

3.20

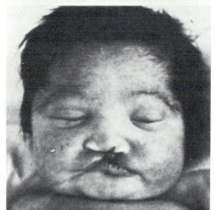

3.19

3.21

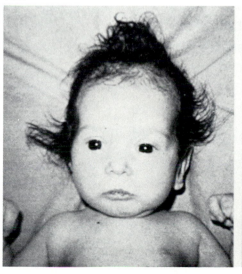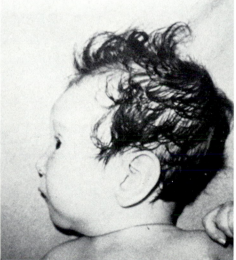

Limbs

Minor anomalies of the limbs include short chubby hands, camptodactyly, cutaneous syndactyly, and, in particular, the existence of a deep crease between the first and second toes.

Malformations

Eleven out of fourteen children had congenital heart disease, including patent ductus arteriosus, atrial or ventricular septal defects.

Other malformations are rare and are mainly digestive (apart from cleft palate and labial cleft, already mentioned).

Mental Retardation

The IQ is less than 50 in those patients who have been evaluated.

Prognosis

Four children died soon after birth. The other patients have been examined at various ages, the oldest being 19.

Somatic development is variable, sometimes normal, and other times retarded. With age, facial dysmorphism is modified by a lessening of the prominence of frontal bosses and roundness of the cheeks.

CYTOGENETIC STUDY

In all cases, trisomy results from a parental rearrangement, most often a translocation. It concerns the p2 region, with the breakpoint varying (depending on the authors' interpretations) between p21 and p25. The second chromosome involved in the parental translocation is variable. Associated monosomy is never important and the phenotype is very similar in all reported cases. The trisomic segment common to all the observations is formed by bands p25 and p26.

DERMATOGLYPHICS

An excess of whorls is noted.

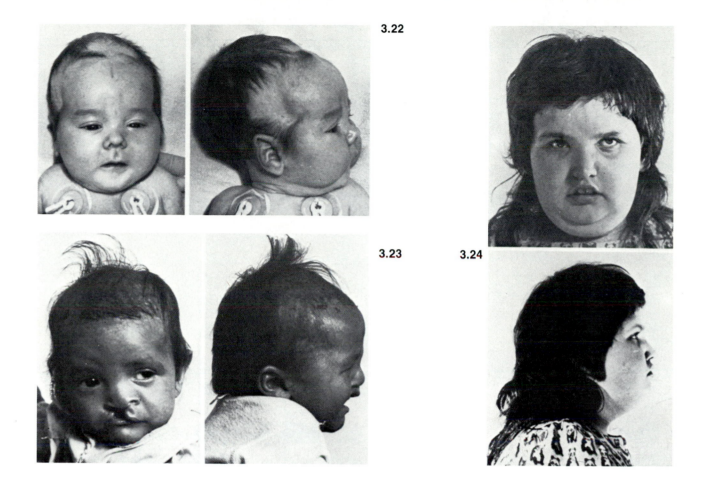

3.22

3.23

3.24

3.25

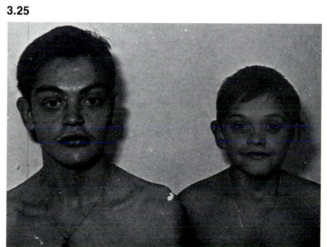

CHROMOSOME 4

Approximately 15 genes have been assigned to chromosome 4, including the MNSs complex, a defector from chromosome 2, and an atypical form of phenylketonuria.

Several syndromes are due to imbalance of chromosome 4, among which are the very classical monosomy 4p, or Wolf syndrome, and its counterpart, trisomy 4p. Terminal trisomy 4q had already been known, but the corresponding terminal 4q monosomy has only recently been recognized.

Observations of intercalary deletions of 4q are presently still too disparate to permit characterization of specific syndromes.

A polymorphism of the centromere of 4 is observed in Q-banding.

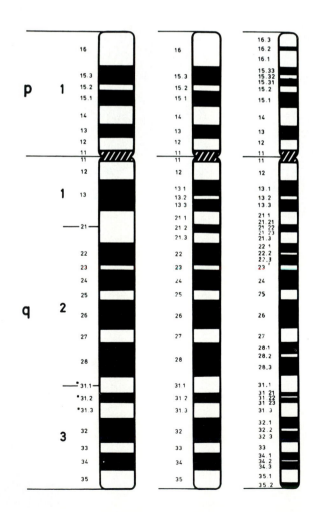

GENE MAP

	PLG	(L)	Plasminogen (17335) F
	DO	(I)	Dombrock blood group (11060) F
(4q)	SF		Stoltzfus blood group (11180) F
(4q)	TYS		Sclerotylosis (18160) F
	ASOD	(P)	Anterior segment mesenchymal dysgenesis (10725) F
4p12-4q13	PGM2		Phosphoglucomutase-2 (17200) S
(4q11-4q13)	DG11	(P)	Dentinogenesis imperfecta-1 (12549) F
4q11-4q13	GC		Group-specific component (13920) F, Fc
(4q11-4q13)	ALB		Albumin (10360) F (linked to GC), A
(4q11-4q13)	AFP	(L)	Alpha-fetoprotein (10415) H
4pter-4q12	PEPS		Peptidase S (17025) S
4pter-4q21	PPAT	(P)	Phosphoribosylpyrophosphate amidotransferase (17245) S
	DHPR	(P)	Quinoid dihydropteridine reductase (26161) S
4q28-4q31 (EM)	MN		MN blood group (11130) F, Fc,AAS
	Ss		Ss blood group (11174) F,Fc, AAS

MORPHOLOGY AND BANDING

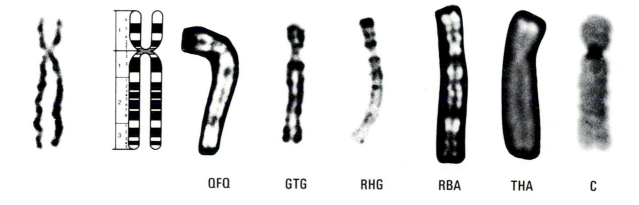

QFQ   GTG   RHG   RBA   THA   C

4p MONOSOMY

Severe growth retardation
Severe mental deficiency
Microcephaly
"Greek warrior helmet"

REFERENCES

BERGER R.: Délétion du bras court du chromosome 4. *Nouv. Presse méd.*, 1:809–810, 1972.

CHAVIN-COLIN F., TURLEAU C., LIMAL J.M., GROUCHY J. de: Anneau du chromosome 4. II. Sans dysmorphie faciale. *Ann. Génét.*, 20:105–109, 1977.

GERMAN J., LEJEUNE J., MACINTYRE M.N., GROUCHY J. de: Chromosomal autoradiography in the cri du chat syndrome. *Cytogenetics*, 3:347–352, 1964.

GUTHRIE R.D., AASE J.M., ASPER A.C., SMITH D.W.: The 4p– syndrome. A clinically recognizable chromosomal deletion syndrome. *Amer. J. Dis. Child.*, 122:421–426, 1971.

LAZJUK G.I., LURIE I.W., OSTROWSKAJA T.I., KIRILLOVA I.A., NEDZVED M.K., CHERSTVOY E.D., SILYAEVA N.F.: The Wolf-Hirschhorn syndrome. II. Pathologic anatomy. *Clin. Genet.*, 18:6–12, 1980.

LEAO J.C., BARGMANN G.L., NEU R.L., KAJII T., GARDNER L.I.: New syndrome associated with partial deletion of short arm of chromosome 4. *JAMA*, 202:434–437, 1967.

LURIE I.W., LAZJUK G.I., USSOVA Y.L., PRESSMAN E.B., GUREVICH D.B.: The Wolf-Hirschhorn syndrome. *Clin. Genet.*, 17:375–384, 1980.

MILLER O.J., BREG W.R., WARBUTON O., MILLER D.A., DE CAPOA A., ALLLDERDICE P.W., DAVIS J., KLINGER H.D., McCILVRAY E., ALLEN F.H.: Partial deletion of the short arm of chromosome n° 4 (4p–). Clinical studies in five unrelated patients. *J. Pediat.*, 77:792–801, 1970.

RETHORE M.O.: Chromosome deletions and ring chromosome syndromes. In N.C. MYRIANTHOPOULOS, ed: *Handbook of Clinical Neurology. Vol. 31, part II, Congenital malformations of the brain and skull.* Amsterdam, North Holland publ., pp. 549–620, 1977.

WOLF U., PORSCH R., BAITSCH H., REINWEIN M.: Deletion on short arms of a B chromosome without «cri du chat» syndrome. *Lancet*, i:769.

WOLF U., REINWEIN H., PORSCH R., SCHROTER R., BAITSCH H.: Defizienz an den kurzen Armen eines Chromosoms n° 4. *Humangenetik*, 1:397–413, 1965.

YOUNG R.S.K., ZALNERAITIS E.L.: Neurological and neuropathological findings in ring chromosome 4. *J. med. Genet.*, 17:487–490, 1980.

In 1963 Lejeune et al. described the "cri du chat" syndrome as resulting from partial deletion of the short arm of a B chromosome, and arbitrarily decided that this chromosome was chromosome number 5.

The following year German et al. (1964) demonstrated that it was possible to distinguish between chromosomes 4 and 5 by autoradiography after isotopic labeling.

In 1965 Wolf et al. described a child, carrier of a partial deletion of the short arm of a B chromosome, whose phenotype did not correspond to the "cri du chat" syndrome. Autoradiography confirmed that the abnormal chromosome was a 4. The 4p monosomy syndrome characterized after this first observation is relatively frequent (Leao et al., 1967; Miller et al., 1970; Guthrie et al., 1971; Berger, 1972). Nearly 120 cases have been reported in the literature (Rethoré, 1977; Lurie et al., 1980; Lazjuk et al., 1980).

GENERALIZATIONS

The deletion arises *de novo* in 90 percent of the cases. A parental mosaicism or translocation is responsible for 10 percent of the cases.

- Sex ratio: 1M/2F
- Mean parental ages in *de novo* cases:
    maternal: 27.2 years
    paternal: 31 years
- Pregnancy: postmaturity in 40 percent of the cases; prematurity in 25 percent of the cases
- Mean birth weight: 2,000 g
- Mean crown-heel length at birth: 44 cm
- Mean head circumference at birth: 30.7 cm

PHENOTYPE

Growth retardation, distinct at birth, accentuates with age. The trunk is long and the limbs are slender. Facial dysmorphism is very characteristic.

Craniofacial Dysmorphism

Microcephaly is constant and very pronounced. It is accompanied by dolichocephaly, with protruding frontal bossing, and median defect of the scalp (in one case out of ten). The forehead is high and, when the infant cries, marked by deep wrinkles that attest to the hypertonia of the subcutaneous muscles. The glabella is large with an occasional hemangioma. The eyebrows, sparse toward the interior, accentuate this impression of broadness. The nose is noteworthy: the edges, rectilinear and

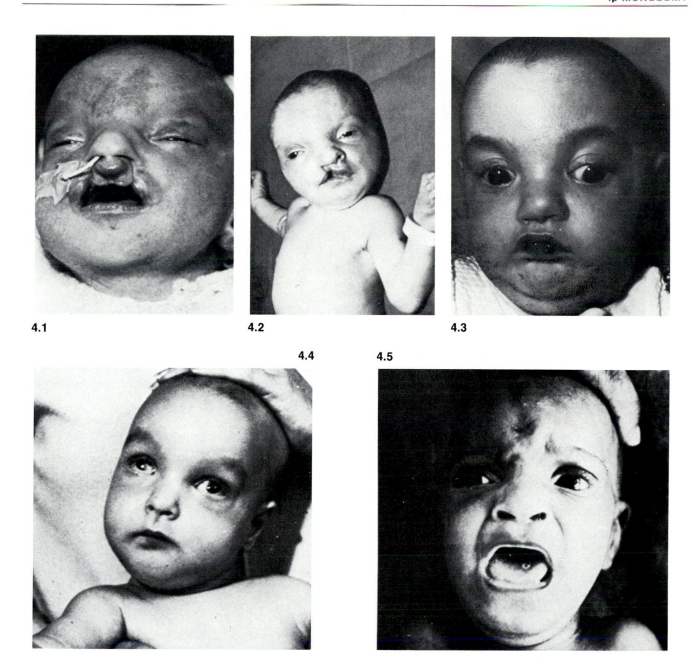

4.1

4.2

4.3

4.4

4.5

parallel, prolong the eyebrow line; the nose bridge is as wide as the apex, which is square, flattened, and occasionally beaked. This aspect evokes a "Greek warrior helmet."

Palpebral fissures are horizontal or slanted downward and outward. Hypertelorism is present and often bilateral epicanthus as well. The eyeballs appear to protrude because of hypoplasia of the eye sockets. The following are often noted: ptosis or a unilateral retraction of the upper eyelid, Marcus Gunn's sign in some cases, strabismus, nystagmus, myopia, and an obliteration of the lacrimal ducts. More severe ocular malformations are observed in one-third of the cases: iridoschisis, pupilar ectopia, cataract, etc.

The upper lip is short. Harelip and especially cleft palate are frequent. The philtrum is very characteristic, deep and narrow, with strongly marked pillars that give the impression of emerging from the nostrils. The nares are triangular.

The mouth is downturned.

The chin is small and receding.

The ears are normally set. The helix is flat, the antitragus protruding, the lobe adherent. A preauricular dimple is frequently noted.

Neck, Thorax, and Abdomen

The neck is long and slender.

The trunk is elongated.

Skeletal deformations are frequent: scoliosis, supernumerary ribs and vertebral malformations, spina bifida and a bifid sacrum, dislocation of the hip. A coccygeal dimple is often present.

Limbs

The limbs are slender with dimples on the elbows and knees.

The fingers are long and thin with pointed tips and narrow, convex fingernails without cuticula. There are supernumerary flexion creases on the first and second phalanges. The thumbs are proximally implanted.

The toes are very often defectively implanted; the big toe is very long.

Genitalia

In males, hypospadias (first degree) is almost constantly found. The testicles are often ectopic. In females the genital organs are usually normal, though some malformations have been noted (absence of the uterus and of the vagina).

Malformations

As has been seen, defects in closure are frequent: defects of the scalp, harelip or cleft palate, coloboma, preauricular dimple, dimple of the coccygeal fissure, and hypospadias.

In more than half the cases there is cardiac malformation, mainly of the septum: interventricular septum defect, atrial septum defect, patent ductus arteriosus, and others.

Central nervous system anomalies have been observed, but their frequency is not well known.

Renal anomalies have been reported in some cases.

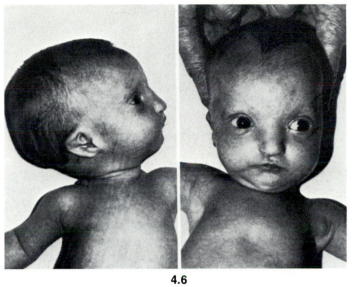

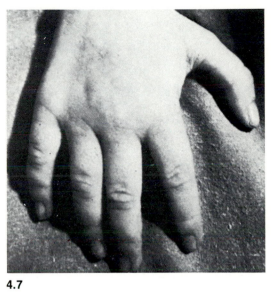

4.6

4.7

4.8

4.10

4.9

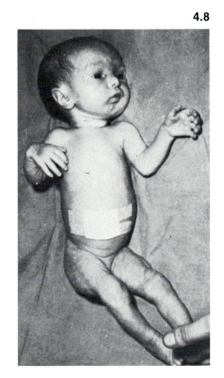

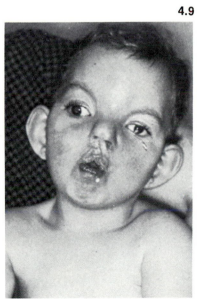

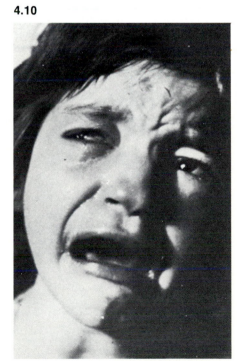

Mental Retardation

Mental retardation is very pronounced and is probably more severe in this than in most other chromosomal diseases. The IQ is below 20. The children have poor gesticulation and are indifferent to their environment. Seizures are habitual, with modification of the EEG.

Prognosis

The life expectancy of the patients is still not well known. Death can occur at variable ages in childhood. Several individuals in their twenties are known.

CYTOGENETIC STUDY

The deletion seems to affect the distal portion of 4p—more precisely the 4p16 band. The size of the deleted fragment would seem to be much less variable than in the "cri du chat" syndrome.

Ring chromosomes 4 represent a rather peculiar situation. Fifteen cases have been reported (see Young and Zalneraitis, 1980). The corresponding phenotype was analyzed by Chavin-Colin et al. (1977). In one-third of the cases, the ring is accompanied by loss of band p16 and the typical 4p-phenotype. In the other cases, the clinical syndrome is variable. The most notable sign is the existence of severe malformations of the limbs (radial aplasia, or club foot) observed in several cases.

DERMATOGLYPHICS

Dermal ridges remain hypoplastic in the infant for some time. Palmar creases are normal in three-fourths of the cases. In one-fourth of the cases there is a single transverse palmar crease.

The axial triradius is in t or t′ and terminates in 11 or 13.

Thenar and hypothenar patterns are rare.

The subdigital main line A often terminates in 1, a very rare occurrence in normal individuals. The mean transversality index is 23.

The digital patterns show an excess of arches.

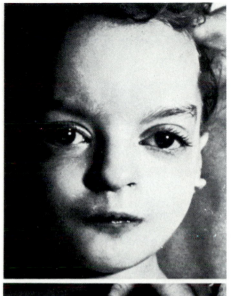

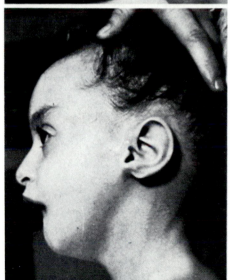

4.11

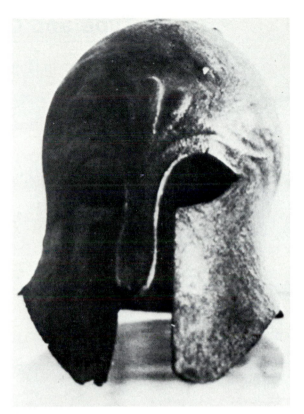

4.12
4.13

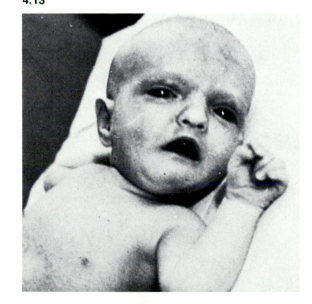

Type and Countertype

Symptoms in Common

Mental retardation
Microcephaly
Skeletal malformations
Hypotonia

Symptoms in Opposition

	Monosomy	*Trisomy*
Forehead	Prominent	Flat
Glabella	Aplastic	Prominent
Nasal bones	Prominent	Hypoplastic
Nasal apex	Square	Round
Chin	Receding	Protruding
Neck	Long	Short
Dermatoglyphics	Excess of arches	Excess of whorls

4p TRISOMY

Aplasia of the nasal bones
"Boxer nose" in adults

REFERENCES

ANDRE M.J., AURIAS A., BERRANGER P. de, GILLOT F., LEFRANC G., LEJEUNE J.: Trisomie 4p de novo par isochromosome 4p. *Ann. Génét.*, **19**:127–131, 1976.

CRANE J., SUJANSKY E., SMITH A.: 4p trisomy syndrome: report of 4 additional cases and segregation analysis of 21 families with different translocations. *Am. J. med. Genet.*, **4**:219–229, 1979.

DALLAPICCOLA B., MASTROIACOVO P., GANDINI E.: Centric fission of chromosome n° 4 in the mother of two patients with trisomy 4p. *Hum. Genet.*, **31**:121–125, 1976.

DALLAPICCOLA B., MASTROIACOVO P., MONTALI E., SOMMER A.: Trisomy 4p: five new observations and overview. *Clin. Genet.*, **12**:344–356, 1977.

GIRAUD F., MATTEI J.F., MATTEI M.G., AYME S., BERNARD R.: La trisomie 4p. A propos de 3 observations. *Humangenetik*, **30**:89–108, 1975.

GUSTAVSON K.H., FINLEY S.C., FINLEY W.H., JALLING B.: A 4-5/21-22 chromosomal translocation associated with multiple congenital anomalies. *Acta paediat. scand.*, **53**:172–181, 1964.

MORTIMER J.G., CHEWINGS W.E., GARDNER R.J.M.: A further report on a kindred with cases of 4p trisomy and monosomy. *Hum. Hered.*, **30**:58–61, 1980.

RETHORE M.O., DUTRILLAUX B., GIOVANELLI G., FORABOSCO A., DALLAPICCOLA B., LEJEUNE J.: La trisomie 4p. *Ann. Génét.*, **17**:125–128, 1974.

RETHORE M.O.: Syndromes involving chromosomes 4, 9, and 12. *In* J. J. YUNIS, ed: *New Chromosomal Syndromes.* New York, Academic Press, pp. 119–183, 1977.

WILSON M.G., TOWNER J.W., COFFIN G.S., FORSMAN I.: Inherited pericentric inversion of chromosome n° 4. *Amer. J. hum. Genet.*, **22**:679–689, 1970.

The first observation of cytologically proven 4p trisomy was reported by Wilson et al. in 1970. The phenotypic concordance suggests that an earlier observation by Gustavson et al. (1964) also concerned 4p trisomy. The clinical syndrome was delineated in 1974 by Rethoré et al. based on seven cases. Nearly 40 cases are currently known (Giraud et al., 1975; Rethoré, 1977; Dallapiccola et al., 1977; Crane et al., 1979; Mortimer et al., 1980).

GENERALIZATIONS

In almost every case trisomy results from a parental translocation or pericentric inversion. Exceptionally, it may arise *de novo* as a consequence of a more or less complex rearrangement.

- Sex ratio: 1M/1F
- Mean birth weight: 2,680 g
- Mean crown-heel length at birth: 47.7 cm

PHENOTYPE

Craniofacial Dysmorphism

In very young infants the cranium is small and round, occasionally asymmetrical. The face is puffy and has a moonlike appearance. The forehead is low and flat and covered with abundant down. The nose is the most characteristic feature of the dysmorphism: owing to aplasia of the nasal bones, it is reduced to its apex, which is small and spherical. This has been verified in several cases by X-ray films. The protruding glabella converges with the orbital ridges and overhangs the nose. The eyebrows are very dense toward the interior and are confluent.

As the child grows, the form of the nose is modified and lengthened. The well-defined bridge is continuous with the glabella but without a nasofrontal angle. The round and fleshy tip may persist, giving a very characteristic "boxer nose" aspect.

The palpebral fissures are horizontal or slanted downward and outward. Hypertelorism is noted, and occasionally blepharophimosis. The eyeballs are small. Strabismus is frequent.

The mouth is broadly fissured. The upper lip is long and slightly protuberant, with a poorly defined philtrum. The palate is high-arched. The teeth are decayed and poorly inserted, with occasional protrusion of the upper incisors.

The chin, massive and round, contrasts with the angles of the jawbone, which are slight.

The ears are low-set. The external ear is large, the anthelix protruding, and the helix strongly folded in its upper horizontal portion.

Neck, Thorax, and Abdomen

The neck is short. The hairline is low. The nipples are abnormally widespaced. The spinal column is often deformed by scoliosis.

Limbs

Irreducible flexion of the digits, bifid thumb, supernumerary digits, clubhand, varus of the feet, hallux valgus, club and "rockerbottom" feet have been reported.

Genitalia

Shortened penis, hypospadias, and cryptorchidism are reported in the male. The genital organs in the female are normal in appearance.

Malformations

Malformations, which are very diverse, mainly affect the skeleton: scoliosis (already noted); costal hypoplasia; anomalies of the vertebrae, the hips, the iliac alae, and the sacrum; and significantly retarded ossification.

More rarely: cardiopathy, harelip, atresia of the larynx, ocular disorders, cerebral, renal, and digestive malformations.

Mental Retardation

Mental deficiency is fairly marked. The IQ is about 50 at two years of age. It subsequently deteriorates to 30 or even 10 in older subjects. An autistic attitude, a state of stupor, and seizures are sometimes observed.

Prognosis

Nutritional growth retardation usually persists, and stature is consistently below normal.

Prognosis for life is relatively harsh. More than one-third of the patients die in childhood, mainly as a result of pneumopathies.

CYTOGENETIC STUDY

A parental rearrangement is responsible for the trisomy in nine out of ten cases. In half of these cases, this involves translocation of 4p onto an acrocentric, usually a 22. These cases may be considered as pure 4p trisomies. The other cases involve reciprocal translocations or pericentric inversions leading to associated monosomy or trisomy.

Two observations merit special attention due to the rareness in humans of the mechanisms in question. One of them involves the de novo formation of an isochromosome i(4p) associated with a translocation t(1;4) (André et al., 1976). The other involves a centromeric fission found in the mother of two children who were 4p trisomic through malsegregation of the small acrocentric corresponding to 4p (Dallapiccola et al., 1976). These two rearrangements concern the centromere of chromosome 4. It is possible that this structure has specific properties.

DERMATOGLYPHICS

The dermatoglyphics are not truly characteristic. The following can be noted:

- High frequency of a single transverse palmar crease, or its equivalent
- Inconsistent axial triradius high in t'
- Decrease in the index of transversality
- Excess of whorls on the digital pads

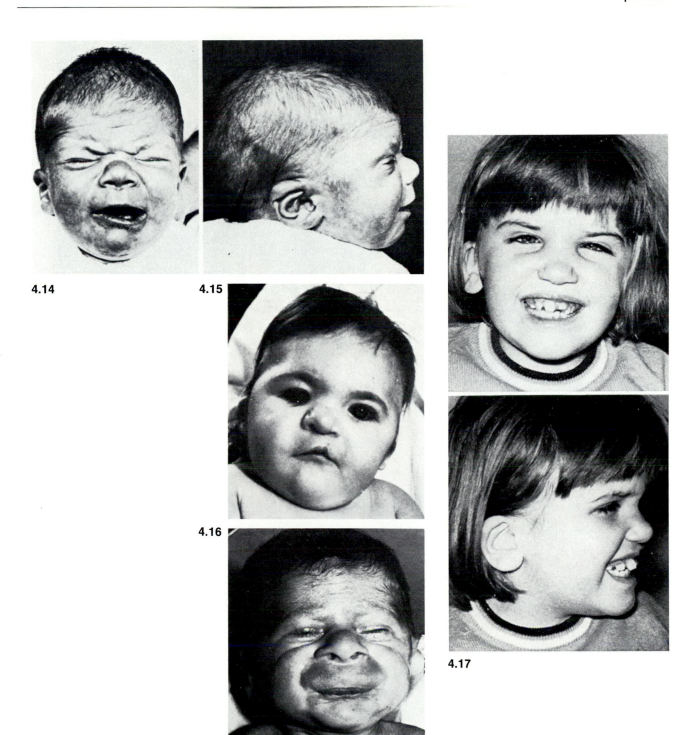

4.14

4.15

4.16

4.17

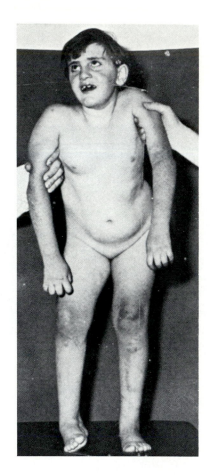

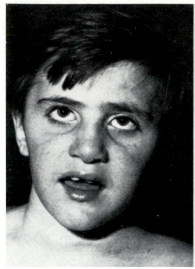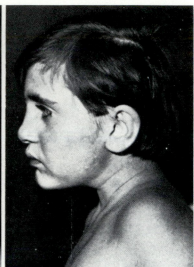

4.18

4.19

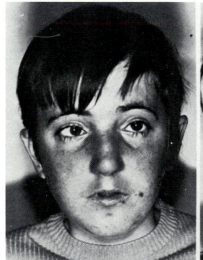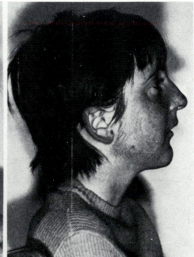

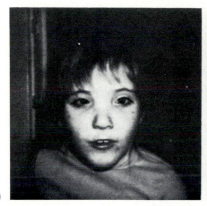 

4.20

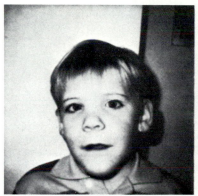 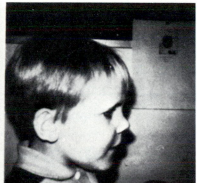

4.21

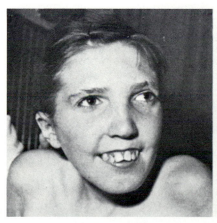 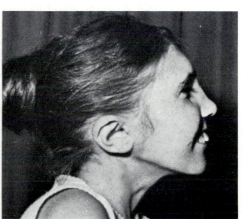

4.22

4q2 & 3 TRISOMY

Absence of nasal bridge
"Cul de poule" or pursed mouth

REFERENCES

ANDRLE M., ERLACH A., RETT A.: Partial trisomy 4q in two unrelated cases. *Hum. Genet.,* **49**:179–183, 1979.

BONFANTE A., STELLA M., ROSSI G.: Partial trisomy 4q: two cases resulting from a familial translocation t(4;18) (q27;p11). *Hum. Genet.,* **52**:85–90, 1979.

DUTRILLAUX B., LAURENT C., FORABOSCO A., NOEL B., SUERINC E., BIEMONT M.C., COTTON J.B.: La trisomie 4q partielle: à propos de trois observations. *Ann. Génét.,* **18**:21–27, 1975.

RETHORE M.O.: Syndromes involving chromosomes 4, 9, and 12. *In* J. J. YUNIS, ed: *New Chromosomal Syndromes,* New York, Academic Press, pp. 119–183, 1977.

STOLL C., ROTH M.P.: Partial 4q duplication due to inherited der(13),t(4;13) (q26;q34)mat in a girl with a deficiency of factor X. *Hum. Genet.,* **53**:303–304, 1980.

SURANA R.B., CONEN P.E.: Partial trisomy 4 resulting from a 4/18 reciprocal translocation. *Ann. Génét.,* **15**:191–193, 1972.

YUNIS E., GIRALDO A., ZUNIGA R., EGEL H., RAMIREZ E.: Partial trisomy 4q. *Ann. Génét.,* **20**:243–248, 1977.

The first cytologically demonstrated case was reported in 1972 by Surana and Conen. The 4q trisomy syndrome was delineated in 1975 by Dutrillaux et al. based on four observations in the literature and three personal observations. Some 29 cases are now known (ref. in Rethoré et al., 1977; Yunis et al., 1977; Bonfante et al., 1979; Andrle et al., 1979; Stoll and Roth, 1980).

GENERALIZATIONS

In the majority of cases trisomy is a consequence of parental translocation.

- Sex ratio: 1M/1F
- Maternal age: advanced in two cases arising *de novo*
- Mean birth weight: 2,700 g
- Mean crown-heel length at birth: 47 cm
- Mean cranial circumference at birth: 33 cm

PHENOTYPE

Craniofacial Dysmorphism

Microcephaly is constant and is often accompanied by brachycephaly. The forehead is receding, with protrusion of the metopic suture.

The palpebral fissures are narrow and slanted slightly downward and outward. Ocular anomalies are often noted: blepharophimosis, epicanthus, ptosis, microphthalmia.

The nasal bridge is shallow or totally absent. The nose is long and extends the forehead line.

The mouth is the most characteristic feature: the upper lip is very short but protruding, with pronounced philtrum pillars and sometimes a median raphe. It overlaps the lower lip, which rises in its median portion, giving the appearance of a "cul de poule" or pursed mouth.

The chin is small and slightly receding.

The ears are characteristic: they are large, round, and low-set with posterior rotation; the helix is strongly folded; the anthelix is protruding and continuous with a rim on the antitragus; the tragus and the lobe are poorly developed.

Neck, Thorax, and Abdomen

The neck is short, sometimes webbed or crowded with cutaneous folds.
The nipples are wide-spaced.
Hernias or diastasis of the musculi recti abdomini have been reported.

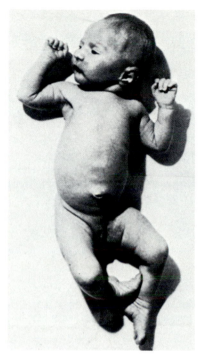

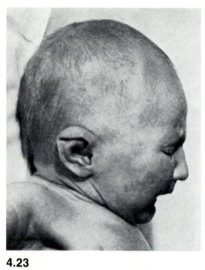

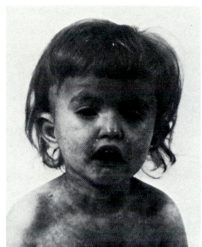

4.23

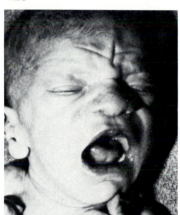

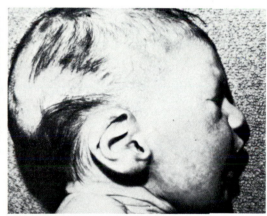

4.24

4.25

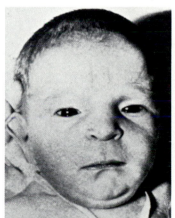

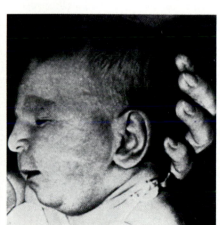

Limbs

Anomalies of the hands have been noted in several cases: digitalization of the thumb, bifid thumb, absence of the thumb, and bifid index.

Genitalia

Cryptorchidism is present in all of the males.

Malformations

Cardiac malformations are those most frequently found. They occur in half of the cases, and involve coarctation of the aorta, atrial septal defect, and Fallot's trilogy. Kidney deformities are also found frequently, including horseshoe kidney, renal hypoplasia, dilatation of the ureter, vesico-ureteral reflux, and hydronephrosis.

Mental Retardation

Mental retardation is severe, with an IQ of less than 50.

Prognosis

Statural retardation is very clear-cut, but ponderal retardation is less so. One-fourth of the patients died during their first year. The age at examination was variable, with one of the patients observed at age 42.

CYTOGENETIC STUDY

In most cases, trisomy is the result of a parental translocation. In several cases, this occurs with 18q and in one case, with 18p.

According to observations, trisomy involves either the two regions, 4q2 and 4q3, or only the latter. The phenotype would seem to be determined mainly by trisomy for the 4q3 region. Survival does not appear to depend on the size of the trisomy.

DERMATOGLYPHICS

These do not appear to be characteristic. A single transverse palmar crease has been reported in several cases.

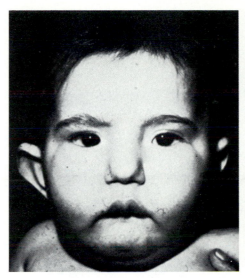 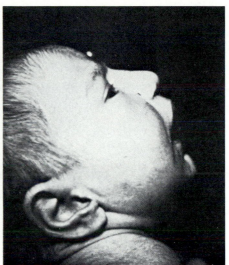

4.26

4.27

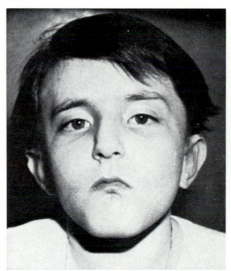 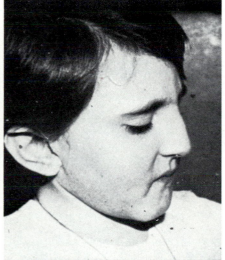

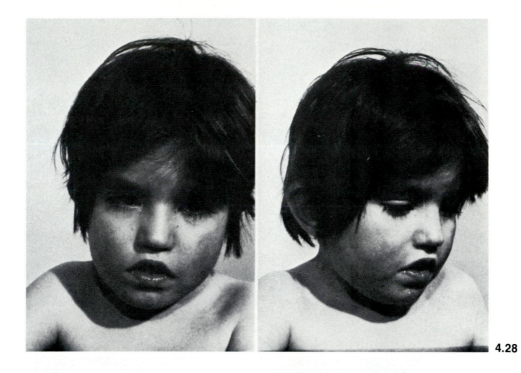

4.28

4.29

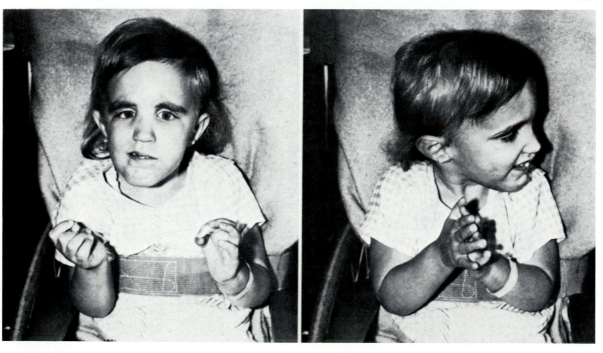

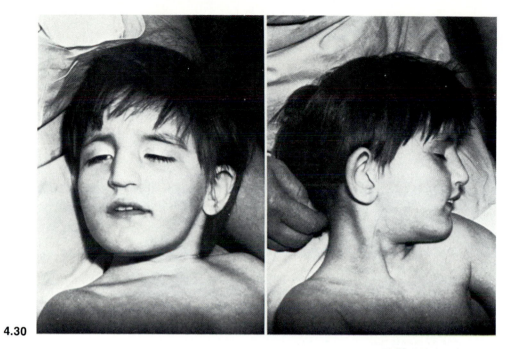

4.30

4.32

4.31

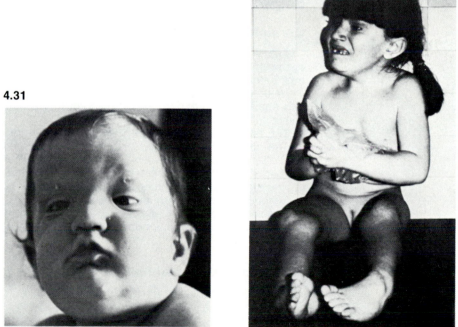

4q3 MONOSOMY

Abnormal shape of the skull
Short nose with flat bridge
Cleft lip/palate
Micrognathia
Oropharyngeal incoordination

REFERENCES

GOLBUS M.S., CONTE F.A., DAENTL D.L.: Deletion from the long arm of chromosome 4 (46,XX,4q−) associated with congenital anomalies. *J. med. Genet.,* **10**:83–85, 1973.

MITCHELL J.A., PACKMAN S., LOUGHMAN W.D., FINEMAN R.M., ZACKAI E., PATIL S.R., EMANUEL B., BARTLEY J.A., HANSON J.W.: Deletions of different segments of the long arm of chromosome 4. *Am. J. med. Genet.,* **8**:73–89, 1981.

OCKEY C.H., FELDMAN G.V., MACAULAY M.E., DELANEY M.J.: A large deletion of the long arm of chromosome n° 4 in a child with limb abnormalities. *Arch. Dis. Childh.,* **42**:428–434, 1967.

RETHORE M.O., COUTURIER J., MSELATI J.C., COCHOIS B., LAVAUD J., LEJEUNE J.: Monosomie 4q32.1→4qter survenue *de novo* chez un nouveau-né multimalformé. *Ann. Génét.,* **22**:214–216, 1979.

The first observation of a 4q monosomy (identified by isotopic banding) was that of Okey et al. (1967). The first observation by G-banding was that of Golbus et al. (1973). At present, 12 cases are known (ref. in Rethoré et al., 1979; Mitchell et al., 1981).

GENERALIZATIONS

Chromosomal rearrangement is nearly always *de novo*.

- Sex ratio: 5M/7F
- Mean parental ages (*de novo* cases)
    maternal age: 26.3 years
    paternal age: 29.6 years
- Mean birth weight: 2,880 g
- Mean crown-heel length at birth: 48 cm

PHENOTYPE

Craniofacial Dysmorphism

The skull is nearly always abnormal in shape, though no characteristic anomalies are seen. Microcephaly, occipital prominence, cranial asymmetry, or large sutures may be observed.

Palpebral fissures are either horizontal or slant upward and outward. Hypertelorism, epicanthus, with dysplasia of the internal canthus, and strabismus are occasionally seen.

The bridge of the nose may be flattened "en coup de hache."

The nose is short and snubbed, with anteverted nares.

Cleft palate is nearly always present and, in half of the cases, is combined with a cleft lip. The form of the nose is modified in such cases.

The mouth is small with an everted lower lip.

Severe microretrognathia is present.

The ears are often low-set. Their morphology may be normal, but usually they have a very characteristic appearance with a pointed helix which gives the impression of faunlike ears.

Neck, Thorax, and Abdomen

In general, the neck is normal, though it may be thickened by cutaneous folds.

The thorax may be deformed, either funnel or barrel chest.

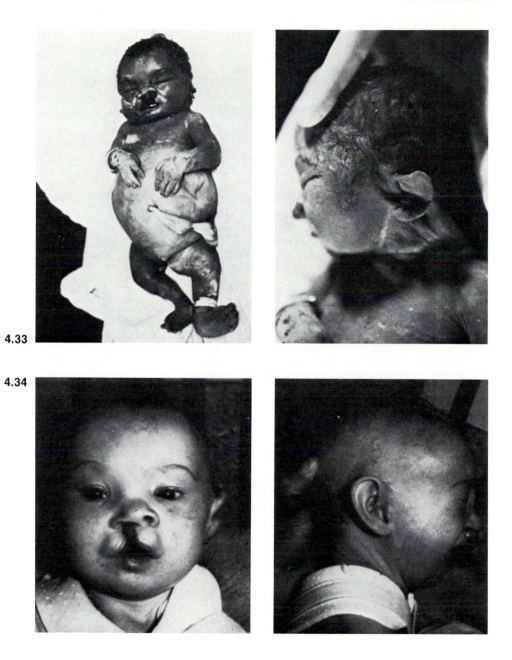

4.33

4.34

Limbs

Clinodactyly of the Vth with absence of one or two flexion creases is nearly always found. In addition, deformities of the other fingers, cutaneous syndactylies, high implantation of the thumb, an absence of a metacarpus or a phalange, and duplication of a fingernail may be observed.

The toes are almost always poorly implanted.

Genitalia

External genitals are usually normal.

Complications

Congenital heart disease is very frequent (in 10 out of 12 patients). It may include Fallot's tetralogy, ventricular or atrial septal defect, and patent ductus arteriosus.

Kidney anomalies are also frequently found.

Severe anomalies of the radial axis have been reported (Ockey et al., 1967).

Prognosis

The neonatal period is marked by serious respiratory troubles, such as edemas and laryngeal hypotonia, leading to tracheobronchial inspiration, as well as cyanosis outbursts during feeding. Growth retardation is little pronounced.

Seizures are seen in several cases.

More than half of the children died at birth or shortly thereafter of heart failure or oropharyngeal incoordination. The oldest patient was still alive at 14½. Another had a nonverbal IQ of 60 at age 5–6, but language learning was severely retarded.

CYTOGENETIC STUDY

The breakpoint is in 4q31 in almost all cases. The deletion is nearly always *de novo*. In only one case does monosomy result from a parental translocation.

DERMATOGLYPHICS

A single transverse palmar crease is observed in half of the cases.

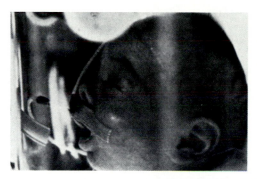
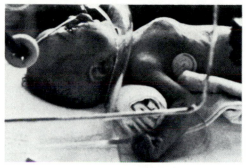

4.35

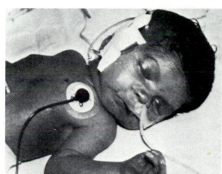
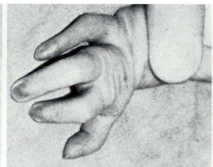
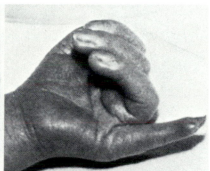

4.36

4.37

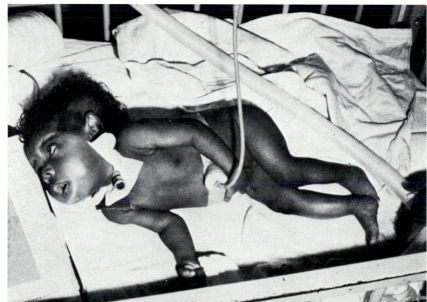

CHROMOSOME 5

Few gene assignments are known for chromosome 5. Among them are hexosaminidase, the deficiency responsible for Sandhoff's disease, and arylsulfatase B, the deficiency that causes mucopolysaccharidosis, type VI, or Maroteaux-Lamy disease.

The "cri du chat" syndrome due to 5p monosomy is the most frequent deletion observed in man. 5p trisomy is a very striking counterpart to this. 5q3 trisomy is a more recently characterized syndrome.

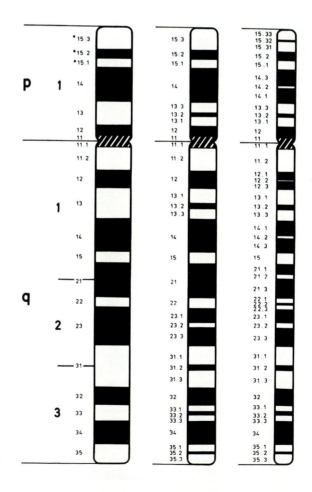

GENE MAP

5p	AVRR	(L)	Antiviral state repressor regulator (10747) (S)
5q11-5q13	HEXB		Hexosaminidase B (14265) S
5q15-5qter	DTS		Diphtheria toxin sensitivity (12615) S
	ARSB		Arylsulfatase B (25320) S
	LARS	(P)	Leucyl-tRNA synthetase (15135) S
	LDLR	(L)	Low density lipoprotein receptor (14389)S (on 5 or 21 or both)
5q35	CHR		Chromate resistance (11884) S
5q23-5q35	EMTR	(P)	Emetine resistance (13062) S
	BAR	(P)	Beta-adrenergic receptor (10969) S

MORPHOLOGY AND BANDING

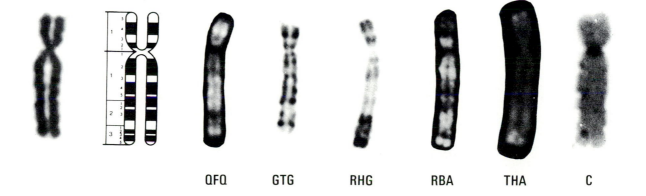

QFQ   GTG   RHG   RBA   THA   C

5p MONOSOMY ("CRI DU CHAT" SYNDROME)

In the Infant

Characteristic cry
Microcephaly
Moonlike face
Hypertelorism

In the Older Child and Adult

Narrow face
Small mandible, effacement of the angles of the jawbone

REFERENCES

BERGER R.: Déficience du bras court du chromosome 5. Maladie du cri du chat. *Nouv. Presse méd.,* **1**:873–875, 1972.

BREG W.R., STEELE M.W., MILLER O.J. et al.: The cri du chat syndrome in adolescents and adults: clinical findings in 13 older patients with partial deletion of the short arm of chromosome n° 5 (5p-). *J. Pediat.,* **77**:782–791, 1970.

LEJEUNE J., LAFOURCADE J., BERGER R., VIALATTE J., BOESWILLWALD M., SERINGE P., TURPIN R.: Trois cas de délétion partielle du bras court d'un chromosome 5. *C.R. Acad. Sci. (Paris),* **257**:3098–3102, 1963.

LEJEUNE J., LAFOURCADE J., GROUCHY J. de, BERGER R., GAUTIER M., SALMON C., TURPIN R.: Délétion partielle du bras court du chromosome 5. Individualisation d'un nouvel état morbide. *Sem. Hôp. Paris,* **18**:1069–1079, 1964.

LUCHSINGER R., DUBOIS G., VASSELA F., JOSS E., CLOOR R., WIESMAN U.: Spektranalyse des «miauens» bei Cri du chat syndrom. *Folia phoniat. (Basel),* **19**:27–33, 1967.

NIEBUHR E.: The cri du chat syndrome. Epidemiology, cytogenetics, and clinical features. *Hum. Genet.,* **44**:227–275, 1978.

NIEBUHR E.: Anthropometry in the cri du chat syndrome. *Clin. Genet.,* **16**:82–95, 1979.

MILLER O.J., WARBURTON D., BREG W.R.: Deletion of group B chromosomes. *Birth Defects, Orig. Art. Ser.,* **V-5**:100–105, 1969.

PITT D., BRASCH J., WONG J.: Another "Cri-du-Chat." *Med. J. Australia:* 606–607, 1966.

PONCET E., LAFOURCADE J., ZHA J., AUTIER C.: Les troubles de la voix et les malformations laryngées dans la maladie par aberration chromosomique dite la «maladie du cri du chat». *Ann. Oto-Laryngol.,* **82**:10, 1965.

RETHORE M.O.: Chromosome deletions and ring chromosome syndromes. *In* N.C. MYRIANTHOPOULOS, ed: *Handbook of Clinical Neurology, Vol. 31, Part II, Congenital Malformations of the Brain and Skull.* Amsterdam, North Holland publ., pp. 549–620, 1977.

The delineation of the 5p monosomy or "cri du chat" syndrome by Lejeune et al. (1963, 1964) was based on three cases. The first description was confirmed by numerous authors (Pitt et al., 1966; Miller et al., 1969; Berger, 1972). At present more than 330 cases have been reported in the literature (Rethoré, 1977; Niebuhr, 1978). Isolated cases have long ceased to be the object of publication.

GENERALIZATIONS

The following data were collected by Niebuhr (1978).

- Frequency: the frequency of the disease is not precisely known. The best estimate seems to be on the order of 1 in 50,000 births. A dizygotic twin pregnancy has been reported in four cases. No case of monozygotic twins has been reported.
- Sex ratio: 7F/5M
- Mean maternal age: 26.9 years
- Mean paternal age: 30.8
- Mean birth weight: 2,610 g
- Mean crown-heel length at birth: 47.4 cm
- Mean head circumference at birth: 31.7 cm

PHENOTYPE

The most remarkable feature is that for which the disease is named: the shrill and plaintive nature of the cry, unmistakably calling to mind the mewing of a kitten. In the infant, this feature in itself allows diagnosis. The craniofacial dysmorphism is also very suggestive.

Craniofacial Dysmorphism

Craniofacial dysmorphism develops with age. In the infant, microcephaly, moonlike face, hypertelorism, and micrognathia are striking.

5.1

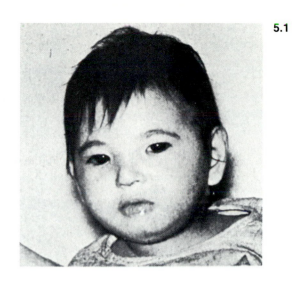

5.2

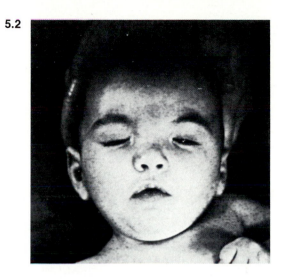

5.3

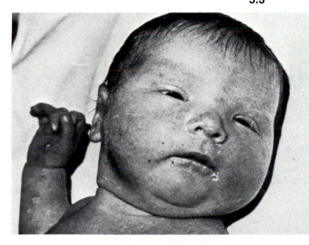

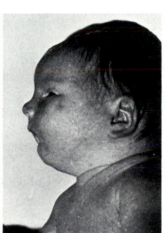

5.4

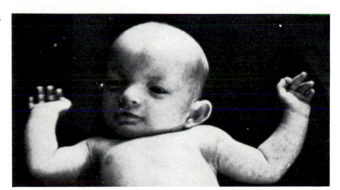

WARD P.H., ENGEL E., NANCE W.E.: The larynx in the cri du chat syndrome. *Laryngoscope (St. Louis)*, **78**:1716–1733, 1968.

The cranium is small and spherical. The metopic suture sometimes protrudes.

Palpebral fissures, normal in length, are very slightly slanted downward and outward.

Hypertelorism appears to be constantly present and may be extensive. Niebuhr (1979) contests the radiologic existence of this hypertelorism. Nevertheless, it remains a very useful clinical sign. It is accompanied by bilateral epicanthal folds and often by divergent or, more rarely, convergent, strabismus.

The nasal bridge is very wide and flat.

The upper lip and mouth are normal.

The mandible is small, with poorly defined angles. The chin may be slightly receding.

The ears are small but normal in morphology; they are somewhat low-set. A pretragal tag and narrow auditory canals are sometimes observed.

With age, the dysmorphism is appreciably modified: microcephaly persists but the face lengthens, hypertelorism and epicanthus lessen, and the mandibular aplasia becomes more evident. The teeth are decayed and poorly inserted. The palate is high-arched.

Neck, Thorax, and Abdomen

In general, no anomalies are observed, apart from a slightly increased frequency of hernias.

Limbs

The limbs are normal.

Genitalia

These are usually normal. Reports of hypospadias, cryptorchidism, hypertrophy of the clitoris, and delayed puberty are exceptional.

The Cry and Anomalies of the Larynx

The very distinctive cry, high-pitched and plaintive, is striking at birth. Its resemblance to the mewing of a kitten has been studied by means of acoustic recordings; the tracings are identical. The phenomenon entails a cry emitted on exhalation, without the inhalation noise suggestive of laryngeal stridor. It is a monotonous regular vocal emission of low variation and high timbre (Luchsinger et al., 1967).

In a few cases, laryngoscopy revealed a generally hypoplastic or "undifferentiated" larynx, without significant anatomical disorder (Poncet et al., 1965; Ward et al., 1968).

Malformations

Malformations are relatively rare and nonspecific apart from laryngeal anomalies. The following may be noted:

- Cerebral anomalies: diffuse cerebral atrophy, ventricular dilatation, cerebellar atrophy, hydrocephaly
- Ocular anomalies: optic atrophy, cataract
- Diverse cardiac malformations
- Renal and abdominal malformations

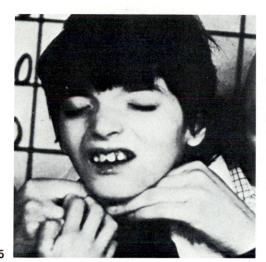
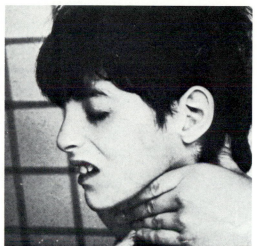

5.5

5.6

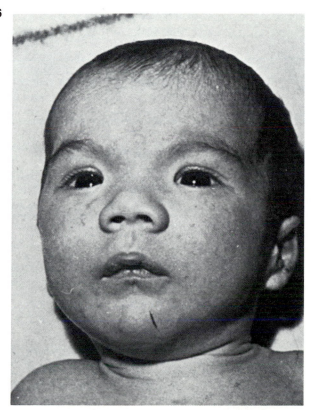

- Diverse skeletal malformations, such as scoliosis, clubfoot, costal and vertebral anomalies

Mental Retardation

Hypotonia is constant; the archaic reflexes persist for an abnormally long time or are weakened. With age, hypertonia of the limbs, with strong reflexes, and spastic gait appear. The EEG may be abnormal.

Accomplishments are delayed: sitting posture after 2 years, independent walking after 4 years. Certain patients remain bedridden in a fetal position.

Mental deficiency is severe. Language is nonexistent or reduced to a few words. The IQ is most often below 20.

Prognosis

Fatality is low. Many patients have attained adult age. They remain below normal in size and weight.

CYTOGENETIC STUDY

In a majority of cases, the deletion appears *de novo*. The dimension of the deleted segment varies from patient to patient and has no evident correlation with phenotypic variation. The major symptoms would seem to be due to the loss of a very small segment in the 5p14p15 bands. Mosaicisms, unbalanced *de novo* translocations, or ring chromosomes may occasionally be observed.

In one-fifth of the published cases there is parental rearrangement (the actual frequency is probably lower because of a "selection" of published cases). Almost every case involves a translocation, more often maternal than paternal, which is responsible for the occurrence of multiple cases of 5p monosomy and trisomy in a given family. Exceptionally, a pericentric inversion, parental mosaicism, or even a much more complex rearrangement leading to "aneusomie de recombinaison" can be observed.

DERMATOGLYPHICS

Dermatoglyphic anomalies, especially of the flexion creases, constitute an important element in the diagnosis. The most common pattern is the presence of a distal flexion crease interrupted at the perpendicular of the second interdigital space. The proximal flexion crease is normal. Also present is a single transverse palmar crease or its equivalent.

The axial triradius is in t'. A high frequency of patterns on the thenar eminence has been noted. The distribution of digital patterns is normal.

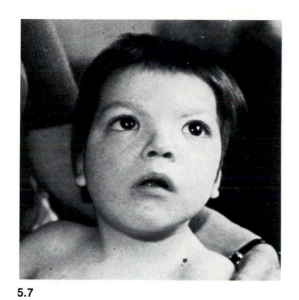

5.7

5.8

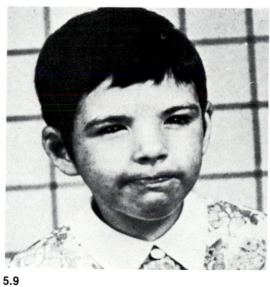

5.9

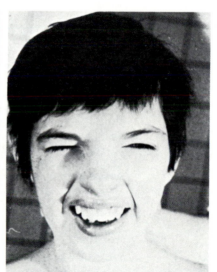

5p TRISOMY

Hypotelorism
Large mandible

REFERENCES

LEJEUNE J., LAFOURCADE J., BERGER R., RETHORE M.O.: Maladie du cri du chat et sa réciproque. *Ann. Génét.*, **8**:11–15, 1965.

LESCHOT N.J., LIM K.S.: «Complete» trisomy 5p; *de novo* translocation t(2;5) (q36;p11) with isochromosome 5p. *Hum. Genet.*, **46**:271–278, 1979.

YUNIS E., SILVA R., EGEL H., ZUNIGA R., TORRES de CABALLERO O.M., RAMIREZ E., POVEDA de RUIZ H.: Partial trisomy 5p. *Hum. Genet.*, **43**:231–237, 1978.

The first case of 5p trisomy was described in 1965 by Lejeune et al. More than 20 observations are known. They result from malsegregation of a parental rearrangement and are observed in families together with 5p monosomies (Yunis et al., 1978; Leschot and Lim, 1979).

GENERALIZATIONS

- Sex ratio: 4F/1M
- Mean birth weight: 3,150 g

PHENOTYPE

The phenotype of 5p trisomy is not very suggestive and usually does not allow diagnosis. In most cases the patients were diagnosed because of the discovery of a "cri du chat" syndrome in their siblings, cousins, or family. For this reason several were not identified until adolescence. In the infant no anomaly of the cry has been reported. In other, more rare cases, the trisomy involves all of 5p and is accompanied by a more suggestive syndrome.

Craniofacial Dysmorphism

When trisomy is limited to p14 and p15, the phenotype is best described by countertypical comparison with those of the "cri du chat" syndrome.
The cranial circumference is normal. The forehead is flat.
Distinct hypotelorism is present. The bridge of the nose is narrow and protruding. The palpebral fissures are horizontal.
The chin is well defined and the angles of the jawbone are consistently protruding. In adolescents the development of the mandible lends a massive character to the face, which contrasts with the narrow face of 5p monosomy patients.
The ears are normal.
When trisomy concerns all or nearly all of 5p, the following may be observed:

- Macrocephaly
- Slanting upward and outward of palpebral fissures
- A flattening of the bridge of the nose
- Low set ears

Neck, Thorax, Abdomen, Limbs, and Genitalia

No particular malformation has been reported.

Mental Retardation

Mental retardation seems slightly less severe than in the "cri du chat" syndrome. IQs are between 20 and 80.

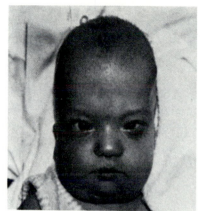

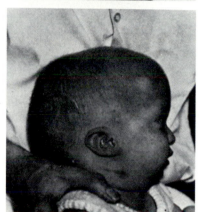

5.10

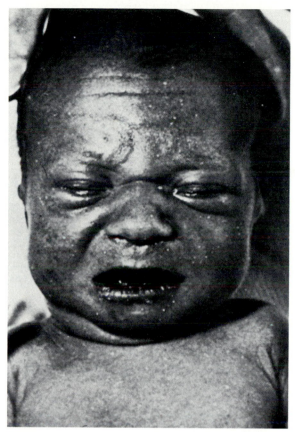

5.11

5.12

5.13

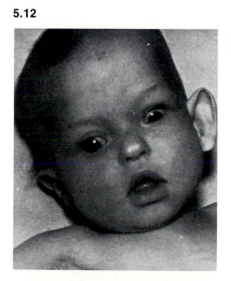

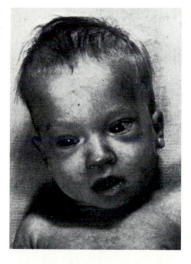

Prognosis

There is no growth retardation, except in a case of complete 5p trisomy. The panniculus adiposus is rather enlarged.

Lifespan seems normal.

CYTOGENETIC STUDY

Most of the known 5p trisomies issue from parental translocations. In one family, a pericentric inversion with "aneusomie de recombinaison" is involved.

In several cases, the trisomy concerns all of 5p. One of these cases is due to a *de novo* translocation, with formation of 5p isochromosome (Leschot and Lim, 1979).

DERMATOGLYPHICS

The flexion creases are normal.

The axial triradius is in t, exceptionally in t'.

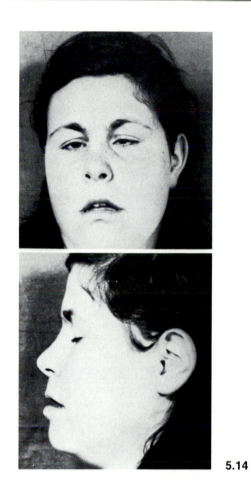

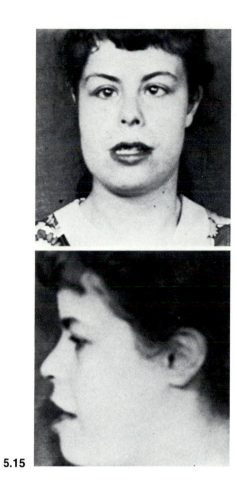

5.14

5.15

5.16

5.17

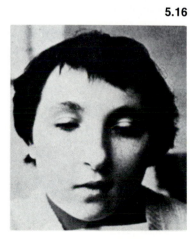

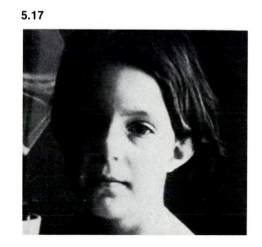

5q3 TRISOMY

Microcephaly
Large eyes
Small, carp-like mouth

REFERENCES

FERGUSON-SMITH M.A., NEWMAN B.F., ELLIS P.M., THOMSON D.M.G.: Assignment by deletion of human red cell acid phosphatase gene locus to the short arm of chromosome 2. *Nature New Biol.*, **243**:271–274, 1973.

JALBERT P., JALBERT H., SELE B., MOURIQUAND C., MALKA J., BOUCHARLAT J., PISON H.: Partial trisomy for the long arms of chromosome n° 5 due to insertion and further aneusomie de recombinaison. *J. med. Genet.*, **12**:418–422, 1975.

KESSEL E., PFEIFFER R.A.: Tandem duplication (5q13→22) in a mentally deficient girl. *Hum. Genet.*, **52**:217–220, 1979.

RODEWALD A., ZANKL M., GLEY E.O., ZANG K.D.: Partial trisomy 5q: three different phenotypes depending on different duplication segments. *Hum. Genet.*, **55**:191–198, 1980.

Since the first observation of Ferguson-Smith et al. (1973), 15 cases of partial 5q trisomy have been reported and analyzed by Rodewald et al. (1980). Thirteen of these involved a more or less extensive trisomy of the 5q3 region and will be considered here. Two others involved a proximal interstitial trisomy, but, due to their small number, they will not be considered here (Jalbert et al., 1975; Kessel and Pfeiffer, 1979).

GENERALIZATIONS

- Sex ratio: 4M/9F
- Mean birth weight: 2,600 g
- Mean crown-heel length at birth: 45 cm
- Mean head circumference at birth: 32.6 cm

PHENOTYPE

Dysmorphism is rather minor. A low birth weight is initially observed followed by a retarded growth rate. Psychomotor retardation is severe.

Craniofacial Dysmorphism

Microcephaly is always present.
Hypertelorism, epicanthus, and strabismus are often seen.
Eyebrows are fine and are sparse at the external extremities.
The eyes are large, with the external part of the upper lid slightly downturned.
The nose is somewhat flat and wide in the middle.
The nares are normally oriented. The nasal wings are slightly "spread". The philtrum is long and smooth.
The mouth is small with very thin lips. Anomalies of the tooth enamel and cavities are frequent. The lower lip is horizontal, the upper lip forms a regular half circle, but has no Cupid's bow, resembling a carp's mouth.
The ears are low-set, poorly folded, and detached.

Neck, Thorax, and Abdomen

Frequently, hernias are found.

Limbs

Anomalies of the fingers and toes have been reported, including duplication of the thumb, and retroversion of the last phalanx.

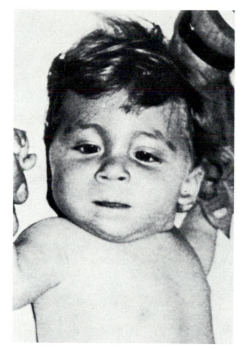

5.18

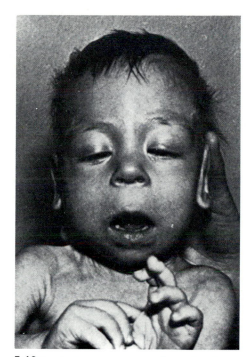

5.19

5.20

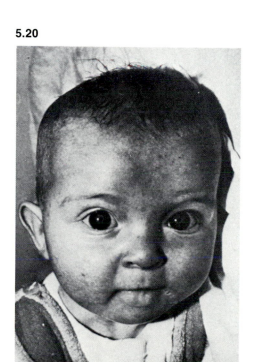

5.21

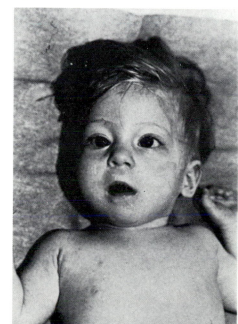

Malformations

Cardiac abnormalities, including ventricular and atrial septal defect, defect in conduction, and pulmonary hypertension are frequently present.

Mental Retardation

This is usually quite severe, but would seem to be more moderate if the breakpoint is in q34.

Prognosis

The patients were seen at various ages. A number of early deaths have been observed in their families among children having a phenotype compatible with trisomy 5q3.

CYTOGENETIC STUDY

Trisomy is always the result of a parental translocation. The breakpoint in 5q varies from q31 to q34. Craniofacial dysmorphism does not show great variability as a function of the size of the trisomic segment, and congenital heart disease may be seen no matter what the size of the latter. Mental retardation would seem to be less severe in the more distal trisomies.

DERMATOGLYPHICS

Dermatoglyphics are not characteristic.

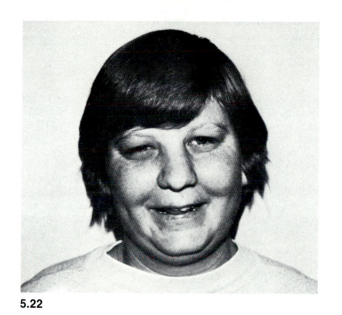

5.22

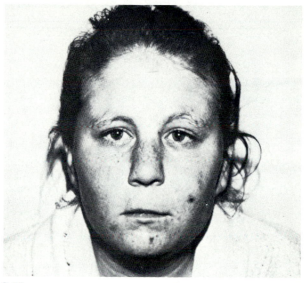

5.23

5.24

5.25

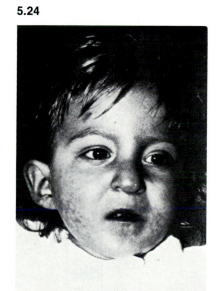

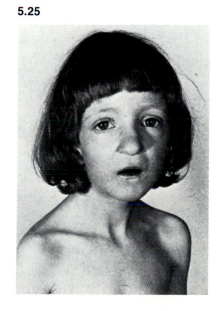

CHROMOSOME 6

REFERENCE

MADAN K., BRUINSMA A.H.: C-band polymorphism in human chromosome no 6. *Clin.
Genet.*, **15**:193–197, 1979.

Chromosome 6 carries the major histocompatibility complex (MHC) locus. Among loci linked to HLA is the gene coding for 21-hydroxylase, whose deficiency causes congenital adrenal hyperplasia. This very close linkage allows, in certain families, a prenatal diagnosis of the disease. Hemochromatosis, some complement components deficiencies, some forms of congenital heart defect, and one form of spinocerebellar ataxia are also part of the HLA linkage group.

Two highly characteristic cytogenetic syndromes, trisomy 6p2 and trisomy 6q2, are known to exist.

A polymorphism of chromosome 6, characterized by the presence of supplementary heterochromatin in the juxtacentromeric region of the short arm, has been found (Madan et al., 1979).

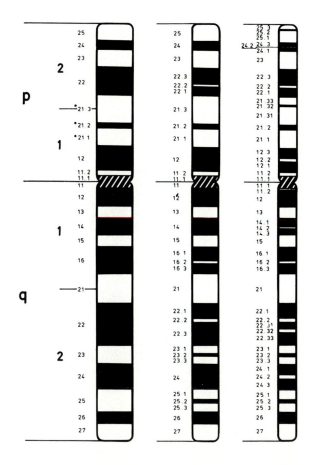

GENE MAP

6p2105-6p2300 (about 6cM from centromere) order: Cen. -D-B-C-A**	MHC		MAJOR HISTOCOMPATIBILITY COMPLEX (MHC)F, S HLA-A (14280) F HLA-B (14283) F HLA-C (14284) F HLA-D (14285) F
		(P)	HLA-DR (14286) F
On B side of D		(P)	HLA-DR2 (14287) F
No crossover with HLA-B	C2		Complement component 2 (21700) F
Not close to HLA	C3BR	(P)	Complement component C3b, receptor for C3BR (12062) S
Not close to HLA	C3DR	(P)	Complement component C3d, receptor for C3DR (12065) S
	C4F		Complement component 4F, or C4B (12080) F
	C4S		Complement component 4S, or C4A (12081) F
	C4BP	(L)	Complement component 4 binding protein (12083) H
No crossover with C2	BF		Properdin factor B (Bf) (13847) F
Near HLA-A end	MLRW	(P)	Mixed lymphocyte reaction, weak (15786) F
Centromeric to DR		(P)	HLA-SB (14289) F, Ch
Near HLA-D/DR	PLT1		Primed lymphocyte test 1 (17668) F
	RWS	(L)	Ragweed sensitivity (17945) F
Near HLA-B?	IHG	(P)	Immune response to synthetic polypeptide HGAL (14695) F
Near HLA-B?	ITG	(P)	Immune response to synthetic polypeptide TGAL (complementing loci) (14696) F
	IS	(P)	Immune suppression (IS) (14685) H
Close to LHA-D/DR	DC1		Ia (immune response antigens) of DC2 specificity (14688)
	HFE		Hemochromatosis (23520) LD, F
	NDF	(P)	Neutrophil differentiation factor (20270) LD
	CAH1		Congenital adrenal hyperplasia due to 21-hydroxylase deficiency (20191) F
	NEU1	(L)	Neuraminidase 1 (16205) H
	RNTMI	(P)	Initiator methionine tRNA (18062) REa
	FEA	(L)	F9 embryonic antigen (13701) H

	ASH		Asymmetric septal hypertrophy (19260) F
	ASD2	(P)	Atrial septal defect, secundum type (10880) F
6p21-6p22 (about 3cM from HLA)	GLO1		Glyoxalase I (13875) F, S
6p21-6q15	ME1		Malic enzyme, soluble (ME1) (15425) S
?between GLO1 and PGM3	SCA1		Spinocerebellar ataxia 1 (16440) F
6q12 or 6q13	PGM3		Phosphoglucomutase 3 (17210) S, F, OT
6q21	SOD2		Superoxide dismutase 2, mitochondrial (14746) S
	HC	(L)	Hypercholesterolemia (14400) F
	PG	(L)	Pepsinogen (16970) F, H
	PDB	(L)	Paget disease of bone (16725) F
	LAP	(L)	Laryngeal adductor paralysis (15027) F
	PLA	(P)	Plasminogen activator (17337) S
BEVI	BEVI		Baboon M7 virus replication (10918) S
	S5	(P)	Surface antigen 5 (18551) S
	RMBC	(P)	Monkey RBC receptor (15805) S
	ADCP1	(I)	Adenosine deaminase complexing protein 1 (10271) S
6pter-6p23	PRL		Prolactin (17674) REa
6p23-6p25	HAF	(P)	Hageman factor (23400) D
6p	SPC	(P)	SALIVARY PROTEIN COMPLEX (16872–16881) F
6q12-6q21	CGA	(P)	Chorionic gonadotropin, alpha chain (11885) REa
6q22-6q24	MYB		Onc gene: avian myeloblastosis virus (18999) S
	IDDM	(L)	Insulin dependent diabetes mellitus (22210) F, LD
	MDI	(L)	Manic depressive illness (12548) F
	ASSP	(P)	Argininosuccinate synthetase pseudogene (10784) REa

MORPHOLOGY AND BANDING

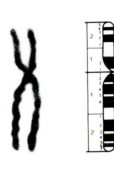

QFQ   GTG   RHG   RBA   THA   C

6p2 TRISOMY

Blepharophimosis
Bulbous nose
Small mouth

REFERENCES

BERNHEIM A., BERGER R., VAUGIER G., THIEFFRY J.C., MATET Y.: Partial trisomy 6p. *Hum. Genet.*, **48**:13–16, 1979.

COTE G.B., PAPADAKOU-LAGOYANNI S., SBYRAKIS S.: Partial trisomy 6p with karyotype 46,XY,der(22)t(6;22) (p22;q13)mat. *J. med. Genet.*, **15**:479–481, 1978.

PEARSON G., MANN J.D., BENSEN J., BULL R.W.: Inversion duplication of chromosome 6 with trisomic codominant expression of HLA antigens. *Am. J. hum. Genet.*, **31**:29–34, 1979.

ROSI G., VENTI G., MIGLIORINI BRUSCHELLI G., DONTI E., BOCCHINI V., ARMELLINI R.: Trisomy 6p22→6pter due to familial t(6;13) (p22;q34 or 33) translocation. *Hum. Genet.*, **51**:67–72, 1979.

THERKELSEN A.J., KLINGE T., HENNINGSEN K., MIKKELSEN M., SCHMIDT G.: A family with a presumptive C/F translocation in five generations. *Ann. Génét.*, **14**:13–21, 1971.

THERKELSEN A.J.: Personal communication, 1975.

TURLEAU C., CHAVIN-COLIN F., GROUCHY J. de: La trisomie 6p partielle. *Ann. Génét.*, **21**:88–91, 1978.

Since the first familial observation by Therkelsen et al. (1971, 1975), 11 cases of 6p2 trisomy have been published (Turleau et al., 1978; Côté et al., 1978; Bernheim et al., 1979; Pearson et al., 1979; Rosi et al., 1979).

GENERALIZATIONS

- Sex ratio: 3M/2F
- Pregnancy: frequent prematurity
- Mean birth weight: 2,050 g
- Mean crown-heel length at birth: 44.6 cm
- Mean head circumference at birth: 31.5 cm

PHENOTYPE

The phenotype is highly specific and allows clinical diagnosis. The neonatal period is marked by serious feeding difficulties which are often lethal.

Craniofacial Dysmorphism

The forehead may be high and bossing, with a flat or prominent occiput. Hair is fine and thin. Skin is dry.

The appearance of the eyes is the most striking characteristic of facial dysmorphism. They appear abnormally close together with a prominent bridge of the nose. Palpebral fissures are short and narrow, contracted by blepharophimosis and blepharoptosis. Strabismus, with or without nystagmus, has been reported in half of the cases. The eyes are often closed. Lashes are long, and very curved. Eyebrows are sparse. An opthalmological examination may disclose microphthalmia or a cataract.

The nose appears to be short, with tiny nares. The philtrum is long and well-defined.

The mouth is small, with thin lips; the fleshy part of the upper lip is rarely visible, but the Cupid's bow is clearly defined.

The chin is small and pointed.

The ears may be low set and deformed, with a simple helix, an absence of antitragus, and a poorly developed lobe.

The overall set of dysmorphic signs lend a rather ironic look to the face.

Neck, Thorax, and Abdomen

The neck is short. The thorax often appears to be too long compared to the limbs. One case of omphalocele has been reported. A sacral dimple is nearly always present.

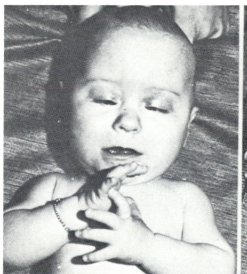
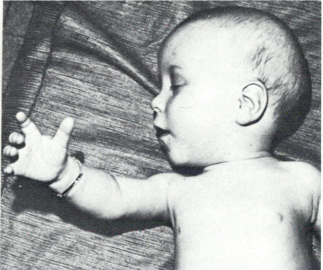

6.1

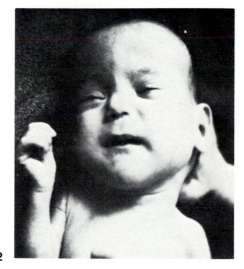

6.2

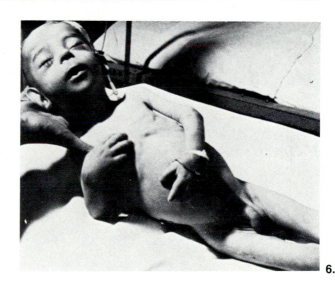

6.3

6.4

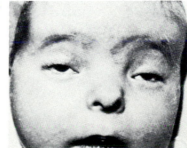

Limbs

Syndactylies and poor positioning of the toes may be observed.

Genitalia

These are usually normal.

Malformations

Inner organ malformations are rare, and have been seen mainly in two cases of t(6;20) (Therkelsen et al., 1975; Brauning et al., 1977). They may have been due to the associated monosomy 20, and involved congenital heart disease and kidney anomalies, such as small kidneys and proteinuria. In another case, due to a pericentric inversion of 6, a patent ductus arteriosus, as well as an absence of the uterus and arhinencephaly were observed (Pearson et al., 1979).

Mental Retardation

Mental retardation appears to be considerable, but is difficult to evaluate because of the young age of the children.

Prognosis

Growth retardation is always very severe. Most infants die very young, mainly due to respiratory troubles or feeding difficulties.

CYTOGENETIC STUDY

The breakpoint varies between p21 and p23. The size of the trisomic segment does not appear to modify the phenotype.

DERMATOGLYPHICS

These do not seem to be characteristic.

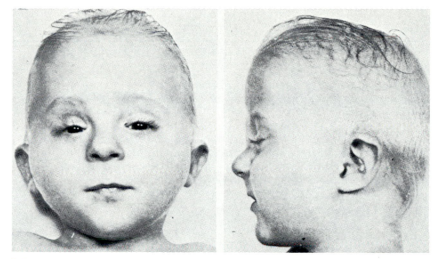

6.5

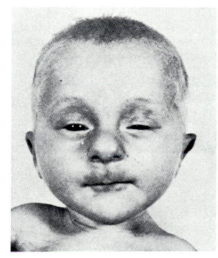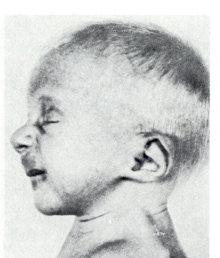

6.6

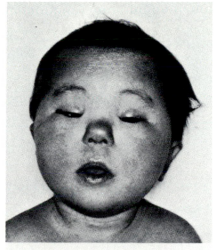

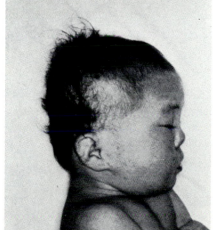

6.7

6q2 TRISOMY

Flattened face
Palpebral fissures slanting downward and outward
Downturned corners of the mouth
Short, webbed neck

REFERENCES

GROUCHY J. de, EMERIT I., AICARDI J.: Trisomie partielle pour le bras long d'un C (?6) par translocation t(6p+; Cqs+). *Ann. Génét.*, **12**:133–137, 1969.

NEU R.L., GALLIEN J.U., STEINBERG-WARREN N., WYNN R.J., BANNERMANN R.M.: An infant with trisomy 6q21→6qter. *Ann. Génét.*, **24**:167–169, 1981.

TURLEAU C., GROUCHY J. de: Trisomy 6qter. *Clin. Genet.*, **19**:202–206, 1981.

The first case of distal 6q trisomy was reported by Grouchy et al. (1969) and was later confirmed by banding. Ten cases are presently known. Nine were analyzed by Turleau et al. (1981) and the tenth case was reported by Neu et al. (1981).

GENERALIZATIONS

- Sex ratio: 2F/1M
- Mean birth weight: 2,570 g

PHENOTYPE

Diagnosis is often made late, as dysmorphism in the very young child is relatively minor. This becomes more apparent with age, finally turning into a "caricature." Smallness of stature, and the very characteristic appearance of the neck have at times led to a diagnosis of Turner's syndrome.

Craniofacial Dysmorphism

The skull may be small and abnormal in shape due to brachycephaly or acrocephaly, with flattening of the occiput and prominence of frontal bossing. The hairline is set high on the forehead. The face is flat; the features appear to be pulled downward.

Palpebral fissures slant downward and outward. Hypertelorism is often present. Eyebrows are thin and arched.

The nose is wide.

The philtral margins are hardly defined.

The mouth is small, with thin lips and downturned corners. It has been described as a "carp's mouth". The upper lip, in the form of an inverted 'V' reveals the gums. Tooth decay and other dental anomalies have been reported.

The angles of the jaw are effaced.

The ears are low set, and usually normal in shape.

Neck, Thorax, and Abdomen

The neck is strikingly short and wide. An anterior webbing (very different from the pterygium colli observed in Turner's syndrome) is almost constantly present.

The hairline is low set upon the neck. There may be a funnel deformity of the chest, with widely spaced nipples. Deformities of the spine are regularly found in older children.

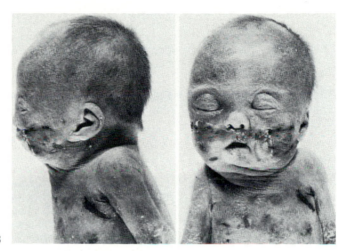

6.8

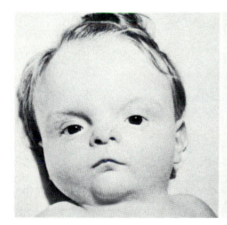

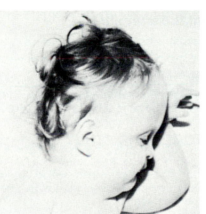

6.9

6.10

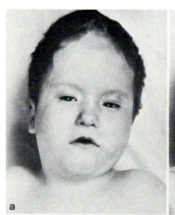

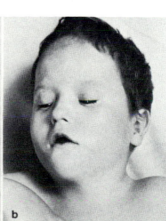

Limbs

Osteoarticular and muscular malformations are always present, leading to faulty posture. Frequently, poor implantation of the toes and cutaneous syndactyly are encountered.

Genitalia

Puberty is delayed in girls. Ovarian biopsies of two patients showed normal ovaries. In males, the scrotum is flat and empty. Hypospadias is regularly found.

Malformations

The previously mentioned osteoarticular and muscular malformations are constantly encountered and may be severe. They may be in the form of severe contractures of the major joints requiring surgery, or retractions of the fingers and toes. Later on, rapidly evolving scoliosis is observed in some of the older patients.

Inner organ malformations are rare. Only one patient, who died at the age of 6½ months, showed severe cerebral, digestive, and cardiac malformations. An autopsy of one stillborn infant showed multiple internal malformations.

Mental Retardation

Mental retardation is very severe.

Prognosis

Growth retardation is severe. In the absence of serious malformations, survival appears to be almost normal. The oldest patient was seen at age 23.

CYTOGENETIC STUDY

Trisomy results, in all known cases, from a parental rearrangement, either a translocation or a pericentric inversion. The breakpoint varies between 6q21 and q26, but is most often in q25. The phenotype does not seem to be dependent upon the size of the trisomic segment.

DERMATOGLYPHICS

These are not characteristic.

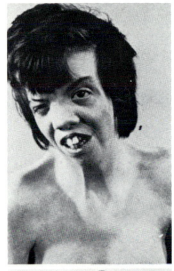

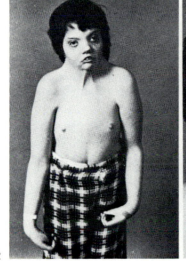

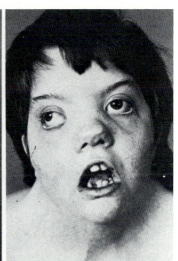

6.12

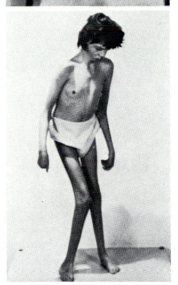

6.13

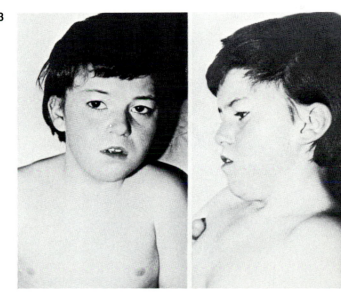

6.11

r(6)

REFERENCES

CARNEVALE A., BLANCO B., CASTILLO J., DEL CASTILLO V., DOMINGUEZ D.: Ring chromosome 6 in a child with minimal abnormalities. *Am. J. med. Genet.*, **4**:271–277, 1979.

KINI K.R., Van DYKE D.L., WEISS L., LOGAN M.S.: Ring chromosome 6: report and review of literature. *Hum. Genet.*, **50**:145–149, 1979.

MOORE C.M., HELLER R.H., THOMAS G.H.: Developmental abnormalities associated with a ring chromosome 6. *J. med. Genet.*, **10**:299–303, 1973.

Since the first observation by Moore et al. (1973), eight cases of r(6) have been reported (see Kini et al., 1979; Carnevale et al., 1979).

The phenotype is not very characteristic. Features most often encountered are:

- A low birth weight
- Moderate to severe mental retardation
- A delay in growth, with retardation of bone age
- Microcephaly
- A flat or broad nasal bridge, bilateral epicanthus, microphthalmy, low set, poorly folded ears
- Micrognathia
- Shortness of the neck

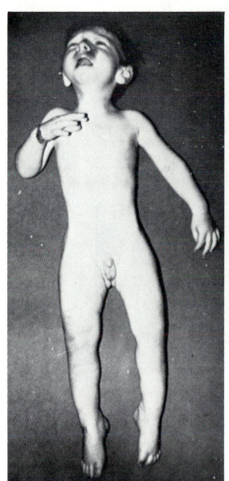

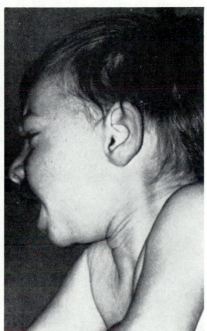

6.14

6.15

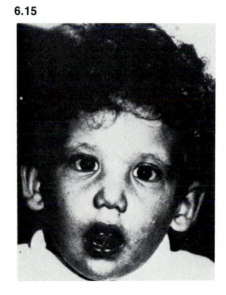

6.16

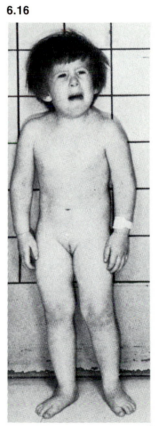

CHROMOSOME 7

Gene assignments for chromosome 7 include the family of genes that controls histone synthesis, as well as several collagen genes. Among hereditary diseases, the following should be noted: type VII mucopolysaccharidosis, due to β-glucuronidase deficiency, type VII Ehlers Danlos disease, Meige's syndrome, argininosuccinuria, and possibly Marfan's syndrome and osteogenesis imperfecta (one or several forms).

Cytogenetic syndromes include trisomy and monosomy 7p, which affect the closure of the cranial sutures and contrast in type and countertype in a striking manner; terminal and interstitial monosomies and trisomies for the long arm.

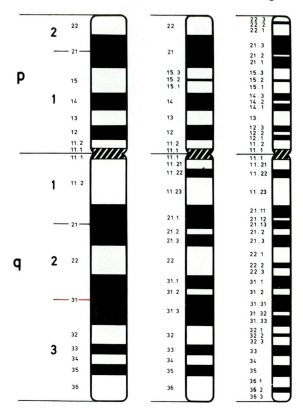

GENE MAP

7p13 (or 3p21.1)	GS	(L)	Greig-polysyndactyly craniofacial dysmorphism syndrome (17570) Ch
	S7	(P)	Surface antigen 7 (18552) S
7pter-7q22	ASL	(P)	Argininosuccinate lyase (20790) S

7p14-7cen	BLVR	(P)	Biliverdin reductase (10975) S
	UP		Uridine phosphorylase (19174) S
7p22-7q22	MDH2		Malate dehydrogenase, mitochondrial (15410) S
7p12-7q22	EGFR		Epidermal growth factor, cell surface receptor for (same as S6) (13155) S
(l)7p13-7cen, or 7cen-7q21 (?7q22) ?linked to Jk	GUSB		Beta-glucuronidase (25322) S, D
	COL1A2		Collagen I (alpha 2) (12016) S, REa
	COL3A1	(P)	Collagen III (alpha 1) (12018) S
	HADH	(P)	Hydroxyacyl CoA dehydrogenase (14345) S
7q22-7qter	GP230	(P)	Neutrophil migration (granulocyte glycoprotein; GP130; formerly neutrophil chemotactic response, NCR) (16282) D
7q21-7q36	H		HISTONE GENE CLUSTER (H1, H2A, H3, H4) (14271–14275) A
	NHCP1	(P)	Nonhistone chromosomal protein-1 (11887) S
	GCF1	(P)	Growth rate controlling factor-1 (13922) S
	DIA2	(L)	Diaphorase-2 (13560) S
	PSP	(P)	Phosphoserine phosphatase (17248) S

MORPHOLOGY AND BANDING

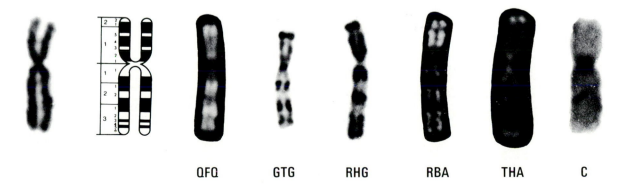

QFQ   GTG   RHG   RBA   THA   C

7q TRISOMIES

7q3 Trisomy

Minor facial dysmorphism

REFERENCES

CARPENTIER S., RETHORE M.O., LEJEUNE J.: Trisomie partielle 7q par translocation familiale t(7;12) (q22;q24). *Ann. Génét.,* **15**:283–286, 1972.

GRACE E., SUTHERLAND G.R., BAIN A.D.: Familial insertional translocation. *Lancet* **ii**:231, 1972.

GRACE E., SUTHERLAND G.R., STARK G.D.: Partial trisomy of 7q resulting from a familial translocation. *Ann. Génét.,* **16**:51–54, 1973.

LEJEUNE J., RETHORE M.O., BERGER R., ABONYI D., DUTRILLAUX B., SEE G.: Trisomie C partielle par translocation familiale t(Cq + ;Cq −). *Ann. Génét.,* **11**:171–175, 1968.

SCHINZEL A., TONZ O.: Partial trisomy 7q and probable partial monosomy of 5p in the son of a mother with a reciprocal translocation between 5p and 7q. *Hum. Genet.,* **53**:121–124, 1979.

SCHMID M., WOLF J., NESTLER H., KRONE W.: Partial trisomy for the long arm of chromosome 7 due to familial balanced translocation. *Hum. Genet.,* **49**:283–289, 1979.

TURLEAU C., ROSSIER A., MONTIS G. de, ROUBIN M., CHAVIN-COLIN F., GROUCHY J. de: Trisomie partielle 7q. Un ou deux syndromes? A propos d'une nouvelle observation. *Ann. Génét.,* **19**:37–42, 1976.

The first two observations of partial 7q trisomy appeared independently of one another (Grace et al., 1972, 1973; Lejeune et al., 1968 and Carpentier et al., 1972). Sixteen cases are now known. Thirteen involve terminal 7q3 trisomy (Turleau et al., 1976; Schmid et al., 1979; Schinzel and Tonz, 1979). Three others concern an intercalary trisomy and will be considered separately.

GENERALIZATIONS

- Sex ratio: 1M/1F
- Mean birth weight: 2,450 g
- Mean crown-heel length at birth: 48.8 cm
- Mean head circumference at birth: 34.8 cm

PHENOTYPE

The children often have delayed development, sometimes significant tonus disorders, either hypertonia or hypotonia, and a relatively minor facial dysmorphism.

Craniofacial Dysmorphism

The head circumference is normal, but the cranium is deformed by frontal and parietal bossing and occasional protrusion of the occiput.

The form of the face is not characteristic. It sometimes seems small compared to the cranium.

The palpebral fissures are most often slightly slanted downward and outward. They are narrow and slightly almond-shaped. Hypertelorism and epicanthus are inconstant. Strabismus and iridoschisis have been reported.

The nose is small and pointed.

The upper lip, sometimes a bit long, overhangs the lower lip. The latter can be everted. Microretrognathia, and cleft palate in several cases, are present.

The ears are low-set and posteriorly rotated.

Neck, Thorax, and Abdomen

The neck is short in all cases.

The thorax and abdomen are usually normal, with the occasional exception of costal or vertebral anomalies.

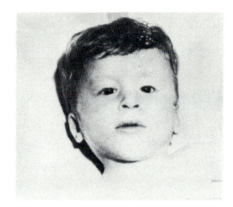

7.1

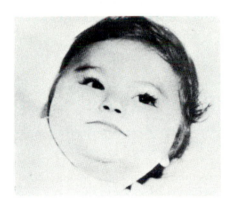

7.2

7.3

7.4

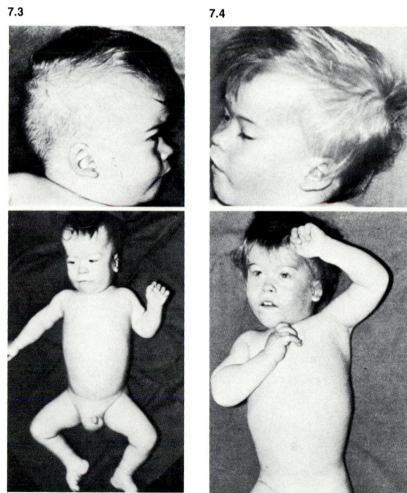

Limbs

Only the following have been reported, each in a single patient: unilateral dislocation of the hip; large thumbs, more proximal than is normal; and abnormally convex fingernails.

Genitalia

These are usually normal.

Malformations

Malformations are more frequently encountered when the trisomy includes the 7q31 band, and involves the brain, heart, and kidneys.

Prognosis

This would appear to depend on the breakpoint. Early death was observed in three out of four cases involving the 7q31 band. The other patients were observed at various ages (from 12 months to 23 years).

CYTOGENETIC STUDY

Apart from one case where the breakpoint was in q22 (Carpentier et al., 1972), trisomy involved the 7q3 region with breakpoints varying between q31 and q33. It nearly always results from a parental rearrangement, a translocation, or, in one case, a pericentric inversion.

DERMATOGLYPHICS

The dermatoglyphics are not characteristic. A unique transverse palmar crease or its equivalent can be observed.

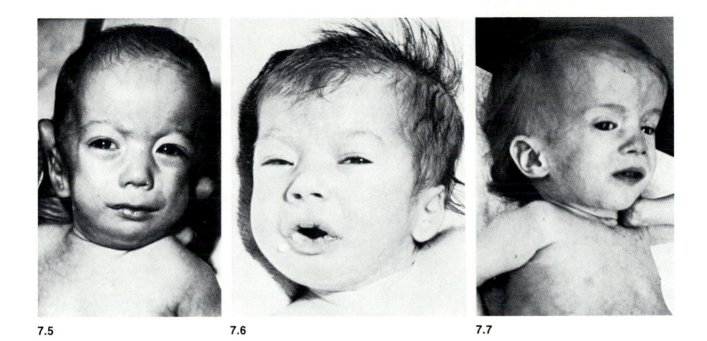

7.5　　　　　　　　7.6　　　　　　　　7.7

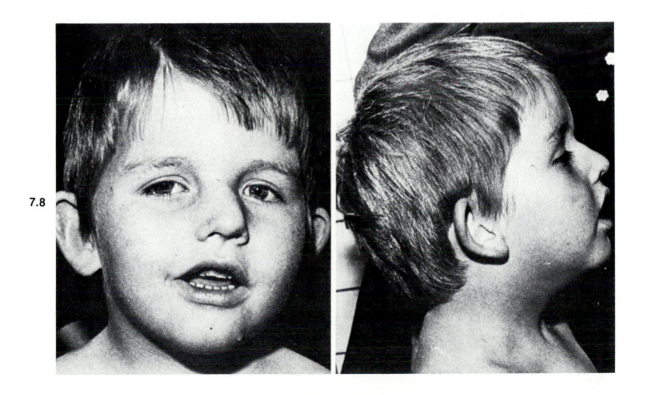

7.8

7q2 Trisomy

REFERENCES

BERGER R., DERRE J., ORTIZ M.A.: Les trisomies partielles du bras long du chromosome 7. *Nouv. Presse méd. Fr.*, **3**:1801–1804, 1974.

SERVILLE F., BROUSTET A., SANDLER B., BOURDEAU M.J., LELOUP M.: Trisomie 7q partielle. *Ann. Génét.*, **18**:67–70, 1975.

Three cases involve an intercalary 7q2 trisomy by insertion (Grace et al., 1973; Serville et al., 1975; Berger et al., 1974). Breakpoints are located in q22 and q31 or q32. Main clinical features are the following:

- Normal staturo-ponderal development
- Extensive hypotonia
- Narrow palpebral fissures, with epicanthus
- Micrognathia, and cleft palate

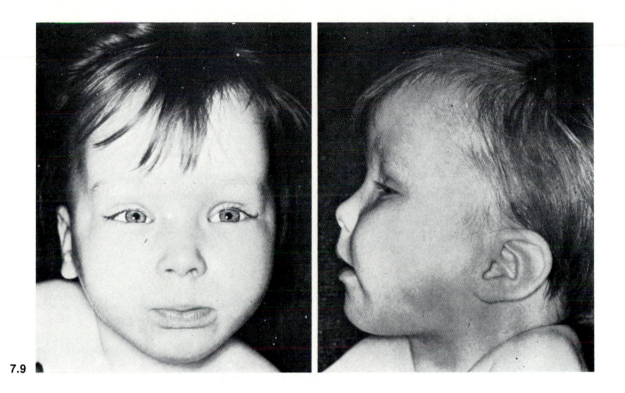

7.9

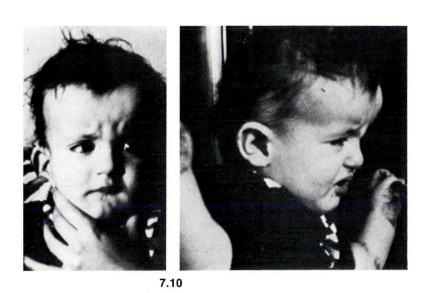

7.10

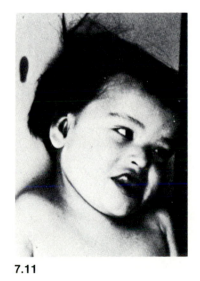

7.11

7q MONOSOMIES

7q3 Monosomy

Microbrachycephaly
Palpebral fissures slanting upward and outward
Thick subcutaneous tissue

REFERENCES

BERNSTEIN R., DAWSON B., MORCOM G., WAGNER J., JENKINS T.: Two unrelated children with distal long arm deletion of chromosome 7: clinical features, cytogenetic and gene marker studies. *Clin. Genet.,* **17**:228–237, 1980.

FRIEDRICH U., ØSTERBALLE O., STEN-BJERG S., JORGENSEN J.: A girl with kary-otype 46,XX,del(7) (pter→q32:). *Hum. Genet.,* **51**:231–235, 1979.

GROUCHY J. de, VESLOT J., BONNETTE J., ROIDOT J.: A case of ?6p- chromosomal aberration. *Amer. J. Dis. Child.,* **115**:93–99, 1968.

GROUCHY J. de, TURLEAU C.: Tentative lo-calization of a Hageman (factor XII) locus on 7q, probably the 7q35 band. *Humangenetik,* **24**:197–200, 1974.

NIELSEN K.B., EGEDE E., MOURIDSEN I., MOHR J.: Familial partial 7q monosomy re-sulting from segregation of an insertional chromosome rearrangement. *J. med. Genet.,* **16**:461–469, 1979.

TURLEAU C., GROUCHY J. de, PERIGNON F., LENOIR G.: Monosomie 7qter. *Ann. Gé-nét.,* **22**:242–244, 1979.

Since the first observation (Grouchy et al., 1969; Grouchy and Turleau, 1974), a num-ber of cases of 7q deletion have been described. The phenotype described here corresponds to terminal deletions which involve all or part of the 7q3 region. Fourteen such deletions have been found (see Turleau et al., 1979; Friedrich et al., 1979; Bern-stein et al., 1980).

One familial observation (3 cases) reported by Nielsen et al. (1979) involves a q32→q34 intercalary deletion and will be considered separately at the end of the chapter.

GENERALIZATIONS

- Sex ratio: 4F/3M
- Mean birth weight: 2,230 g
- Mean crown-heel length at birth: 44.5 cm
- Mean head circumference at birth: 28.6 cm

PHENOTYPE

The most important feature of the phenotype is microcephaly.

Craniofacial Dysmorphism

Microcephaly is very severe and is accompanied by brachycephaly with marked flattening of the occiput. The face is wide and flat, with heavy cheeks.

Palpebral fissures slant upward and outward, and strabismus is often present.

The nose is bulbous with a fleshy, flattened tip.

The mouth, though usually normal in size, may in some cases be rather large. Cleft harelip is often encountered.

The chin is small and round. The presence of a thick subcutaneous tissue, as well as the shortness of the neck, often give the impression of a double chin.

The ears are almost always large and simple.

Neck, Thorax, and Abdomen

A slight pectus excavatum, as well as abnormally widely-spaced nipples, diastasis recti, and hernias, may be observed.

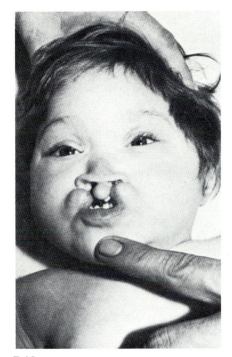

7.12

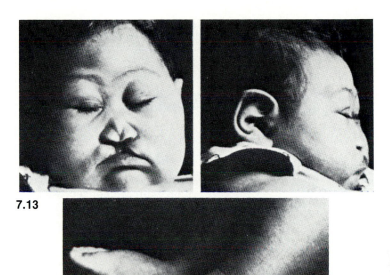

7.13

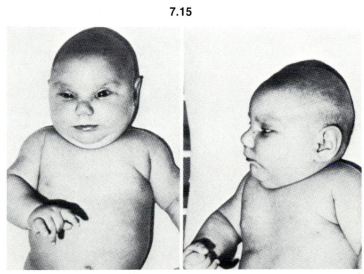

7.14

7.15

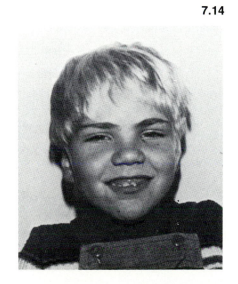

Limbs

Syndactyly, symbrachydactyly, clinodactyly of the Vth, and overlapping of the toes may be present. The back of the feet and hands may be swollen because of the very peculiar thickening of the subcutaneous tissue.

Genitalia

Micropenis, with or without cryptorchidism, may be noted.

Malformations

Inner organ malformations are rare. In one case, a serious malformation of the brain, with arhinencephaly and atrophy of the optic nerve, was observed.

Mental Retardation

Mental retardation is constantly present and is severe. Abnormal muscle tone is also regularly found, and often combines axial hypotonia with peripheral hypertonia. Seizures occur frequently.

Prognosis

Septicemia, and respiratory, urinary, and ear infections are frequently found.

In the case of *de novo* 7q3 deletion, survival does not appear to be affected; only one child died of septicemia.

CYTOGENETIC STUDY

The majority of cases occur *de novo,* with the breakpoint usually being in q32, and in rare cases, more distal in q35. A minority of cases are due to malsegregation of a parental rearrangement.

DERMATOGLYPHICS

Dermatoglyphics are not characteristic. A single transverse palmar crease is frequently found.

q32→q34 DELETION

One familial observation concerned three monosomic children who are 7q32→q34 monosomic due to malsegregation of a parental insertion (Nielsen et al., 1979).

Certain clinical features in this case are present in terminal monosomy, including hypertelorism, a wide bridge of the nose, a fleshy nose tip, a large mouth, urinary infections, and seizures. Neither microcephaly nor harelip are observed. Birth weight was normal, as well as postnatal growth. Mental retardation is moderate in two out of the three patients.

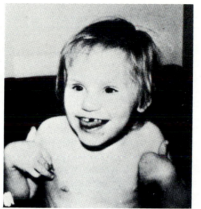

7.16

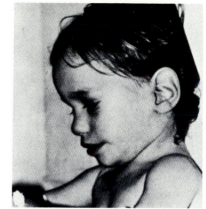

7.17

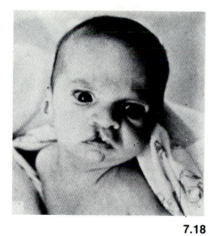

7.18

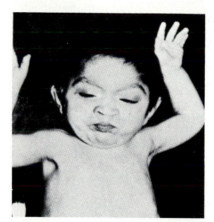

7.19

7.21

7.20

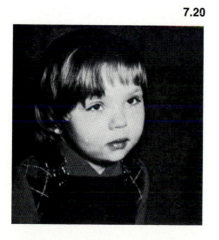

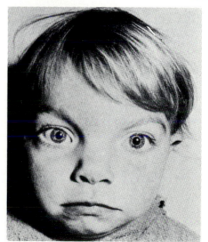

7q2 Monosomy

REFERENCES

AYRAUD N., ROVINSKI J., LAMBERT J.C., GALIANA A.: Délétion interstitielle du bras long d'un chromosome 7 chez une enfant lepréchaune. *Ann. Génét.,* **19**:265–268, 1976.

FRANCESCHINI P., SILENGO M.C., DAVI G.F., SANTORO M.A., PRANDI G., FABRIS C.: Interstitial deletion of the long arm of chromosome 7: 46,XX, del(7) (pter→ q2200::q3200→qter). *Hum. Genet.,* **44**:345– 348, 1978.

HIGGINSON G., WEAVER D.D., MAGENIS R.E., PRESCOTT G.H., HAAG C., HEPBURN D.J.: Interstitial deletion of the long arm of chromosome n° 7 (7q−) in an infant with multiple anomalies. *Clin. Genet.,* **10**:307– 312, 1976.

KLEP-DE PATER J.M., BIJLSMA J.B., BLEE- KER-WAGEMAKERS E.M., DE FRANCE H.F., DE VRIES-EKKERS C.M.A.M.: Two cases with different deletions of the long arm of chromosome 7. *J. med. Genet.,* **16**:151– 154, 1979.

SERUP L.: Interstitial deletion of the long arm of chromosome 7. *Hum. Genet.,* **54**:19– 23, 1980.

Breakpoints are usually found in q21 or q22, and q31 or q32. Five cases have been found (Ayraud et al., 1976; Higginson et al., 1976; Franceschini et al., 1978; Kleb-de Pater et al., 1979; Serup, 1980), all involving girls.

Mean birth weight is 2,210 g, and mean crown-heel length at birth is 45 cm.

These children are brachycephalic, with a narrow but prominent forehead. Aplasia of the supraorbital margin is often found.

Palpebral fissures are small, and slant upward and outward. The nose has a bulbous tip with enlarged alae. The philtrum is short.

The mouth is large, with downturned corners.

Cheeks are heavy, and the chin is small and receding. The ears may be low-set and large; the inner part of the helix is sometimes folded in an abnormal manner; the lobe is underdeveloped.

Minor anomalies such as flaccid skin and wide-spaced nipples have been reported. Hands and feet may be large.

Malformations include two cases of congenital heart disease and one case of arhinencephaly.

The neonatal period is marked by nutritional and respiratory difficulties. Mental and growth retardation are severe. Two children died at an early age.

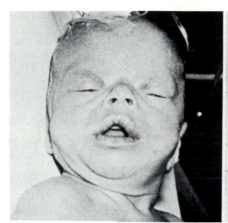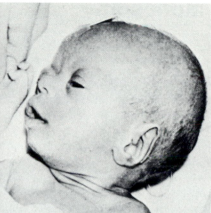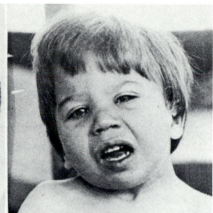

7.22

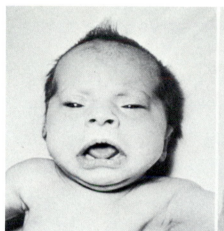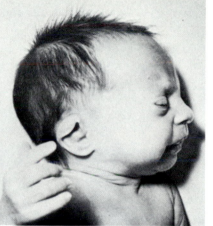

7.23

7.24

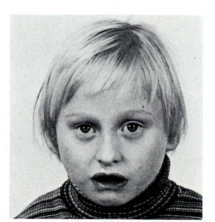

7q1 Monosomy

REFERENCES

CRAWFURD M. d'A., KESSEL I., LIBERMAN M., McKEOWN J.A., MANDALIA P.Y., RIDLER M.A.C.: Partial monosomy 7 with interstitial deletions in two infants with differing congenital abnormalities. *J. med. Genet.*, **16**:453–460, 1979.

SEABRIGHT M., LEWIS G.M.: Interstitial deletion of chromosome 7 detected in three unrelated patients. *Hum. Genet.*, **42**:223–226, 1978.

VALENTINE H., SERGOVITCH F.: A syndrome associated with interstitial deletion of chromosome 7q. *Birth Defects, Orig. Art. Ser. XIII*, 3B:261–262, 1977.

Five cases, including three boys and two girls, are known (Valentine and Sergovitch, 1976; Seabright and Lewis, 1978; Kleb-de Pater et al., 1979; Crawfurd et al., 1979). Breakpoints are usually found in q11 and in q21 or q22.

Mean birth weight is 2,270 g.

The description of the patients is usually rather concise: craniofacial dysmorphism is not very characteristic, and does not usually attract attention. The following should be noted:

- Signs of brain damage, with nutritional troubles, hypotonia, myoclonic seizures and convulsions
- Growth retardation
- Microcephaly, or brachycephaly, with a prominent forehead
- Rather rough facial features, and a gaping mouth
- Diaphragmatic, hiatal, or inguinal hernias
- Mental retardation is variable, ranging from very severe to subnormal

Only one child has congenital heart disease. All children are reported alive.

7.25

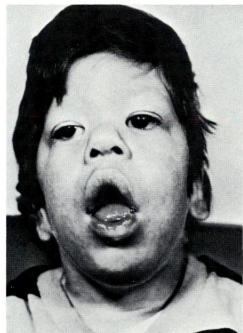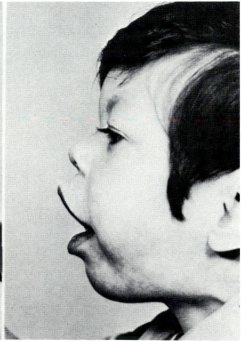

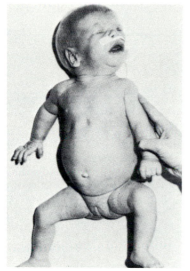

7.26

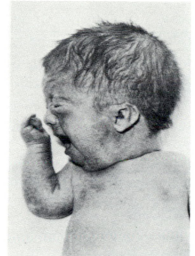

7.27

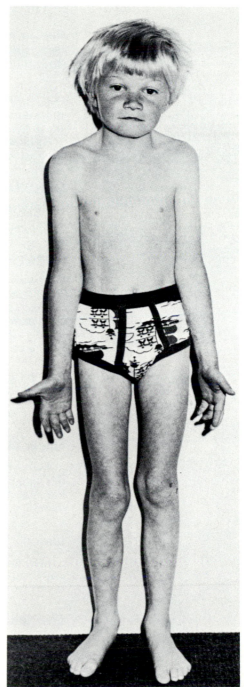

7.28

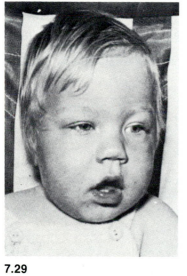

7.29

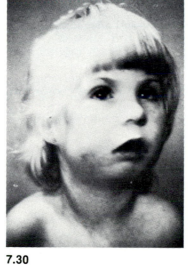

7.30

7p2 TRISOMY

Gaping fontanels and cranial sutures
Hypertelorism
Round cheeks

REFERENCES

BERRY C., HONEYCOMBE J., MACOUN S.J.R.: Two children with partial trisomy for 7p. *J. med. Genet.*, **16**:320–321, 1979.

CARNEVALE A., FRIAS S., DEL CASTILLO V.: Partial trisomy of the short arm of chromosome 7 due to a familial translocation rcp (7;14) (p11;p11). *Clin. Genet.*, **14**:202–206, 1978.

LARSON L.M., WASDAHL W.A., JALAL S.M.: Partial trisomy 7p associated with familial, 7p;22q translocation. *J. med. Genet.*, **14**:258–261, 1977.

MILLER M., KAUFMAN G., REED G., BILEN-KER R., SCHINZEL A.: Familial, balanced insertional translocation of chromosome 7 leading to offspring with deletion and duplication of the inserted segment 7p15→7p21. *Am. J. med. Genet.*, **4**:323–332, 1979.

Five observations of trisomy 7p are known. Four concern the 7p2 region (Larson et al., 1977; Berry et al., 1979; Miller et al., 1979). One of them involves all of 7p (Carnevale et al., 1978).

Several clinical features, and notably, the defect in closure of the cranial sutures, contrast with the symptoms for monosomy 7p2.

GENERALIZATIONS

- Sex ratio: 3M/2F
- Mean birth weight: 2,600 g

PHENOTYPE

The skull is generally elongated, with a prominent occiput. Micro- or macrocephaly may be observed. A gaping of the fontanels and sagittal or metopic sutures, the width of which may attain 6 cm, is the most noteworthy feature.

Hypertelorism is present. Palpebral fissures are short and vary in orientation, but usually slant downward and outward.

The nose is short and slightly beaked; nares are somewhat flattened.

Cheeks are round.

The corners of the mouth are downturned. Lips are fleshy; a high arched palate, cleft palate, and bifid uvula may be observed.

The chin is rather small.

Ears may be low-set and simple.

Neck, Thorax, and Abdomen

In the one case where trisomy involved all of 7p, a slender neck and an absence of, or hypoplasia of, the sterno-cleidomastoids was noted, as well as a dislocation of the humeral heads, a triangular-shaped thorax, and cutaneous dimples opposite the joints.

Limbs

The following anomalies have been noted: camptodactylies, spatular fingers, arachnodactyly, malformations of the wrists, dislocation of the hip, club foot, and rocker bottom feet.

Genitalia

Cryptorchidism and a small penis have been reported in one boy. Hypertrophy of the clitoris and anal imperforation have been noted in a girl.

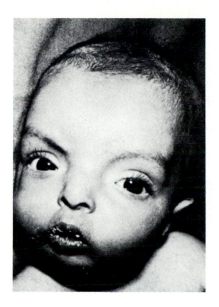 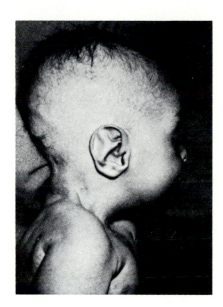

7.31

Malformations

Anomalies of the cranial sutures have been confirmed by X-ray films that, in one case, revealed a deformity of the interorbital region resembling that seen in frontal or fronto-ethmoidal omphaloceles.

Severe congenital heart disease and cerebral malformations are observed.

Mental Retardation

Mental retardation is considerable.

Prognosis

This does not appear to be encouraging. Three children died before 7 months of age; the oldest living child is now 9.

Growth retardation is considerable.

CYTOGENETIC STUDY

In all five cases, the trisomy results from a parental rearrangement, including an intrachromosomal insertion.

Breakpoints are variable, from p11 to p21. A comparison of these breakpoints and the phenotypes suggests that the anomalies in closing of the cranial sutures are associated with trisomy for the 7p21 banding.

DERMATOGLYPHICS

These do not appear to be significant.

7p2 MONOSOMY

Craniosynostosis
Palpebral fissures slanting downward and outward
Hypotelorism
Joint deformities

REFERENCES

DHADIAL R.K., SMITH M.F.: Terminal 7p deletion and 1;7 translocation associated with craniosynostosis. *Hum. Genet.*, **50**:285–289, 1979.

FRIEDRICH U., LYNGBYE T., OSTER J.: A girl with karyotype 46,XX, del(7) (qter→p15). *Humangenetik*, **26**:161–165, 1975.

McPHERSON E., HALL J.G., HICKMAN R.: Chromosome 7 short arm deletion and craniosynostosis. A 7p− syndrome. *Hum. Genet.*, **35**:117–123, 1976.

WILSON M.G., FUJIMOTO A., SHINNO N.W., TOWNER J.W.: Giant satellites or translocation? *Cytogenet. Cell Genet.*, **12**:209–214, 1973.

Several cases of 7p monosomy have been reported, but we have only considered the four comparable observations in which the monosomy involves 7p2 (Wilson et al., 1973; Friedrich et al., 1975; McPhersen et al., 1976; Dhadial and Smith, 1979). We have not considered those observations which involved interstitial deletions or complex rearrangements leading to a different phenotype.

The phenotype for monosomy 7p2 is noteworthy mainly because of the presence of a craniosynostosis which in type and countertype sharply contrasts with the gaping sutures observed in trisomy 7p2.

GENERALIZATIONS

- Sex ratio: 3F/1M
- Mean birth weight: 2,600 g
- Mean crown-heel length at birth: 49.2 cm

PHENOTYPE

Craniofacial Dysmorphism

The skull is very abnormal in form due to premature closure of the cranial sutures. This may be predominant on the sagittal or metopic suture, creating either a turricephaly or a trigonocephaly. The forehead is high and narrow, and the occiput is flat. These deformities lead to effacement of the orbital margins and hypotelorism. Palpebral fissures usually slant downward and outward. The eyes appear to be protruding. Ptosis, epicanthus, and anomalies in ocular mobility may be observed.

The bridge of the nose is flat, and the nose is snubbed with anteverted nares.

The mouth may be small and triangular, with a high-arched palate. Cleft palate and bifid uvula have been reported. Minor retrognathia may be noted. The ears are low-set, simple, and in a posterior rotation.

Limbs

The fingers are long, tapering, or spatular, or with clawlike nails. Flexion/extension contractures, with abnormal positioning of the fingers and thumbs, may be observed, as well as clubfoot or rocker-bottom feet. Impairment of extension may affect the large joints.

Malformations

Cardiac malformations have been observed (aortic stenosis, or moderate ventricular septal defect).

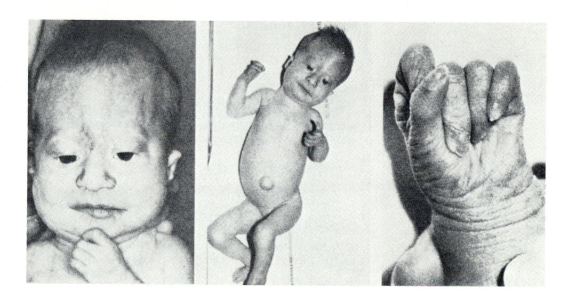

7.32

7.33

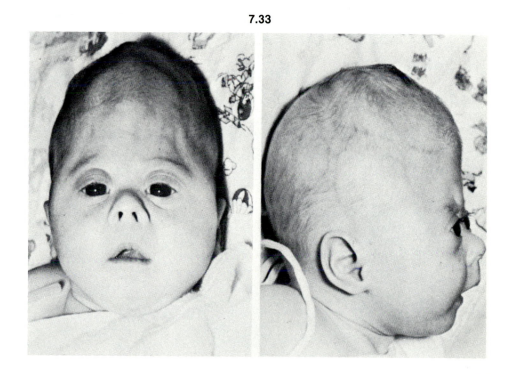

Mental Retardation

Psychomotor retardation is very difficult to evaluate due to the young age of the patients.

Prognosis

These children were observed at several months of age. Their later development is unknown. One of them died from bronchial pneumonia at the age of 10 weeks.
Staturo-ponderal retardation is considerable.

CYTOGENETIC STUDY

In the four cases considered here, the deletion is *de novo*, with a breakpoint in p15.

DERMATOGLYPHICS

A single transverse palmar crease, or its equivalent, is usually found. The two children for whom dermatoglyphics were described showed an excess of arches.

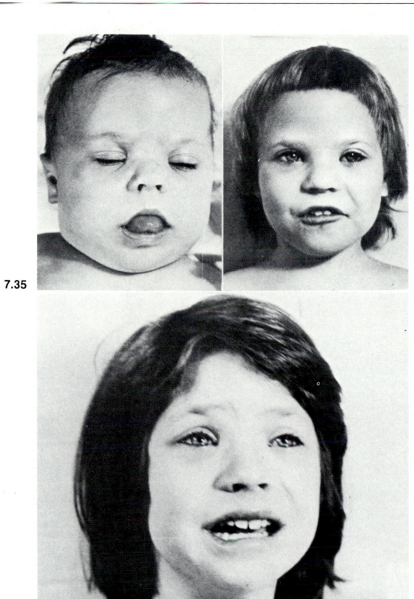

7.35

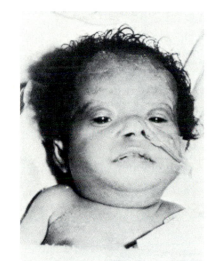

7.34

CHROMOSOME 8

Few gene assignments are known for chromosome 8. Among them are the gene for glutathione reductase, the deficiency of which causes a variety of hemolytic anemia. Glutathione reductase shows a gene dosage effect, and it was through the observations of patients afflicted with chromosomal imbalance that this gene was precisely located in 8p21.

8 trisomy mosaicism is one of the major cytogenetic syndromes. Partial 8q and 8p trisomies are also known, as well as monosomy for 8p.

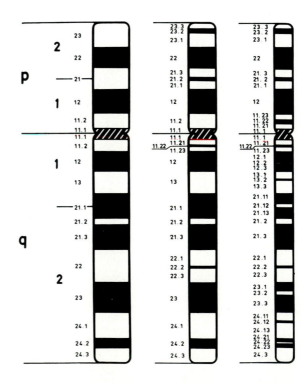

GENE MAP

8p21	FNE	(L)	Fibronectin (13560) S
	GSR		Glutathione reductase (13830) S, D
	CF7E	(P)	Clotting factor VII expression (22750) D
8q23	LGS		Langer-Giedion syndrome (15023) Ch
	GPT2		Glutamate-pyruvate transaminase, soluble liver (13822) S
8q22	MOS		Onc gene: Moloney murine sarcoma virus (19006) S
8q24	MYC		Protooncogene: avian myelocytomatosis virus (19008) A
8q24.3	NLP1	(L)	Neoplastic lymphoproliferation-1 (16184) Ch
?chr 8 or 12	SPH	(L)	Spherocytosis (18290) F

MORPHOLOGY AND BANDING

 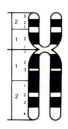      

QFQ   GTG   RHG   RBA   THA   C

8 TRISOMY

Long face
Thick, everted lower lip
Osteoarticular anomalies

REFERENCES

BERRY A.C., MUTTON D.E., LEWIS D.G.M.: Mosaicism and the trisomy 8 syndrome. *Clin. Genet.*, **14**:105–114, 1978.

CASPERSSON T., LINDSTEN J., ZECH L., BUCKTON K.E., PRICE W.H.: Four patients with trisomy 8 identified by the fluorescence and Giemsa banding technique. *J. med. Genet.*, **9**:1–7, 1972.

CHANDLEY A.C., HARGREAVE T.B., FLETCHER J.M., SOOS M., AXWORTHY D., PRICE W.H.: Trisomy 8. Report of a mosaic human male with near-normal phenotype and normal IQ ascertained through infertility. *Hum. Genet.*, **55**:31–38, 1980.

GAFTER U., SHABTAI F., KAHN Y., HALBRECHT I., DJALDETTI M.: Aplastic anemia followed by leukemia in congenital trisomy 8 mosaicism. Ultrastructural studies of polymorphonuclear cells in peripheral blood. *Clin. Genet.*, **9**:134–142, 1976.

GROUCHY J. de, TURLEAU C., LEONARD C.: Etude en fluorescence d'une trisomie C mosaïque probablement 8: 46,XY/47,XY, ?8+. *Ann. Genet.*, **14**:69–72, 1971.

LA CHAPELLE A. de, ICEN A., AULA P., LEISTI J., TURLEAU C., GROUCHY J. de: Mapping of the gene for glutathione reductase on chromosome 8. *Ann. Génét.*, **19**:253–256, 1976.

LA CHAPELLE A. de, VUOPIO P., ICEN A.: Trisomy 8 in the bone marrow associated with high red cell glutathione reductase activity. *Blood*, **47**:815–827, 1976.

PFEIFFER R.A.: Trisomy 8. *In* J.J. YUNIS: *New Chromosomal Syndromes*, New York, Academic Press, pp. 197–217, 1977.

RETHORE M.O., AURIAS A., COUTURIER J., DUTRILLAUX B., PRIEUR M., LEJEUNE J.: Chromosome 8: trisomie complète et trisomies segmentaires. *Ann. Génét.*, **20**:5–11, 1977.

RICCARDI V.M.: Trisomy 8: an international study of 70 patients. *Birth Defects, Orig. Art. Ser., XIII-3C*:171–184, 1977.

RODEWALD A., ZANKL H., WISCHERATH H., BORKOWSKY-FEHR B.: Dermatoglyphic patterns in trisomy 8 syndrome. *Clin. Genet.*, **12**:28–38, 1977.

SINET P.M., BRESSON J.L., COUTURIER J., LAURENT C., PRIEUR M., RETHORE M.O., TAILLEMITE J.L., TOUDIC D., JEROME H., LEJEUNE J.: Localisation probable du gene de la glutathion réductase (EC 1.6.4.2) sur la bande 8p21. *Ann. Génét.*, **20**:13–17, 1977.

The first case of trisomy 8 authenticated by banding was reported by Grouchy et al. in 1971. Trisomy 8 is a relatively frequently encountered syndrome. Seventy cases have been reviewed by Riccardi (1977), Pfeiffer (1977), and Berry et al. (1978).

The syndrome was clearly described by Rethoré et al. (1977).

GENERALIZATIONS

All known observations of trisomy 8 are *de novo*. However, the possibility of transmission from parent to child cannot be excluded, given the possibly moderate impairment of intellectual functions. One patient, who was phenotypically normal, was only detected because she had undergone two spontaneous abortions (Caspersson et al., 1972). In the large majority of cases, the trisomy is in the mosaic stage.

- Sex ratio: 3M/1F
- Mean parental ages:
  maternal: 28.1 years
  paternal: 31.1 years
- Mean birth weight: 3,420 g
- Mean crown-heel length at birth: normal

PHENOTYPE

Trisomy 8 is characterized by very suggestive facial dysmorphism, as well as by osteoarticular anomalies which are often rather severe.

Craniofacial Dysmorphism

The volume of the skull is usually normal. The forehead is high and protruding. Macrocephaly may be present.

The face is elongated and fairly harmonious in appearance, with eversion of the lower lip.

The eyes are normal in shape and orientation, but moderate hypertelorism, ptosis, and strabismus, are often present.

The nose is wide and snubbed.

The upper lip is long, and its fleshy part highly visible. But it is the lower lip that is especially noteworthy: it is thick and everted, and immediately suggests a diagnosis of trisomy 8. Microretrognathia contributes toward the large size of the mouth. A horizontal dimple is present on the chin.

The ears are abnormal in shape. The auricle is large, and the anthelix is prominent. The fold of the helix is well defined in its upper part; in its lower part, it either disappears, or becomes ravelled inside out.

The palate is high-arched.

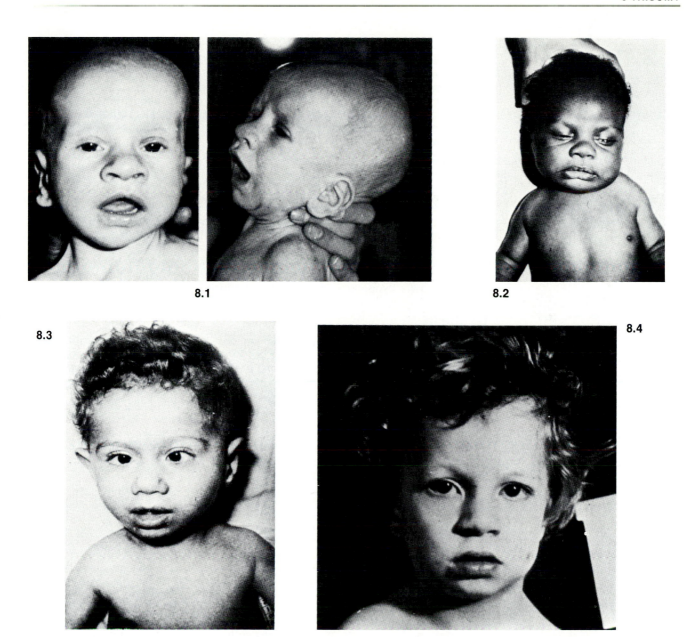

8.1

8.2

8.3

8.4

Neck, Thorax, and Abdomen

The neck is short and wide, and contrasts with the narrowness of the shoulders and the length of the trunk.

Skeletal malformations are encountered frequently and are often quite severe. These include dorsolumbar kyphoscoliosis, abnormal or supernumerary vertebrae, spina bifida, supernumerary ribs, funnel chest, and hypoplastic or narrow pelvis. The nipples are abnormally wide-spaced, and supernumerary nipples may be present. The buttocks are flat (tobacco pouch).

Limbs

Osteoarticular lesions are nearly always found, which include brachydactylies or arachnodactylies, clinodactylies, camptodactylies, knotted joints, clubfoot, and hallux valgus. A bilateral absence of the patella has been reported several times.

Joint anomalies are not always detectable at birth. Contractures appear with age, leading to permanent deformities.

One highly characteristic symptom may be noted especially in very young infants: this is the existence of very deep palmar and plantar skin furrows ("plis capitonnés").

The nails are hypoplastic and convex, and may be entirely absent at birth.

Genitalia

In boys, several cases of cryptorchidism, testicular hypoplasia, and retarded puberty have been reported.

One adult patient with a nearly normal phenotype consulted for a primary sterility due to severe oligospermia. However, we have no precise data on the fertility of trisomy 8 subjects (Chandley et al., 1980).

Malformations

Inner organ malformations are rare, and present no danger to the life of the patient. Congenital heart disease may be observed, but renal and ureteral malformations would appear to be more characteristic.

X-ray films confirm the gravity and variety of the osteoarticular malformations. Bone age may be retarded.

Mental Retardation

Except in a few rare cases, mental deficiency is not very pronounced. IQ varies between 12 and 94, with an average of between 45 and 75. It may even be normal and, when combined with a fairly eumorphic phenotype, may mean that some cases have not been ascertained. In addition, in one-third of the cases, the diagnosis was performed in adulthood.

In general, patients are calm, even-tempered and sociable, but occasionally anxiety-ridden, or given to fits of anger. They suffer from language difficulties. Seizures have been reported.

Prognosis

Physical growth and development are highly variable. Subjects are long-limbed, but height and weight are generally lower than normal in the adult.

In one female patient with trisomy 8 mosaicism, acute leukemia, following medullary aplasia, appeared at the age of 40 (Gafter et al., 1976).

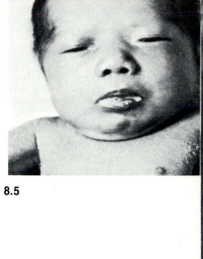
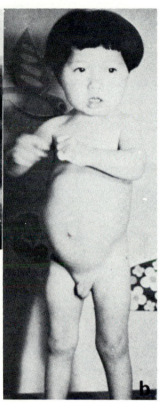

8.5

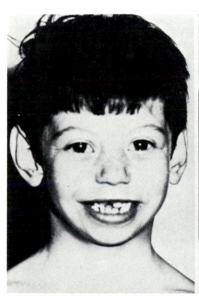
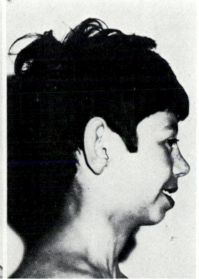

8.6

CYTOGENETIC STUDY

Trisomy 8 is typified by the very high frequency of mosaicism. It is also noteworthy that the relative frequency of the trisomy 8 cell population compared with the normal cell population decreases with age in several observations.

There does not seem to be any correlation between phenotypical impairment and the relative size of the trisomic clone.

DERMATOGLYPHICS

An examination of palms and plantar soles confirms the existence of deep skin furrows ("plis capitonnés") in nursing infants. The principle dermatoglyphic features of the hands are as follows (Rodewald et al., 1977):

- In nearly half of the cases, a single transverse palmar crease
- An axial triradius in t' or t" in two-thirds of the cases
- A very high frequency of thenar and hypothenar patterns
- A decreased index of transversality
- A high frequency of arches on the fingertips

The following may be noted on the feet:

- Deep longitudinal furrows, especially between the first and second toes
- A high intensity of plantar patterns
- An arch on the big toe

LABORATORY FINDINGS

The gene assignment for glutathione reductase to 8p21 and the possibility of measuring the activity of this enzyme mean that this biological assay is an important test when confronted with trisomy 8 (La Chapelle, 1976; Sinet et al., 1977).

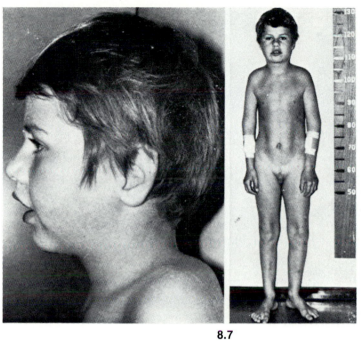

8.7

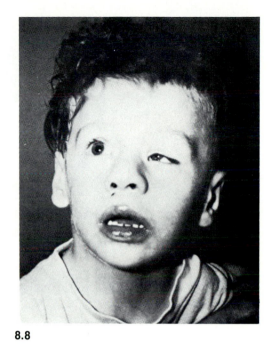

8.8

8.9

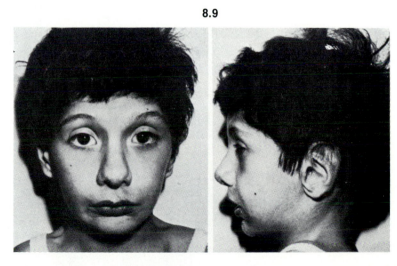

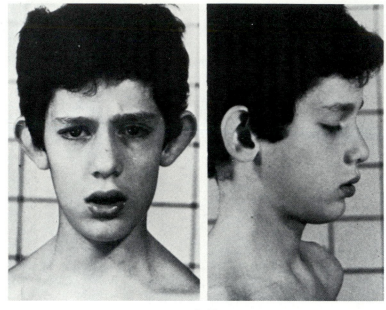

8.10

8.11

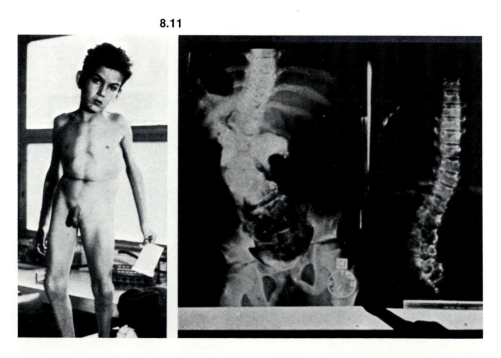

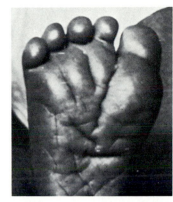

8.12

8.13

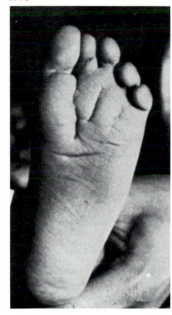

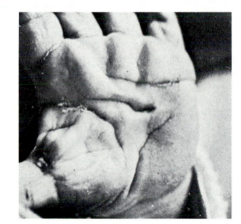

8.14

8.16

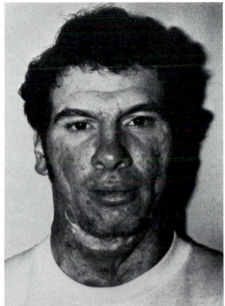

8.15

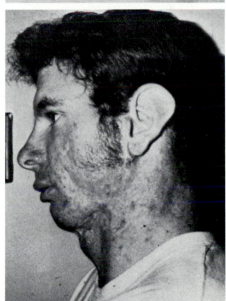

8q2 TRISOMY

Prominent forehead
Blunt nasofrontal angle
Microretrognathia

REFERENCES

ABUELO D., PERL D.P., HENKLE C., RICHARDSON A.: Partial trisomy 8 (trisomy 8q2106→8qter). *J. med. Genet.*, **14**:463–466, 1977.

BALLESTA F., FERNANDEZ E., MILA M.: Translocation t(1;8;15) maternelle et trisomie 8qter chez la fille. *J. Génét. hum.*, **28**:361–366, 1980.

FINEMAN R.M., ABLOW R.C., BREG W.R., WING D., ROSE J.S., ROTHMAN L.G., WARPINSKI J.: Complete and partial trisomy of different segments of chromosome 8: case reports and review. *Clin. Genet.*, **16**:390–398, 1979.

LEJEUNE J., RETHORE M.O., DUTRILLAUX B., MARTIN G.: Translocation 8-22 sans changement de longueur et trisomie partielle 8q. *Exp. Cell Res.*, **74**:293–295, 1972.

RETHORE M.O., AURIAS A., COUTURIER J., DUTRILLAUX B., PRIEUR M., LEJEUNE J.: Chromosome 8: trisomie complète et trisomies segmentaires. *Ann. Génét.*, **20**:5–11, 1977.

The first observation of 8qter trisomy was that of Lejeune et al. (1972). Sixteen cases are currently known (see Rethoré et al., 1977; Fineman et al., 1979; Abuelo et al., 1977; and Ballesta et al., 1980).

GENERALIZATIONS

- Sex ratio: 5M/3F
- Mean birth weight: 2,940 g

PHENOTYPE

Craniofacial Dysmorphism

The forehead is prominent, but may be high or low. Myelomeningocele and a frontal cyst have been observed. Hypertelorism is constantly present and is sometimes extensive. Palpebral fissures usually slant upward and outward. Eyelids often form a double fold. The bridge of the nose is very wide, with a blunt nasofrontal angle. The nose is wide, with a short septum.

The upper lip is short and may be everted. The lower lip appears to droop slightly, and to form a horizontal chin dimple. Buccal deformities are often observed, including bifid tip of the tongue, cleft upper gum, abnormal frenula, bifid uvula, high arched or cleft palate.

Microretrognathia is regularly found.

The ears are often poorly folded, possibly low-set or in posterior rotation, and detached, especially in older children.

Neck, Thorax, and Abdomen

The neck is short and wide.

The thorax is elongated. A depression of the mesosternum, as well as deformities of the spinal column, may be observed.

Limbs

Contractures reduce the play of certain joints. Brachydactylies, clinodactylies, and camptodactylies are also noted. The toes may often overlap. Only one case of deep palmar skin furrows has been reported.

Genitalia

Anomalies such as micropenis, hypospadias, and cryptorchidism have been noted in several cases.

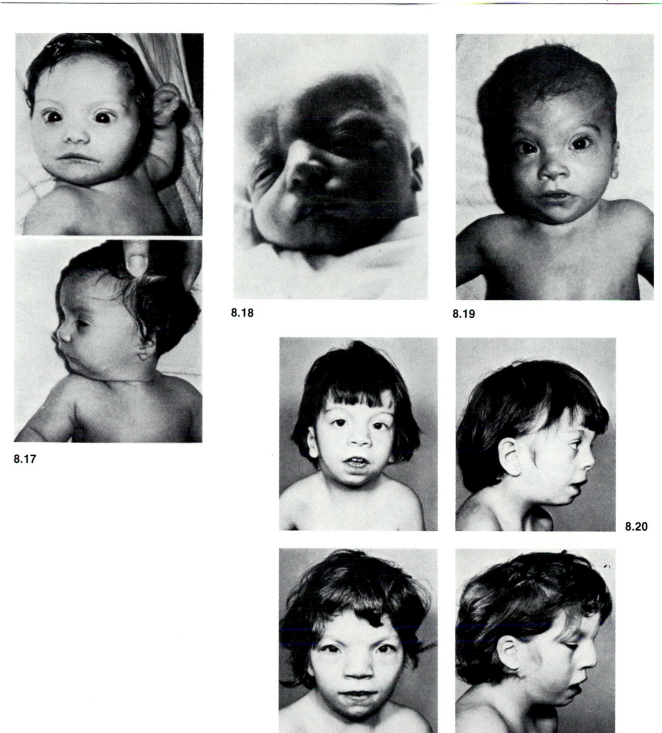

8.17

8.18

8.19

8.20

8.21

Malformations

Osteoarticular malformations have already been mentioned. They seem to appear less regularly than had previously been thought according to the first publications. Cardiac malformations, however, are found much more frequently, and are much more severe than had initially been considered. They are seen in more than half of the cases, and most often involve Fallot's tetralogy.

Among other malformations that occur sporadically are one case of the Treacher Collins syndrome, with agenesis of the thymus and gall bladder, cerebral malformations of the Dandy Walker type, anomalies of the bones of the skull, one case of spina bifida, one case of hydronephrosis with hydroureter, one pilonidal sinus, and a hypertrichosis.

Mental Retardation

This is variable; in several cases it was extremely severe, with an IQ of less than 10.

Prognosis

Somatic development is retarded. The children were observed at various ages; four died before the age of 7 months. Several patients were seen in adulthood, with the oldest aged 29.

CYTOGENETIC STUDY

The trisomy usually results from a parental translocation and, in one case, from a pericentric inversion. Several cases involve close relatives. The breakpoint on 8 varies from q21 to q24.

DERMATOGLYPHICS

There is an absence of deep furrows, and an elevation of the axial triradius in t' or t'', as well as a prolongation of the distal flexion crease in 11.

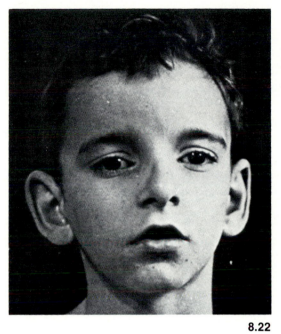
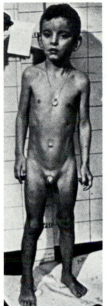
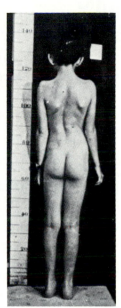
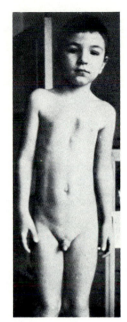

8.22

8.23

8.24

8.25

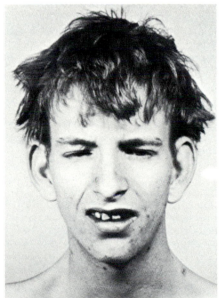
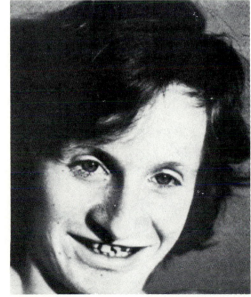

8p TRISOMY

High, prominent forehead
Fleshy, everted lower lip

REFERENCES

CLARK C.E., TELFER M.A., COWELL H.R.: A case of partial trisomy 8p resulting from a maternal balanced translocation. *Am. J. med. Genet.,* **7**:21-25, 1980.

FINEMAN R.M., ABLOW R.C., BREG W.R., WING D., ROSE J.S., ROTHMAN L.G., WARPINSKI J.: Complete and partial trisomy of different segments of chromosome 8: case reports and review. *Clin. Genet.,* **16**:390-398, 1979.

FUNDERBURK S.J., BARRETT C.T., KLISAK I.: Report of a trisomy 8p infant with carrier father. *Ann. Génét.,* **21**:219-222, 1978.

JONES L.A., DENGLER D.R., TAYSI K., SHACKELFORD G.D., HARTMANN A.F.: Partial trisomy of the short arm of chromosome 8 resulting from balanced maternal translocation. *J. med. Genet.,* **17**:232-235, 1980.

LA CHAPELLE A. de, ICEN A., AULA P., LEISTI J., TURLEAU C., GROUCHY J. de: Mapping of the gene for glutathione reductase on chromosome 8. *Ann. Génét.,* **19**:253-256, 1976.

LAZJUK G.I., LURIE I.W., USOVA Y.I., GUREVICH D.B., NEDZVED M.K.: Trisomy 8p due to the 3:1 segregation of the balanced translocation t(8;15)mat. *Hum. Genet.,* **46**:335-339, 1979.

MATTEI J.F., MATTEI M.G., ARDISSONE J.P., COIGNET J., GIRAUD F.: Clinical, enzyme, and cytogenetic investigations in three new cases of trisomy 8p. *Hum. Genet.,* **53**:315-321, 1980.

YANAGISAWA S., HIRAOKA K.: Familial G/C translocation in three relatives associated with severe mental retardation, short stature, unusual dermatoglyphics and other malformations. *J. ment. Defic. Res.,* **15**:136-146, 1971.

YANAGISAWA S.: Partial trisomy: further observation of a familial C/G translocation chromosome identified by the Q-staining method. *J. ment. Defic. Res.,* **17**:28-32, 1973.

The first observation of 8p trisomy was a familial one involving three brothers (Yanagisawa and Hiraoka, 1971; Yanagisawa, 1973). Eighteen cases are presently known (see Mattei et al., 1980; Funderburk et al., 1978; Lazjuk et al., 1979; Fineman et al., 1979; Jones et al., 1980; and Clark et al., 1980).

GENERALIZATIONS

- Sex ratio: 5M/4F
- Mean birth weight: 2,885 g
- Mean crown-heel length at birth: 51.6 cm (for five children)

PHENOTYPE

The infants are rather big at birth, and remain long limbed, although growth retardation is frequently encountered. On the whole, observed anomalies are relatively minor.

Craniofacial Dysmorphism

The forehead is high and prominent, with slight temporal retraction and prominence of the parietals.

In children, the face is wide with chubby cheeks.

The nose is bulbous and short, with very thick nares. The philtrum is long, with undefined pillars. The mouth is large, and the lower lip is generally fleshy and everted. Cleft palate and a high arched palate have been reported in several cases. The chin is small and receding, although the lower part of the face is large. A horizontal chin dimple may be present. The ears are large, sometimes rectangular, and tend to tilt backwards.

Neck, Thorax, and Abdomen

The neck is short with an excess of skin. The thorax may appear to be long when compared to the rest of the body. Kyphoscoliosis may eventually set in. Inguinal hernias are often noted.

Limbs

The hands are often deformed, and the following may be observed: flexion contractures, hyperextensibility of the finger joints, clinodactyly of the Vth, short fingers with a single flexion crease, and hypoplasia of the nails.

Flexion contractures may occur in the lower limbs, with club feet and coxa valga.

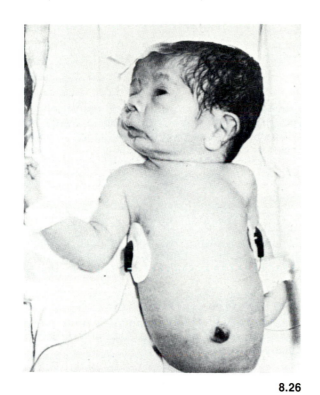

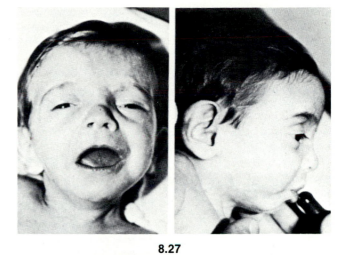

8.27

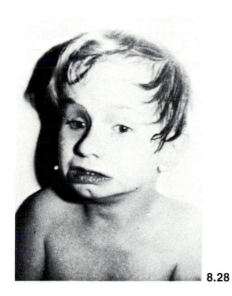

8.26

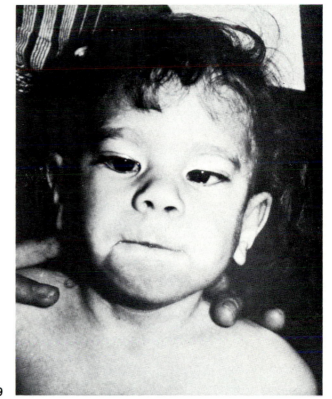

8.28

8.29

Genitalia

The penis is often small, and cryptorchidism may be present.

Malformations

Cardiac malformations, which may at times be rather complex, are observed in about one-third of the cases. Cerebral malformations may also be observed, the most frequent of these being agenesis of the corpus callosum. Other malformations that have been sporadically noted are spina bifida occulta, ureteral stenosis, the absence of the bladder, or a single umbilical artery. X-ray films have shown a delayed bone age, and various osseous dysplasias, such as supernumerary ribs, and anomalies of the phalanges.

Mental Retardation

This is often considerable and is much more severe than with trisomy 8 mosaicism. Several children are unable to walk, and they move about the floor by bracing themselves on their arms.

Prognosis

Two children died very young, one after severe hemorrhaging. The other patients have been observed at various ages, and several are adults. Statural retardation is irregular. One adult female has primary amenorrhea.

CYTOGENETIC STUDY

Trisomy may result from a parental translocation on an acrocentric, in which case it involves all of 8p. In nearly half of the cases, the trisomy occurs *de novo,* and most often is a direct or mirror duplication of nearly all of 8p. The frequency of this rather unusual rearrangement should be emphasized.

In two cases, the trisomy included the most proximal part of the long arm.

DERMATOGLYPHICS

The following may be noted:

- The frequency of a single median palmar crease
- The absence of a digital flexion crease
- An excess of arches
- Deep furrows on the hands have not been mentioned, but deep plantar furrows may be observed in babies.

LABORATORY FINDINGS

The gene for glutathione reductase has been localized in 8p21. A gene dosage effect is observed in trisomies involving this band (La Chapelle et al., 1976).

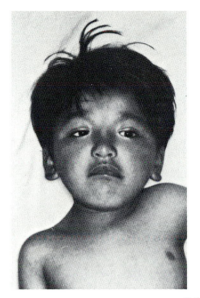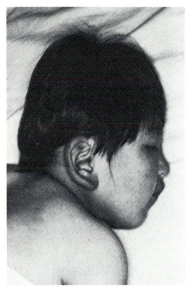
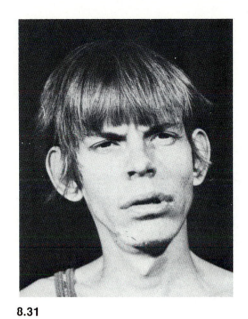

8.30

8.31

8.32

8.33

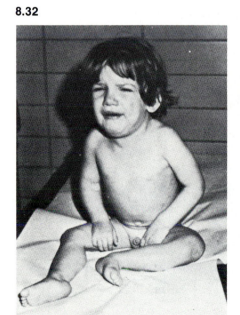
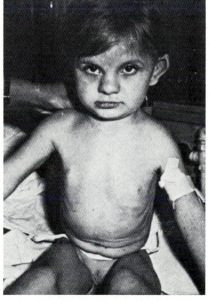

8p2 MONOSOMY

Small skull
Short nose
Small mouth

REFERENCES

LUBS H.A., LUBS M.L.: New cytogenetic technics applied to a series of children with mental retardation. *In* CASPERSSON T., and ZECH L.: *Nobel Symposia 23. Chromosome identification-technique and applications in biology and medicine.* N.Y. and London, Academic Press, pp. 241–250, 1973.

REISS J.A., BRENES P.M., CHAMBERLIN J., MAGENIS R.E., LOVRIEN E.W.: The 8p−syndrome. *Hum. Genet.*, **47**:135–140, 1979.

RODEWALD A., STENGEL-RUTKOWSKI S., SCHULZ P., CLEVE H.: New chromosomal malformation syndromes. I. Partial monosomy 8p. An attempt to establish a new chromosome deletion syndrome. *Eur. J. Pediat.*, **125**:45–58, 1977.

Since the first observation (Lubs and Lubs, 1973), eight observations of 8p2 monosomy have been reported (see Rodewald et al., 1977; Reiss et al., 1980).

GENERALIZATIONS

- Sex ratio: 5M/3F
- Mean birth weight: 2,600 g
- Mean crown-heel length at birth: 45.4 cm (4 cases)

PHENOTYPE

These children are only slightly dysmorphic.

Craniofacial Dysmorphism

The skull is generally small, high and narrow, with a protrusion of the occiput. A slight temporal retraction may be present, and the forehead may be receding.

The nasofrontal angle is relatively undefined. The nose is short, rectilinear, and fairly wide in the middle.

The mouth is small and sometimes slightly triangular. The palate may be high-arched, the teeth poorly implanted, the gonions well-defined, and the chin slightly receding, giving a "pointed" appearance.

The ears are normally implanted, but are sometimes detached and other times tilted backward.

Neck, Thorax, and Abdomen

In babies, the neck is short and wide.
The thorax is wide, with somewhat wide-spaced nipples.

Limbs

Limbs are normal. In several babies, an excess of subcutaneous tissue has been noted on the back of the feet and hands. In one observation, deep furrows were noted on the feet and hands.

Genitalia

Balanic hypospadias and cryptorchidism, as well as testicular hypoplasia in the adolescent, may be noted.

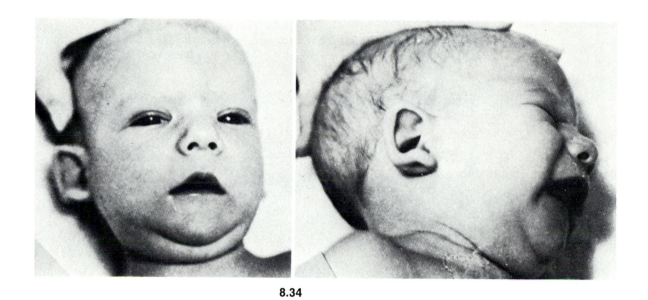

8.34

8.35

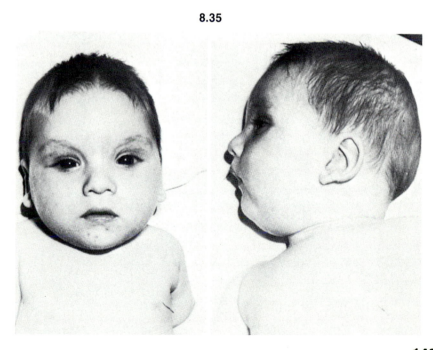

Malformations

Cardiac malformations are found in six out of eight cases, including pulmonary stenosis, with or without ventricular septal defect, or a more complex congenital heart disease.

Mental Retardation

Mental retardation is highly variable, but language difficulties are always present.

Prognosis

Children have been observed either in infancy or in adolescence. One child died at 20 months of bronchopneumonia.

CYTOGENETIC STUDY

In six out of eight cases, deletion is *de novo,* with the breakpoint located in p21 or p22.

DERMATOGLYPHICS

These are not in any way characteristic. A single transverse palmar crease, or its equivalent, may be observed.

LABORATORY FINDINGS

In one case, where deletion includes p21, the activity of glutathione reductase is reduced by half. In another case, where the breakpoint is in p22, it is normal, thus confirming the assignment of the corresponding gene to p21.

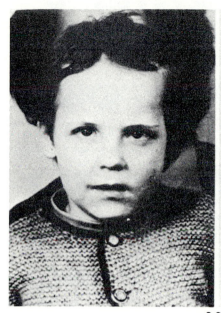
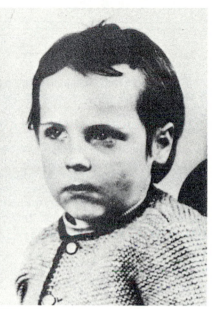

8.36
8.37

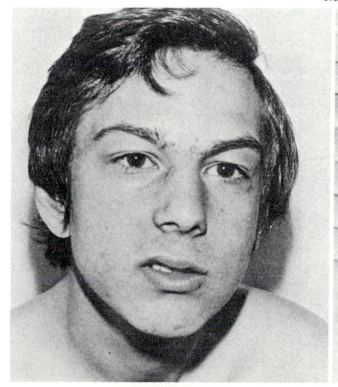
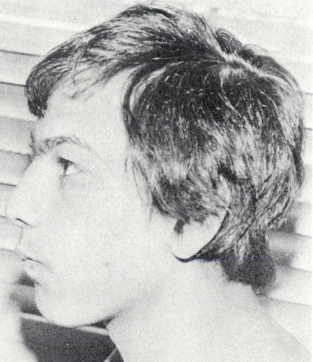

CHROMOSOME 9

REFERENCES

BOUE J., TAILLEMITE J.L., HAZEL-MASSIEUX P., LEONARD C., BOUE A.: Association of pericentric inversion of chromosome 9 and reproductive failure in ten unrelated families. *Humangenetik*, **30**:217–224, 1975.

CHANY C., FINAZ C., WEIL D., VIGNAL M., NGUYEN VAN CONG., GROUCHY J. de.: Investigations on the chromosomal localizations of the human and chimpanzee interferon genes. *Ann. Génét.*, **23**:201–207, 1980.

FERGUSON-SMITH M.A., AITKEN D.A., TURLEAU C., GROUCHY J. de.: Localisation of the human *ABO:Np-1:AK-1* linkage group by regional assignment of AK-1 to 9q34. *Humangenetik*, **34**:35–43, 1976.

LUBS H.A., RUDDLE F.H.: Chromosome polymorphism in American negro and white populations. *Nature*, **233**:134–136, 1971.

Chromosome 9 is noteworthy because of its gene assignments and cytogenetic pathology. Mapping the ABO-NP-AK1 linkage group was made possible by the assignment of AK1, through cellular hybridization, to chromosome 9, and through gene dosage effect to 9q34 (Ferguson–Smith et al., 1976). Other important assignments are those of genes coding for the fibroblast and leukocyte interferons (Chany et al., 1980) and that of GALT, which shows gene dosage effect in certain aneusomies of 9p and whose deficiency causes galactosemia.

Cytogenetic syndromes also are highly important: 9p trisomy, the first partial trisomy to have been identified before the use of banding techniques, and probably one of the most frequently found; 9p monosomy, which offers a striking counterpart to 9p trisomy; 9q3 trisomy, which also constitutes a well-characterized syndrome; and complete 9 trisomy, in mosaic. Several cases of r(9) should also be considered.

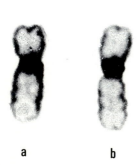

a   b

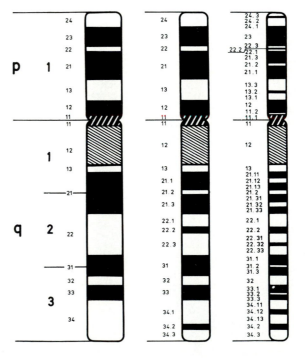

GENE MAP

9p (distal to IFL)	IFF		Interferon-F (fibroblast interferon) (14764) REa
9p21-9pter	IFL		LEUKOCYTE INTERFERON GENE FAMILY (Interferon-L) (14766) REa
9p13-9p24	AK3		Adenylate kinase-3, mitochondrial (10303) S
9p13-9p22	ACO1		Aconitase, soluble (10088) S
9p13-9p21	GALT		Galactose-1-phosphate uridyltransferase (23040) S, D
9qh	DNCM	(P)	Cytoplasmic membrane DNA (12633) A
9q33	FS		Fragile site 9q33 (13660)
(9q34)	ABO		ABO blood group (11030) F
9q34	AK1		Adenylate kinase-1, soluble (10300) F, S, D
Linked to ABO definitely or provisionally. Linkage group assigned to No. 9 by assignment of AK1	NPS1		Nail-patella syndrome (16120) F
	ORM		Orosomucoid (13860) F
	DBH	(L)	Dopamine-beta-hyroxylase (22336) F
	WS1	(L)	Waardenburg syndrome-1 (19350) F
	FPGS	(P)	Folylpolyglutamate synthetase (13651) S
9q34-9qter	ASS		Argininosuccinate synthetase (21570) S, D
	ABL		Onc gene: Abelson strain, murine leukemia virus (18998) S, Ch
	IGE3	(P)	Immunoglobulin epsilon heavy chain pseudogene (14721) A

MORPHOLOGY AND BANDING

 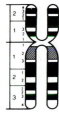       

QFQ   GTG   RHG   RBA   THA   C

147

9p TRISOMY

Brachycephaly
Bulbous nose
Unilateral grin
Worried look

REFERENCES

BACCICHETTI C., LENZINI E., TEMPERANI P., PALLOTTA R., GIORGI P.L., TARANTINO E., MENGARDA G., DORDI B.: Partial trisomy 9: clinical and cytogenetic correlations. *Ann. Génét.*, **22**:199–204, 1979.

CENTERWALL W.R., MAYESKI C.A., CHA C.C.: Trisomy 9q−. A variant of the 9p trisomy syndrome. *Humangenetik*, **29**:91–98, 1975.

LEWANDOWSKI R.C., YUNIS J.J., LEHRKE R., O'LEARY J., SWAIMAN K.F., SANCHEZ O.: Trisomy for the distal one half of the short arm of chromosome 9 [9(p21pter)+] and the 9p+ syndrome. *Amer. J. Dis. Child.* **130**:663–667, 1976.

LURIE I.W., LAZJUK G.I., GUREVICH D.B., USOEV S.S.: Genetics of the +p9 syndrome. *Hum. Genet.*, **32**:23–33, 1976.

RETHORE M.O., LARGET-PIET L., ABONYI D., BOESWILLWALD M., BERGER R., CARPENTIER S., CRUVEILLER J., DUTRILLAUX B., LAFOURCADE J., PENNEAU M., LEJEUNE J.: Sur quatre cas de trisomie pour le bras court du chromosome 9. Individualisation d'une nouvelle entité morbide. *Ann. Génét.*, **13**:217–232, 1970.

RETHORE M.O.: Syndromes involving chromosomes 4, 9, and 12. *In* J.J. YUNIS, ed: *New Chromosomal Syndromes*. New York, Academic Press, pp 119–183, 1977.

SCHINZEL A.: Trisomy 9p, a chromosome aberration with distinct radiologic findings. *Radiology*, **130**:125–133, 1979.

SPARKES R.S., SPARKES M.C., FUNDERBURK S.J., MOEDJONO S.: Expression of GALT in 9p chromosome alterations: assignment of GALT locus to 9cen←9p22. *Ann. hum. Genet.*, **43**:343–347, 1980.

Rethoré et al. delineated the 9p trisomy syndrome as early as 1970, based on four observations. These authors identified chromosome 9 by its secondary constriction, and this identification was further confirmed by banding techniques. One hundred cases are now known from the literature (Rethoré et al., 1977; Lurie et al., 1976). Individual cases are no longer reported.

GENERALIZATIONS

- Frequency: This has not yet been precisely determined, but would seem to be relatively high.
- Sex ratio: 2F/1M
- Parental ages in *de novo* cases:
    mean maternal age: 29 years
    mean paternal age: 30 years
- Pregnancy: normal duration
- Mean birth weight: 2,900 g

PHENOTYPE

In certain cases, 9p trisomy can be associated with partial trisomy or monosomy for another chromosome, since it often results from malsegregation of a reciprocal familial translocation. Described here is the phenotype for 9p trisomy when it occurs alone, or *pure 9p trisomy*. The cytogenetic variants will be discussed in more detail below.

Craniofacial Dysmorphism

Moderate microcephaly with brachycephaly is present. In the infant the occiput is flat. The anterior fontanel is widely opened and expands forward into the metopic suture. In the older child and adult, brachycephaly is less apparent.

The eyes are small and deeply set in their orbits. The palpebral fissures are horizontal or slightly slanted downward and outward. A moderate hypertelorism is present. The pupils are excentric, near the internal edge of the iris. Strabismus is frequent.

The nose is full, primarily at the tip. The nasal ridge thickens in its middle portion. The nares open downward. The membranous nasal septum is thin and protruding.

The upper lip is short. At rest it sometimes exposes the teeth, but it is primarily when the mouth opens that a very characteristic asymmetric grin appears. The mouth is downturned with the lower lip everted.

The chin is round, occasionally protruding with a horizontal dimple.

The ears are large, detached, and normally set.

In the child and adolescent the knit brows, slanted palpebral fissures, and mouth lend an anxious look which is particularly characteristic.

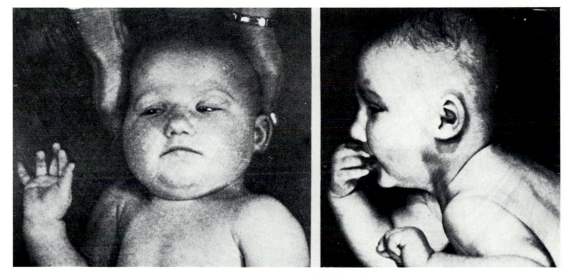

9.1

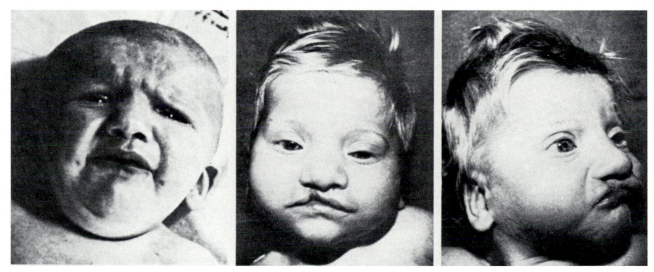

9.2

9.3

Neck, Thorax, and Abdomen

The neck is short, sometimes with rudimentary pterygium colli.

The thorax is wide and deformed into a funnel shape. The nipples are widespaced.

Subacromial and coccygeal dimples, diastasis of the recti musculi, umbilical hernia, and abdominal striae atrophicae cutis are frequently observed. At adolescence the periscapular muscle masses dissolve and lead to defective posture of the spinal column.

Limbs

The hands are very characteristic. The palms are too long compared to the fingers. A single palmar crease is almost constantly encountered. Above all, brachy-mesophalangy is noted, although the fingers seem to be of normal length.

There may even be a fusion of the two flexion creases of the fifth digit. Clinodactyly, a proximal implantation of the thumb, and anomalies of interphalangeal articulations may also be present.

The feet are often malformed.

The nails are dysplastic, sometimes clawlike, especially on the toes.

The hands and feet may be cyanosed. X-ray films of the hands, feet, and pelvis show features which, on the whole, appear to be very characteristic of 9p trisomy (Schinzel, 1979).

Genitalia

There does not seem to be any major anomaly of the external genitalia in the infant. At the age of puberty, hypogonadism in the male and delayed onset of menarch in the female are recorded. In the female, hypotrophy of the labia minora can be observed.

None of the patients who have attained adult age has reproduced.

Malformations

Malformations are rare in cases of pure 9p trisomy.

They are numerous and diverse in cases of associated 9p trisomy.

Harelip has been reported in several cases.

Mental Retardation

Mental retardation is constant but of variable intensity. It is often more pronounced in cases where there is an associated trisomy. The mean IQ is 55 in pure 9p trisomy. The delayed development of language is particularly evident and may attain deaf-mutism.

Psychological tests show evidence of agitation, instability, and disorders in coordination.

Prognosis

Life expectancy is not endangered in pure 9p trisomy. Early deaths are rare, and involve carriers of an associated chromosomal imbalance.

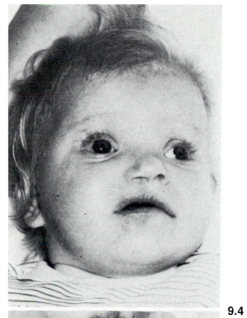

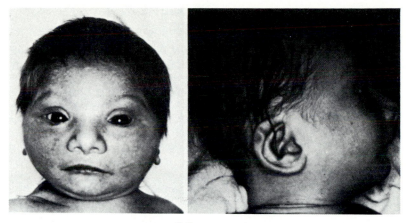

9.5

9.4

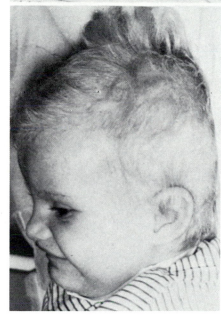

9.6

9.7

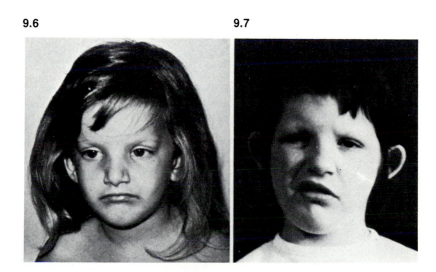

CYTOGENETIC STUDY

The following can be distinguished:

- *Pure 9p trisomy*. This can arise either by breakage of the secondary constriction or by centromeric fusion with an acrocentric chromosome.
- *9q− trisomy*. This trisomy simultaneously involves the short arm and about the proximal half of the long arm. It results from translocation of the distal segment of 9q onto an acrocentric. These translocations preferentially concern acrocentric chromosomes 15 and 22. When the breakpoint on 9q is distal with respect to q13, the trisomy is accompanied by serious malformations that resemble those of complete trisomy 9 (Baccichetti et al., 1979).
- *9p trisomy associated with partial monosomy or trisomy for another chromosome*. This results from a reciprocal translocation. Among the associated trisomies, some involve only the distal half of 9p.

In more than half of the cases of pure 9p trisomy the rearrangement occurs *de novo*. In 9q− trisomy and associated 9p trisomy, malsegregation of a parental rearrangement is always involved.

From a clinical point of view, phenotypic expression is relatively constant. Lower birth weight, lower IQ, and a higher frequency of malformations may be noted in cases of associated 9p trisomy. Trisomies that involve only the distal portion of 9p seem to entail a majority of the clinical features of pure 9p trisomy (Lewandowski et al., 1976; Centerwall et al., 1975); 9q− trisomy is also very similar to pure 9p trisomy, with no particular clinical features when it does not extend beyond 9q13. This seems to indicate that the proximal regions of 9p as well as of 9q are of little importance in phenotypic expression when in a trisomic state.

DERMATOGLYPHICS

Dermatoglyphics are characteristic. Apart from anomalies already mentioned,

- Abnormal palm length
- Single transverse palmar crease or its equivalent
- Brachymesophalangy with fusion of flexion creases
- Clinodactyly

the following are usually observed:

- Axial triradius in t′ position in half of the palms
- Absence or fusion of subdigital triradii b and c in two-thirds of the palms (very rare in the general population)
- Line D terminates in 11 in more than three-fourths of the palms
- Diminished frequency of interdigital patterns
- High frequency of thenar patterns and a decreased frequency of hypothenar patterns
- Excess of arches on the fingertips

LABORATORY FINDINGS

A gene dosage effect for galactose-1-P-uridyl transferase (GALT) is observed when the trisomy involves 9cen p22 (Sparkes et al., 1980).

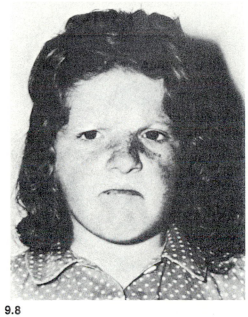

9.8

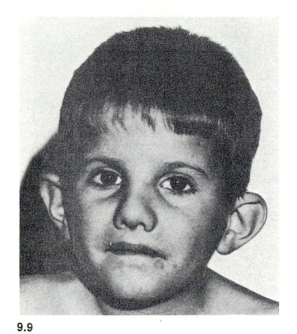

9.9

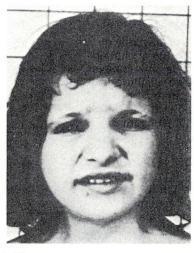

9.10

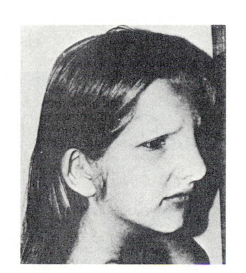

9.11

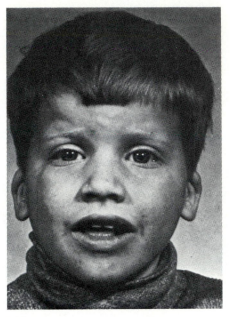

9.12

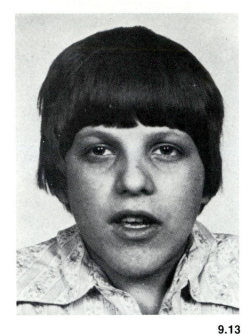

9.13

9.14

9.15

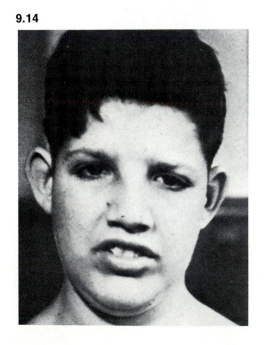

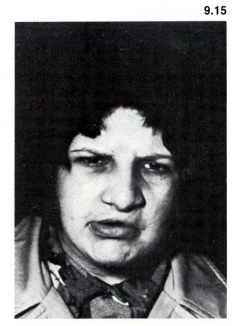

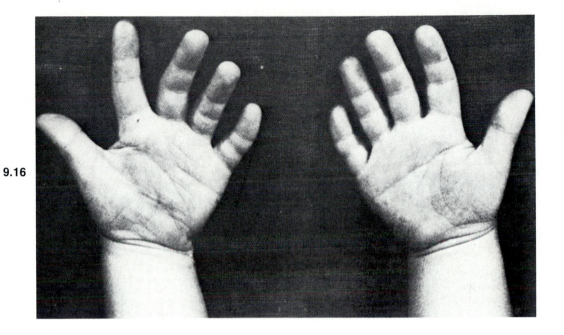

9.16

Trisomy and Monosomy 9p: Type and Countertype

Trisomy	Monosomy
Brachycephaly	Trigonocephaly
Enophthalmos	Exophthalmos
Palpebral fissures slanted downward and outward	Palpebral fissures slanted upward and outward
Nares oriented downward	Anteverted nares
Bulbous nose	Short nose
Short upper lip	Long upper lip
Brachymesophalangia	Dolichomesophalangia
Excess of arches	Excess of whorls

9p TETRASOMY

REFERENCES

GHYMERS D., HERMANN B., DISTECHE C., FREDERIC J.: Tétrasomie partielle du chromosome 9 à l'état de mosaïque chez un enfant porteur de malformations multiples. *Humangenetik,* **20**:273–282, 1973.

MOEDJONO S.J., CRANDALL B.F., SPARKES R.S.: Tetrasomy 9p: confirmation by enzyme analysis. *J. med. Genet.,* **17**:227–242, 1980.

The first observation was that of Ghymers et al., 1973. Six cases are presently known (see Moedjono et al., 1980). Tetrasomy 9p is noteworthy due to the rareness of auto-somal tetrasomies. The phenotype is variable; facial dysmorphism may resemble that of 9p trisomy. Malformations are severe and manifold: hydrocephaly, cardiac, renal, osteoarticular malformations, and harelip.

Mosaicism was demonstrated in three cases, and would seem to lessen the gravity of the prognosis. Homogenous tetrasomies seem to lead to early death.

Cytogenetic mechanisms behind the occurrence of the rearrangement in question are rather unusual (formation of a 9p isochromosome, or of a dicentric chromosome), and might well be associated with the juxtacentromeric secondary constriction of 9q.

A gene dosage effect is demonstrated for galactose-1-uridyl transferase (GALT), whose activity is multiplied twofold compared with normal.

9.17

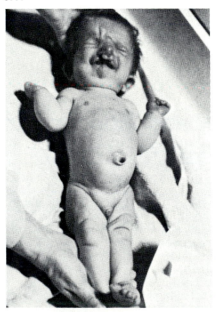
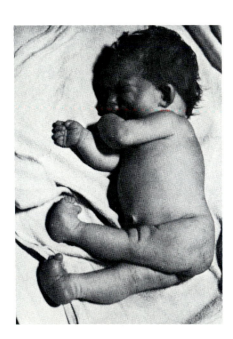

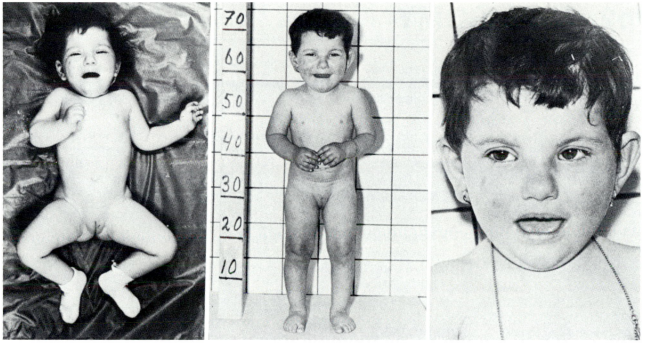

9.18

9.19

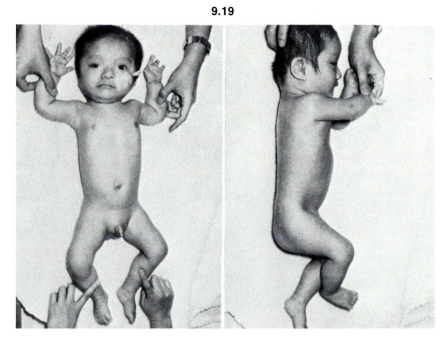

9p2 MONOSOMY

Trigonocephaly
Palpebral fissures slanting upward and outward
Long upper lip

REFERENCES

ALFI O., DONNELL G.N., CRANDALL B.F., DERENCSENYI A., MENON R.: Deletion of the short arm of chromosome 9 (46,9p −): a new deletion syndrome. *Ann. Génét.*, **16**:17–22, 1973.

ALFI O., DONNELL G.N., ALLDERDICE P.W., DERENCSENYI A.: The 9p − syndrome. *Ann. Génét.*, **19**:11–16, 1976.

DEROOVER J., FRYNS J.P., PARLOIR C., HAEGEMAN J., Van den BERGHE H.: Partial monosomy of the short arm of chromosome 9. *Hum. Genet.*, **44**:195–200, 1978.

FUNDERBURK S.J., SPARKES R.S., KLISAK I.: The 9p − syndrome. *J. med. Genet.*, **16**:75–79, 1979.

HOO J.J., PARSLOW M.I., SHAW R.L., VEALE A.M.O.: Complex de novo rearrangement of chromosome 9 with clinical features of monosomy 9p syndrome. *Clin. Genet.*, **16**:151–155, 1979.

NIELSEN J., HOMMA A., CHRISTIANSEN E., RASMUSSEN K., SALDANA-GARCIA P.: The deletion 9p syndrome. A 61 year-old man with deletion of short arm 9. *Clin. Genet.*, **12**:80–84, 1977.

PAVONE L., MOLLICA F., SORGE G., SCIACCA F., D'AGATA A., LAURENCE K.M.: Une nouvelle observation de monosomie partielle 9. *Ann. Génét.*, **21**:186–188, 1978.

RETHORE M.O.: Syndromes involving chromosomes 4, 9, and 12, in J.J. YUNIS, ed: *New Chromosomal Syndromes.* New York, Academic Press, 1977, pp 119–183.

RUTTEN F.J., HUSTINX T.W.J., DUNKTILLEMANS A.A.W., SCHERES J.M.J.C., TJON Y.S.T.: A case of partial 9p monosomy with some unusual clinical features. *Ann. Génét.*, **21**:51–55, 1978.

The 9p monosomy syndrome was characterized by Alfi et al. in 1973, based on six cases. About 20 cases are presently known (Rethoré, 1977; Nielsen et al., 1977; Rutten et al., 1978; Hoo et al., 1979; Funderburk et al., 1979; and Pavone et al., 1979).

GENERALIZATIONS

- Sex ratio: 2F/1M
- Mean parental ages (de novo cases):
  maternal: 29 years
  paternal: 31.2 years
- Mean birth weight: 3,250 g
- Mean crown-heel length at birth: 50 cm

PHENOTYPE

Craniofacial Dysmorphism

Craniofacial dysmorphism is highly pathognomonic, and clinical diagnosis is possible mainly because of trigonocephaly and slanting of the palpebral fissures.

Trigonocephaly is clearly evident. The forehead protrudes above the midface, which appears to be retracted.

Palpebral fissures slant upward and outward and may give the patient a false appearance of 21 trisomy. Minor exophthalmy is present, and bilateral epicanthus is almost always encountered. The eyebrows are very arched.

The bridge of the nose is flattened and wide, with the nose itself relatively short and the nares anteverted.

The upper lip is long. The philtral borders are undefined.

The mouth is small with a high arched palate and slight micrognathia.

Ears are low-set and well attached, but are characterized mainly by aplasia of the lobe. The helix is poorly folded in its vertical portion.

Neck, Thorax, and Abdomen

The neck is short, with an occasional, rather well defined pterygium colli.

The space between the nipples, as well as the anteroposterior diameter of the thorax, are wider than normal.

Genitalia

In girls, hypoplasia of the labia majora and a very distinct hyperplasia of the labia minora are found.

In boys, hypospadias is present in two out of three cases.

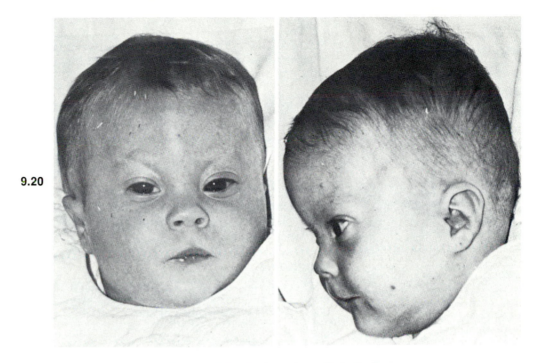

9.20

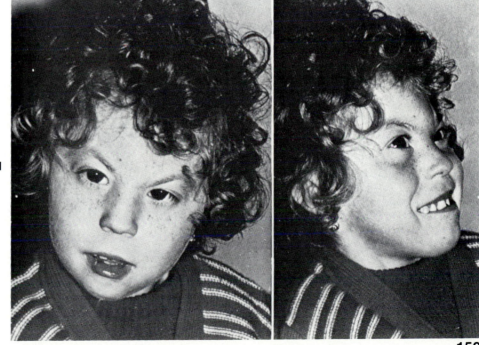

9.21

Limbs

The fingers and toes are long, mainly due to lengthening of the second phalange. A supernumerary flexion crease may even be noted. The nails are square-shaped, wide, and excessively convex.

Malformations

Malformations may be quite severe, especially in the case of associated monosomy. They are usually cardiac malformations but may also include hernias and omphaloceles.

Mental Retardation

Newborns are hypertonic.
The IQ of the patients is between 30 and 60; they tend to be friendly, affectionate, and sociable.

Prognosis

The survival rate of the patients does not seem to be seriously affected. Two carriers of an associated monosomy who were afflicted with severe malformations died in the first few months of life.

CYTOGENETIC STUDY

The deletion is *de novo* in two-thirds of the cases. In the other cases, it results from a parental translocation.
The breakpoint is in p22 in the majority of cases.

DERMATOGLYPHICS

Excess of whorls on the digital pads.
Axial triradius often found in t″.

LABORATORY FINDINGS

In cases where the breakpoint is in p22, GALT activity is normal, thus implying that the locus is proximal with respect to p22.

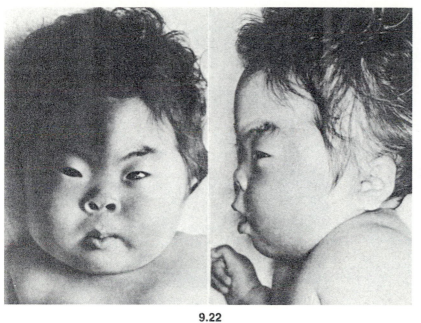

9.22

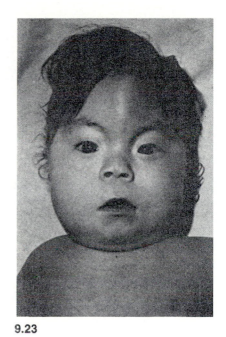

9.23

9.24

9.25

9.26

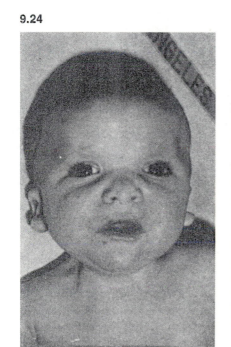

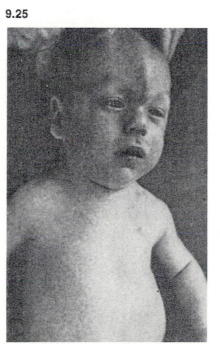

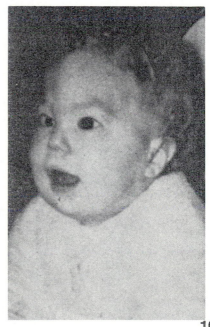

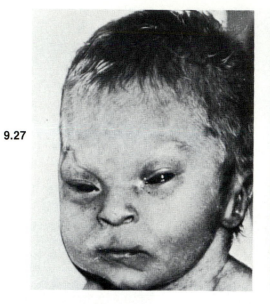

9.27

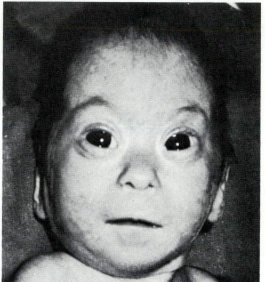

9.28

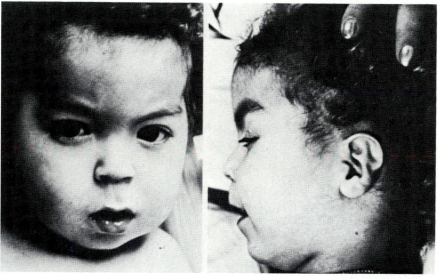

9.29

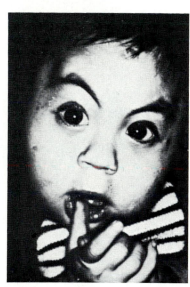

9.30

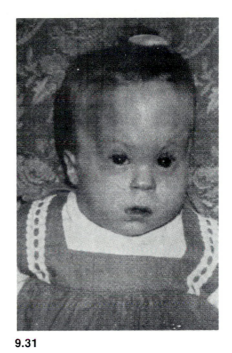

9.31

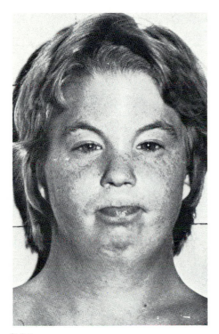

9.32

9.33

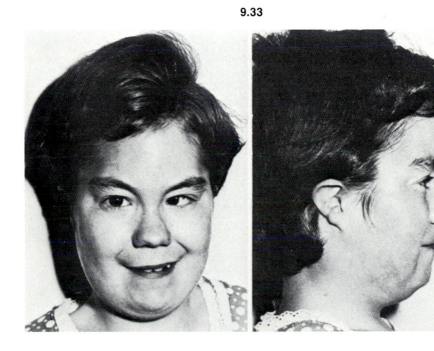

r(9)

REFERENCES

FRYNS J.P., LAMBRECHTS A., JANSSEUNE H., Van den BERGHE H.: Moderate mental retardation and nonspecific dysmorphic syndrome associated with ring chromosome 9. *Hum. Genet.*, **50**:29–32, 1979.

INOUYE T., MATSUDA H., SHIMURA K., HAMAZAKI M., KIKUTA I., IINUMA K., NAKAGOME Y.: A ring chromosome 9 in an infant with malformations. *Hum. Genet.*, **50**:231–235, 1979.

RETHORE M.O.: Syndromes involving chromosomes 4, 9, and 12. *In* J.J. YUNIS, ed: *New Chromosomal Syndromes.* New York: Academic Press, pp 119–183, 1977.

Eight cases of r(9) are known. These were analyzed by Rethoré (1977) and Inouye et al. (1979) (also see Fryns et al., 1979).

PHENOTYPE

As with all ring structures, the phenotype is variable. The principal features of 9p monosomy may be observed, such as:

- Mental retardation
- Microcephaly with trigonocephaly
- Craniofacial dysmorphism, including:

   Slanting upward and outward of the palpebral fissures

   Epicanthus

   Minor exophthalmy

   Very arched eyebrows

   A lengthening of the upper lip, with effacement of Cupid's bow

   Retrognathia

   A protuberant anthelix
- Shortness of the neck
- An excess of whorls on the finger tips

Certain symptoms resembling those of trisomy 9 also may be seen, including skeletal and cardiac anomalies.

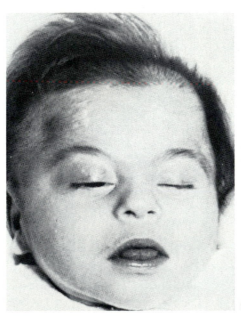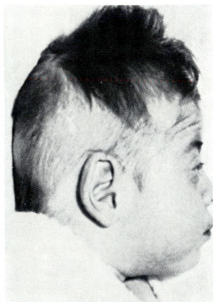

9.34

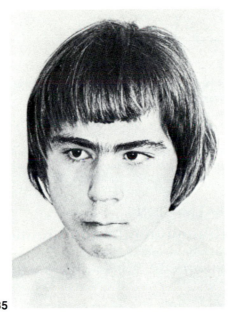

9.35

9.36

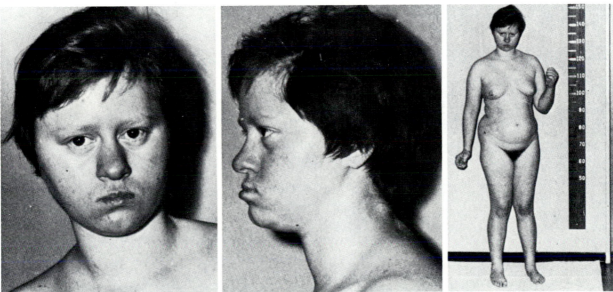

9q3 TRISOMY

Small face
Beaked nose
Microretrognathia
Long fingers, abnormally positioned

REFERENCES

AFTIMOS S.F., HOO J.J., PARSLOW M.I.: Partial trisomy 9q due to maternal 9/17 translocation. *Am. J. Dis. Child.*, **134**:848–850, 1980.

FAED M., ROBERTSON J., BROWN S., SMAIL P.J., MUCKHART R.D.: Pure partial trisomy for the long arm of chromosome 9. *J. med. Genet.*, **13**:239–242, 1976.

MATTEI J.F., MATTEI M.G., ARDISSONE J.P., TARAMASCO H., GIRAUD F.: Pericentric inversion, inv(9) (p22q32), in the father of a child with a duplication-deletion of chromosome 9 and gene dosage effect for adenylate kinase-1. *Clin. Genet.*, **17**:129–136, 1980.

PESCIA G., JOTTERAND-BELLOMO M., CROUSAZ H. de PAYOT M., MARTIN D.: Phénotype de la trisomie 9q distale chez un enfant présentant un chromosome surnuméraire remanié t(X;9). *Ann. Génét.*, **22**:158–162, 1979.

SUBRT I., JANOVSKY M., JODL J.: Partial trisomy 9q – chromosomal syndrome. *Hum. Genet.*, **34**:151–154, 1976.

TURLEAU C., GROUCHY J. de, CHAVIN-COLIN F., ROUBIN M., BRISSAUD P.E., REPESSE G., SAFAR A., BORNICHE P.: Partial trisomy 9q: a new syndrome. *Humangenetik*, **29**:233–241, 1975.

This syndrome was characterized by Turleau et al. (1975) based on two patients. Seven such patients are now known (Faed et al., 1976; Subrt et al., 1976; Mattei et al., 1980; Aftimos et al., 1980; and Pescia et al., 1979).

GENERALIZATIONS

- Sex ratio: 3M/4F
- Mean birth weight: 2,300 g
- Mean crown heel length at birth: 48.3 cm (6 cases)

PHENOTYPE

The dysmorphism of these patients is highly characteristic. Mental and growth retardation are severe.

Craniofacial Dysmorphism

Microcephaly and, in particular, dolichocephaly are present.
The face is small, with frontal bossing. Hair is smooth and thick.
The eyes are deep-set. Palpebral fissures are horizontal. Hypo- or hypertelorism may be visible.
The bridge of the nose is normal, but the nose is beaked, giving the face a birdlike appearance.
The mouth is small, and the upper lip overlaps the lower one. The philtrum is normal, but the palate may be high-arched or cleft.
The chin is small, pointed, and receding.
The ears are large and poorly folded.

Neck, Thorax, and Abdomen

An excess of skin on the neck may be noted.
The thorax is normal.
The pelvis is narrow. Sacral and elbow dimples may be noted.

Limbs

A very pronounced amyotrophy, especially of the buttocks and posterior muscles of the lower limbs, is present. Flexion contractures of the joints are regularly encountered.
The fingers are very long and tapering, with a peculiar position; in flexion, the index finger folds at a right angle above the other fingers, with the thumb folded into the palm.

9.37

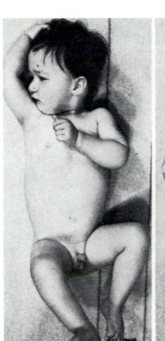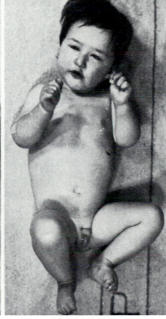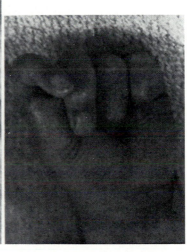

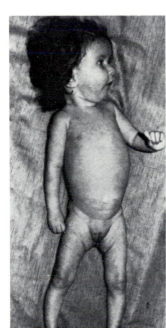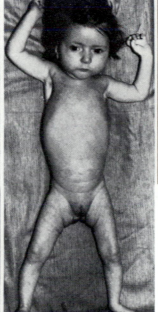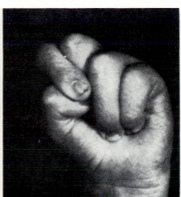

9.38

The toes are also very long, often with abnormal implantation, and, occasionally, hammer toe and rocker-bottom feet.

Genitalia

The genitals may be either normal or hypoplastic.

Malformations

Inner organ malformations basically involve the heart, such as ventricular septal defect and Fallot's tetralogy, but may include urinary (hydroureter) defects. Other malformations also have been reported including deafness, hexadactyly, vertebral anomalies, and supernumerary ribs.

Mental Retardation

Mental retardation is very severe. These children are usually unable to walk alone and are unable to speak; they are agitated and have abnormal, stereotyped movements of the arms and legs.

Prognosis

Two infants died at an early age. The oldest child is 8.

CYTOGENETIC STUDY

Two cases involve nearly the same rearrangement: a tandem duplication arising *de novo*, with its consequence being a partial 9q trisomy involving the segment between the centromere and q33. This type of rearrangement is very rarely observed, and usually concerns the acrocentrics. It is possible that the secondary constriction on 9q enhances its appearance.

Other cases are due to malsegregation of a parental rearrangement (a large pericentric inversion in two cases).

Comparison of the breakpoints indicates that the expression of the clinical syndrome may be attributed to trisomy for q32.

DERMATOGLYPHICS

Dermatoglyphics have no particular characteristics. Supernumerary flexion creases may be observed on the fingers.

LABORATORY FINDINGS

A gene dosage effect for AK1 is observed when trisomy includes 9q34.

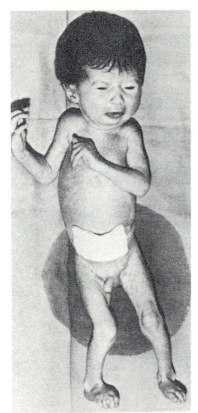

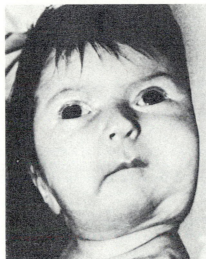

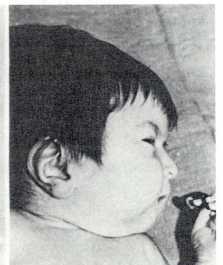

9.40

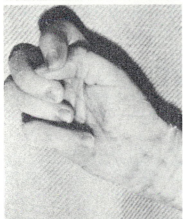

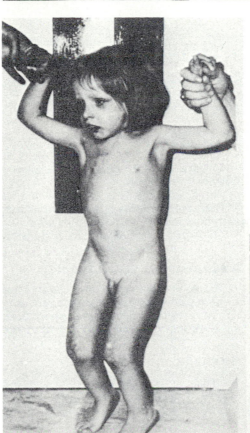

9.39

9.41

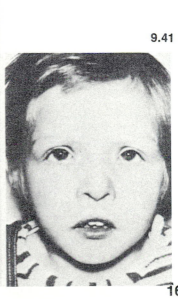

9 TRISOMY

Microcephaly
Enophthalmy
Microretrognathia
Osteoarticular and inner organ anomalies

REFERENCES

AKATSUKA A., NISHIYA O., KITAGAWA T., KAGEYAMA A., INANA I., NAKAGOME Y.: Trisomy 9 mosaicism with punctate mineralization in developing cartilages. *Eur. J. Pediatr.*, **131**:271–275, 1979.

FEINGOLD M., ATKINS L.: A case of trisomy 9. *J. med. Genet.*, **10**:184–187, 1973.

HASLAM R.H.A., BROSKE S.P., MOORE C.M., THOMAS G.H., NEILL C.A.: Trisomy 9 mosaicism with multiple congenital anomalies. *J. med. Genet.*, **10**:180–184, 1973.

MACE S.E., MACINTYRE M.N., TURK K.B., JOHNSON W.E.: The trisomy 9 syndrome: multiple congenital anomalies and unusual pathological findings. *J. Pediatr.*, **92**:446–448, 1978.

The first two observations of 9 trisomy involved mosaicism (Haslam et al., 1973) and an apparently homogenous trisomy (Feingold and Atkins, 1973). Thirteen cases are presently known (see Akatsuka et al., 1970; Mace et al., 1978). Trisomy may be homogenous or in mosaic.

GENERALIZATIONS

- Sex ratio: 1M/1F
- Mean parental ages
    maternal: 26.4 years
    paternal: 31.7 years
- Mean birth weight: 2,435 g
- Mean crown-heel length at birth: 48.2 cm
- Mean head circumference at birth: 32.5 cm

PHENOTYPE

Craniofacial Dysmorphism

Microcephaly with dolichocephaly is present. The forehead is high.

The eyes are sunk deep into the sockets, and palpebral fissures are small, slanting upward and outward.

The base of the nose is wide, with a bulbous tip.

Retromicrognathia is fairly pronounced and the upper lip overlaps the lower one to a great extent. A high arched palate or cleft palate may be observed.

The ears are low-set, round, and flaccid, with a protuberance of the anthelix.

Neck, Thorax, and Abdomen

The neck is somewhat short, with a low hair line.

Highly characteristic osteoarticular anomalies are present. These include dislocation of the hips or limitation of abduction, dislocation of the knees or elbows, deformities of the spinal cord, and rib anomalies.

Genitalia

Cryptorchidism, scrotal hypoplasia, and micropenis are regularly found.

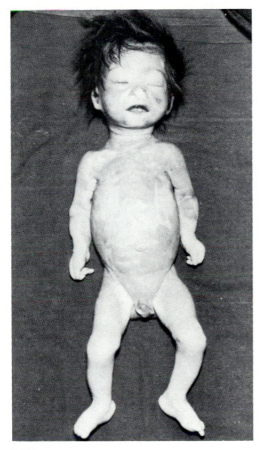

9.42

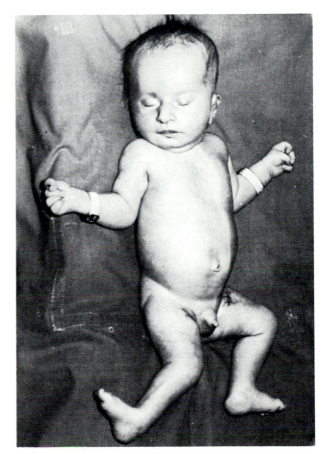

9.43

9.44

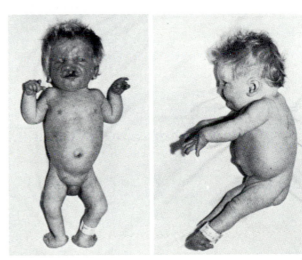

Malformations

Osteoarticular malformations have already been mentioned. Inner organ malformations are nearly always encountered. These usually involve complex congenital heart disease, and cerebral, kidney, and occasionally digestive anomalies.

Mental Retardation

The vital prognosis is most severe. Most infants die during the first four months of life, usually of respiratory infections.

CYTOGENETIC STUDY

In about half of the cases, there is mosaicism. In the other half of the cases, the trisomy is considered as homogenous.

In two cases, trisomy was nearly complete, with the supernumerary chromosome 9 having undergone a deletion of the very distal portion of the long arm. Trisomy occurred *de novo* in one of these cases, and by transmission of a parental rearrangement in the other case.

DERMATOGLYPHICS

Dermatoglyphics are not often described, but single transverse palmar creases, an absence of the b triradius, and deep furrows may be observed.

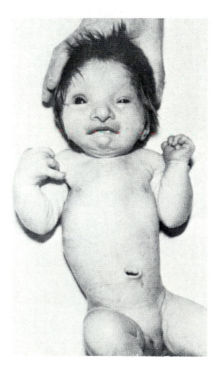

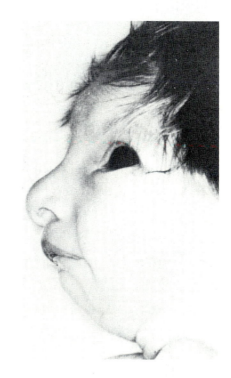

9.45

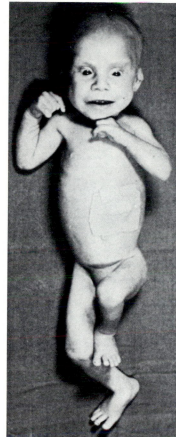

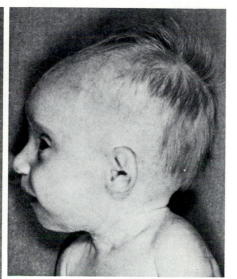

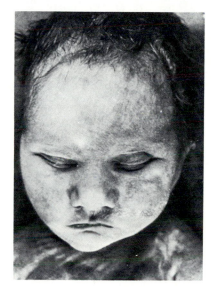

9.47

9.46

9.48

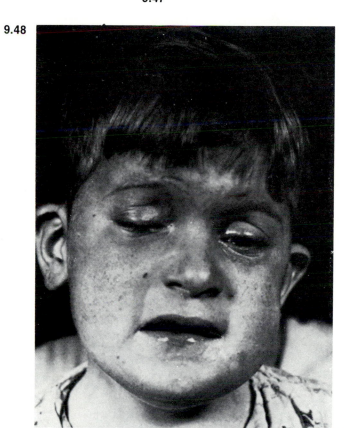

Glutamate oxaloacetate transaminase, soluble (GOT1) and hexokinase 1 (HK1) show a gene dosage effect.

Cytogenetic syndromes mainly include 10p trisomy and 10q trisomy. 10p and 10q monosomies, as well as rings, are rather rare.

On the long arm, two fragile sites are present, probably having no pathological effect. One is in q23, and is folate-dependent, the other is in q25 and is BrdU-dependent.

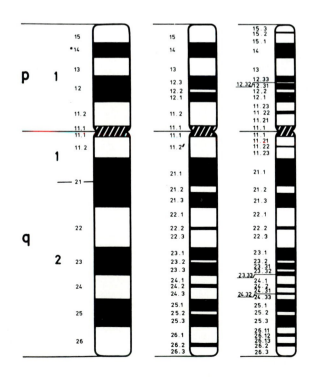

GENE MAP

10p	PFKF	(P)	Phosphofructokinase, fibroblast (or platelet) (17184) S
10p15-10q24	PP		Inorganic pyrophosphatase (17903) S
10p11-10q23	HK1		Hexokinase-1 (14260) S,D
?10q11-10q24	ADK		Adenosine kinase (10275) S, D-EM
10q23	FS		Fragile site 10q23 (folate-dependent) (13654)
10q25	FS		Fragile site 10q25 (BrdU-dependent) (13662)
10q24-10q25	GOT1		Glutamate oxaloacetate transaminase, soluble (13818) S, D,H
	GSAS	(P)	Glutamate-gamma-semialdehyde synthetase (13825) S
	FUSE	(P)	Polykaryocytosis promoter (17475) S
	EMP130	(P)	External membrane protein-130 (13371) S
?10q24-10q25	LIPA	(P)	Lysosomal acid lipase-A (27800) S
	CG	(L)	Chorionic gonadotropin (11885) S (see chr. 5, 18)
	OMR	(P)	Oligomycin resistance (16436) S
	DHFR	(P)	Dihydrofolate reductase (12606) Ch

MORPHOLOGY AND BANDING

 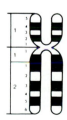       

QFQ	GTG	RHG	RBA	THA	C

175

10q2 TRISOMY

High, protruding forehead
Wide flat face
Narrow palpebral fissures
Ligamentary hyperlaxity

REFERENCES

BACK E., KOSMUTZKY J., SCHUWALD A., HAMEISTER H.: Two cases of partial trisomy 10q in the same family caused by parental direct insertion [ins(15;10) (q15;q24q26)]. *Ann. Génét.*, **22**:195–198, 1979.

BASS H.N., SPARKES R.S., CRANDALL B.F., TANNENBAUM S.M.: Familial partial trisomy 10q(q23→qter) syndrome and paracentric inversion 3(q13q26) in the same patient. *Ann. Génét.*, **21**:74–77, 1978.

CANTU J.M., HERNANDEZ A., NAZARA Z., ROLON A., RAMIREZ L., SANCHEZ-CORONA J., RIVERA H.: Simultaneous trisomy 10q24→qter and monosomy 4p16: an example of epistasis at the chromosome level. *Ann. Génét.*, **24**:41–44, 1981.

DELICADO A., LOPEZ-PAJARES I., VILENTE P., HAWKINS F.: Familial translocation t(10;21) (q22;q22). *Hum. Genet.*, **50**:253–258, 1979.

DUTRILLAUX B., LAURENT C., ROBERT J.M., LEJEUNE J.: Inversion péricentrique, inv(10), chez la mère et aneusomie de recombinaison inv(10), rec (10), chez son fils. *Cytogenet. Cell Genet.*, **12**:245–253, 1973.

GROUCHY J. de, FINAZ C., ROUBIN M., ROY J.: Deux translocations familiales survenues ensemble chez chacune de deux sœurs, l'une équilibrée, l'autre trisomique partielle 10q. *Ann. Génét.*, **15**:85–92, 1972.

KLEP-DE PATER J.M., BIJLSMA J.B., FRANCE H.F. de, LESCHOT N.J., DUIJNDAN-Van den BERGE M., Van HEMEL J.: Partial trisomy 10q. *Hum. Genet.*, **46**:29–40, 1979.

LACA Z., KALICANIN P.: A case of partial trisomy 10q. *J. mental Defic. Res.*, **18**:285–291, 1974.

LAURENT C., BOVIER-LAPIERRE M., DUTRILLAUX B.: Trisomie partielle 10 par translocation familiale t(1;10) (q44;q22). *Humangenetik*, **18**:321–327, 1973.

MIRO R., TEMPLADO C., PONSA M., SERRADELL J., MARINA S., EGOZCUE J.: Balanced translocation (10;13) in a father, ascertained through the study of meiosis in semen, and partial trisomy 10q in his son. *Hum. Genet.*, **53**:179–182, 1980.

PRIEUR M., FORABOSCO A., DUTRILLAUX B., LAURENT C., BERNASCONI S., LEJEUNE J.: La trisomie 10q24→10qter. *Ann. Génét.*, **18**:217–222, 1975.

TALVIK T., MIKKELSAAR A.V., MIKKELSAAR R., KAOSAAR M., TUUR S.: Inherited translocations in two families t(14q+;10q−) and t(13q−;21q+). *Humangenetik*, **19**:215–226, 1973.

YUNIS J.J., LEWANDOWSKI R.C.: Partial duplication 10q and duplication 10p syndromes. In J.J. YUNIS: *New Chromosomal Syndromes*. New York, Academic Press, pp. 219–244, 1977.

The first observation of partial 10q trisomy was reported by Grouchy et al. (1972). Since the trisomy in question resulted from a complex rearrangement, and since the infant was stillborn, this case will not be considered here. The first observations allowing a description of the syndrome are those of Laurent et al. (1973), Dutrillaux et al. (1973), and Talvik et al. (1973). Twenty observations are currently recognized, including several that are familial, bringing the total number of cases to 31 (see reviews by Prieur et al., 1975; Yunis and Lewandowski, 1977; Klep-de Pater et al., 1979; Laca and Kalicanin, 1973; Bass et al., 1978; Back et al., 1979; Delicado et al., 1979; Miro et al., 1980; and Cantù et al., 1981).

The breakpoint is almost always in 10q24.

GENERALIZATIONS

The majority of cases are due to a parental rearrangement:

- Sex ratio: 1M/1F
- Mean birth weight: 2,625 g

PHENOTYPE

Growth retardation is severe. This is especially evident in older children and concerns their weight more than their height.

Craniofacial Dysmorphism

Microcephaly is frequently encountered. Brachycephaly and cranial asymmetry may be observed. The shape of the forehead is characteristic: high and bulging.

The face is round, broad, and flat, especially in the infant; the cheeks may be bloated and flaccid.

The eyebrows are thin and incurved.

The palpebral fissures are small, slanted downward and outward. Blepharophimosis can be observed. Hypertelorism is more or less pronounced; it is always exaggerated in appearance by the smallness of the eyes, a constant feature of the syndrome. Epicanthus inversus and ptosis can also be observed.

The nose is small and often beaked. The nasal bridge is hypoplastic and "pinched."

The upper lip is prominent. It has a rounded shape, and the vermilion is very apparent. The pillars of the philtrum are not strongly marked. The mouth is often open. The palate is high-arched, and a palatal cleft can be present.

The chin is receding and often small.

The ears are well folded, low-set, and posteriorly rotated.

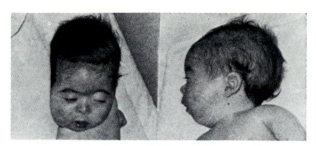

10.1

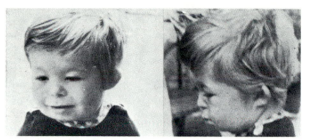

10.2

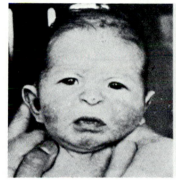

10.3

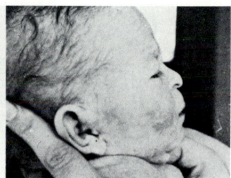

10.4   10.5

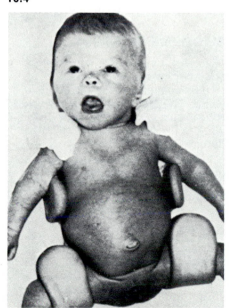

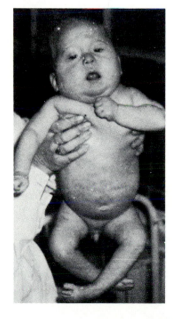

Neck, Thorax, and Abdomen

The neck is short.
Scoliosis is reported in more than half the cases.

Limbs

Ligamentary hyperlaxity may be encountered, allowing voluntary dislocation of diverse articulations.
The fingers are slender and tapering. Clinodactyly of the fifth digit can be seen. The big toe is occasionally hammer-shaped, and the space between it and the second toe is augmented. Syndactyly of the second and third toes has been reported.

Genitalia

Cryptorchidism is frequent in males.

Malformations

Inner organ malformations are rarely found. Besides scoliosis, cardiac and renal malformations may be present, eventually causing death at an early age.

Mental Retardation

Hypotonia is a feature found almost constantly. Mental deficiency is usually very severe. However, an IQ of 70 has been reported in one case.

Prognosis

Survival is essentially a function of internal malformations and respiratory infections that result in death during the first year in about one-third of the patients.

CYTOGENETIC STUDY

Trisomy usually occurs following a parental rearrangement, translocation or pericentric inversion.
Translocation usually occurs onto an acrocentric, that is, without reciprocal exchange. The breakpoint is almost always located at 10q24. This band seems to represent a fragile zone on the chromosome. In several cases, the breakpoint is in q25 or in q22. The phenotype is only slightly influenced by the size of the trisomic segment (Klep-de Pater et al., 1979).

DERMATOGLYPHICS

The dermatoglyphics do not display any characteristic feature. An excess of ulnar loops and a single palmar crease have been reported.

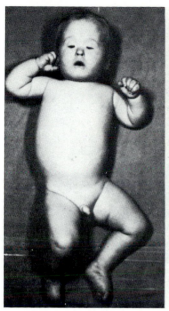
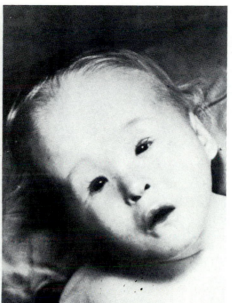

10.6

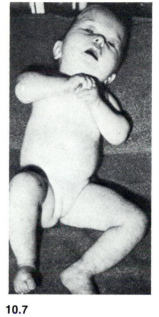

10.7

10.8

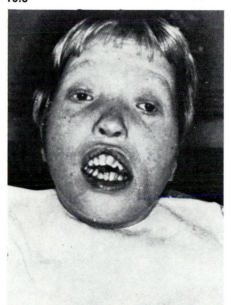

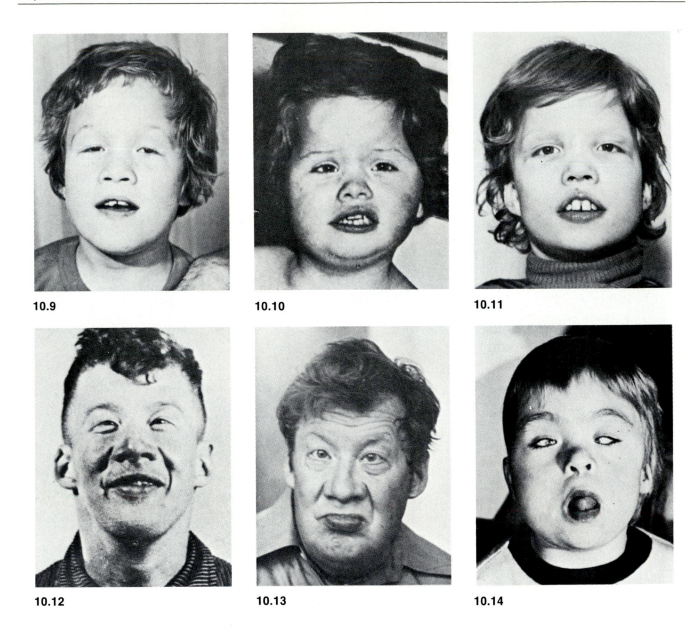

10.9

10.10

10.11

10.12

10.13

10.14

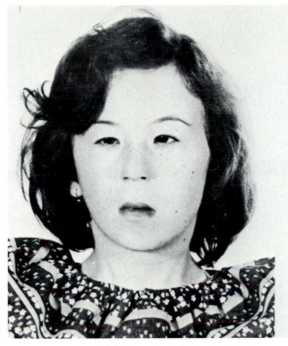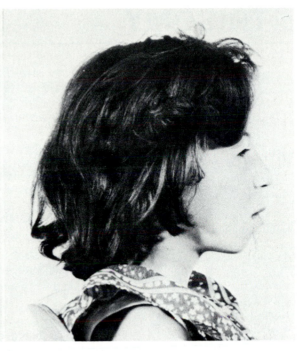

10.15

10.16

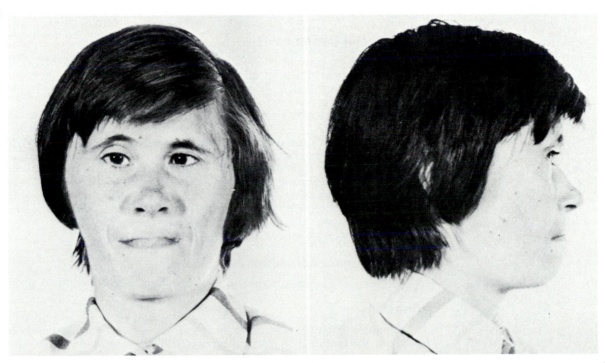

10p TRISOMY

Severe growth retardation
Dolichocephaly
Turtle's beak
Osteoarticular anomalies

REFERENCES

ALLER V., ABRISQUETA J.A., PEREZ-CASTILLO A., DEL MAZO J., MARTIN-LUCAS M.A., TORRES M.L. de.: Trisomy 10p due to a de novo t(10p;13p). *Hum. Genet.,* **46**:129–134, 1979.

FRYNS J.P., DEROOVER J., HAEGEMAN J., VAN DEN BERGHE H.: Partial duplication of the short arm of chromosome 10. *Hum. Genet.,* **47**:217–220, 1979.

HIRSCHHORN K., LUCAS M., WALLACE I.: Precise identification of various chromosomal abnormalities. *Ann. hum. Genet.,* **36**:375–379, 1973.

INSLEY J., RUSHTON D.I., EVERLEY JONES H.W.: An intersexual infant with an extra chromosome. *Ann. Génét.,* **11**:88–94, 1968.

OBRY E., PIUSSAN C., RISBOURG B., DU-TRILLAUX B.: Trisomie partielle (10pter→10q21) et monosomie partielle (21pter→21q21) dues à une translocation réciproque familiale équilibrée (10;21) (q21;q21). *Ann. Génét.,* **23**:216–220, 1980.

RODEWALD A., STENGEL-RUTKOWSKI S.: The dermatoglyphic pattern of the trisomy 10p syndrome. *Clin. Genet.,* **14**:330–337, 1978.

STENGEL-RUTKOWSKI S., MURKEN J.D., RENKENBERGER R., RIECHERT M., SPIESS H., RODEWALD A., STENE J.: New chromosomal dysmorphic syndromes. II. Trisomy 10p. *Europ. J. Pediatr.,* **126**:109–125, 1977.

STENE J., STENGEL-RUTKOWSKI S.: Risk for short arm 10 trisomy. *Hum. Genet.,* **39**:7–13, 1977.

YUNIS J.J., LEWANDOWSKI R.C.: Partial duplication 10q and duplication 10p syndromes. *In* J.J. YUNIS: *New Chromosomal Syndromes,* New York, Academic Press, pp. 219–244, 1977.

The first observation was that of Insley et al., 1968, and Hirschhorn et al., 1974. Twenty-two observations are presently known, several of which are familial, making a total of 29 patients (see Stengel-Rutkowski et al., 1977; Yunis and Lewandowski, 1977; Obry et al., 1980; Aller et al., 1979; Dallapiccolla et al., 1981; and Fryns et al., 1979).

GENERALIZATIONS

- Sex ratio: 1M/1F
- Mean birth weight: 2,670 g
- Mean crown-heel length at birth: 47 cm
- Mean head circumference at birth: 32 cm

PHENOTYPE

Physiological disorders can be observed at birth: apnea, absence of the sucking reflex, persistent neonatal jaundice, and significant livedo.

Growth retardation is a major feature of the syndrome and is accompanied by muscular hypoplasia and hypotonia, as well as defective postures.

Craniofacial Dysmorphism

The infants are generally dolichocephalic. The forehead is high, with frontal bossing and occasional protrusion of the occiput.

The face is narrow and seems small in relation to the cranium. The cheeks are round and sagging.

The hair is backswept, and the hairline may suggest a heart-shaped curve, its tip on the midline.

The palpebral fissures are large and horizontal. The eyebrows are thin, and extend laterally upon the temples.

The nasal bridge is broad, most often protruding.

The mouth is deformed by a harelip (uni- or bilateral) in one-third of the cases. Cleft palate can be observed in the absence of harelip. The fleshy portion of the lips are inverted, producing the appearance of a "turtle's beak."

The chin is round, often small, and poorly defined.

The ears are large, rotated posteriorly, and low-set.

Neck, Thorax, and Abdomen

The thorax and pelvis are narrow. Scoliosis has been observed in several cases.

Limbs

The upper limbs have an abnormal hyperflexed position; the hands flex on the wrists and the forearms on the arms. In one case, considerable ligamental hyperlaxity

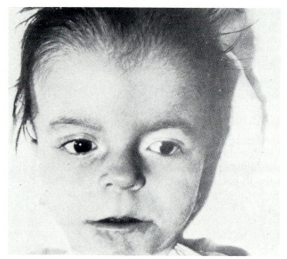

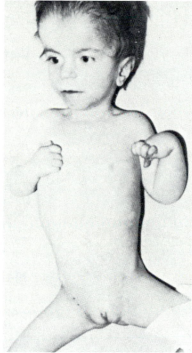

10.17

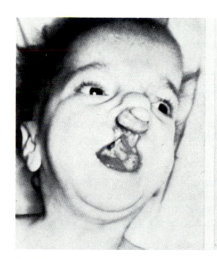

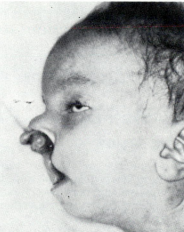

10.18

10.19

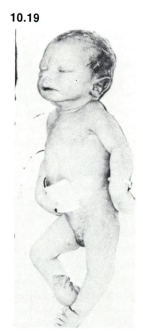

responsible for "contortionist positions" was noted. The fingers flex in the palm. Clinodactyly and camptodactyly can be observed.

The lower limbs are in abduction-flexion. Clubfeet exist in almost all cases.

Genitalia

The genitalia may be hypoplastic in males. One case of associated intersexuality has been reported.

Malformations

Cardiac malformations are present in one-third of the cases: dextrocardia, coarctation of the aorta, and other cardiopathies.

Renal malformations are more specific. Polycystic, aplastic, or double kidneys, anomalies in rotation, and aplasia of the ureter have been reported.

Bone anomalies visible on X-ray films are also characteristic. The diaphyses of the long bones are slender, the metaphyses narrow. Spontaneous fractures may occur. Delay in bone age may be considerable.

Ocular anomalies have been reported: coloboma, nystagmus, microphthalmy, microcornea, and atrophia of the optic nerves.

In several cases hypochromic anemia and thrombocytopenia were noted.

Mental Retardation

In those cases where it could be determined, mental retardation seemed profound, with an IQ of the order of 20.

Hypotonia is often severe.

Prognosis

Lethality seems to be considerable. One-fourth of the infants were stillborn. Half died before 14 months. The oldest child is 12 years old.

In cases of survival, statural, and especially ponderal retardation remains considerable. The children are dolichomorphic and emaciated.

CYTOGENETIC STUDY

The near totality of cases are due to malsegregation of a familial translocation.

The breakpoint on 10p is most often in p11. Thus, trisomy involves all of the short arm. The breakpoint may be more distal. No noteworthy phenotypical difference exists. Serious malformations are observed more often in cases of complete 10p trisomy.

A study of the segregation in 11 families showed that there was a higher risk of occurrence of 10p trisomy (22 per 100), no matter what the sex of the parent carrier of the translocation, as well as a high proportion of balanced carriers in normal descendants (71 per 100) (Stene and Stengel-Rutkowski, 1977).

DERMATOGLYPHICS

The following may be noted:

- An excess of whorls
- Axial triradius in t″
- Hypoplastic dermal ridges
- Abnormal palmar creases

(Rodewald and Stengel-Rutkowski, 1978)

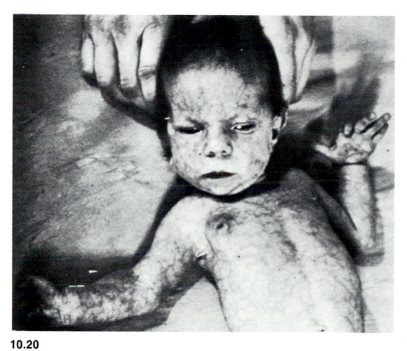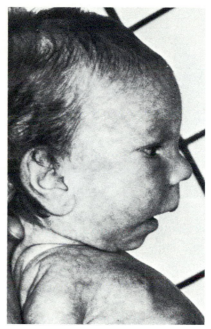

10.20

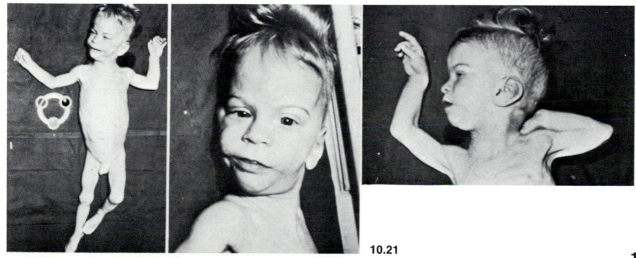

10.21

185

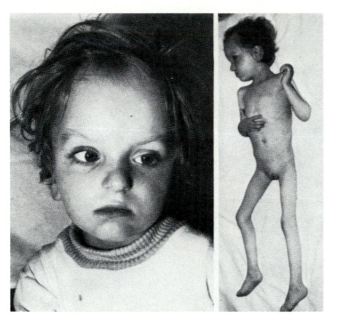

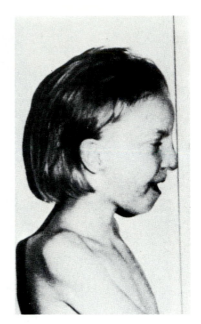

10.22

10.23

10.24

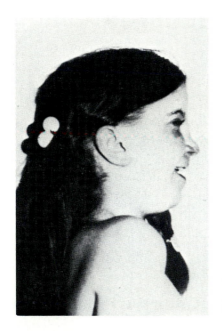
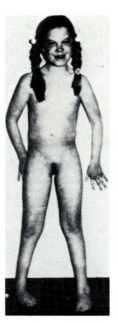

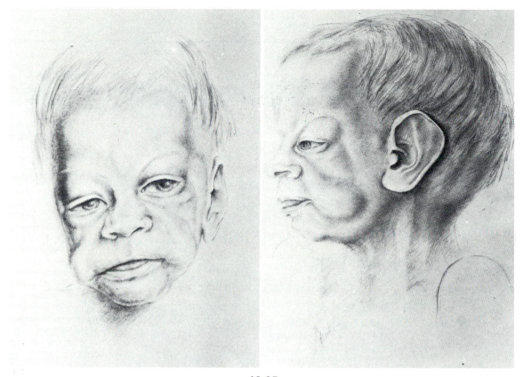

10.25

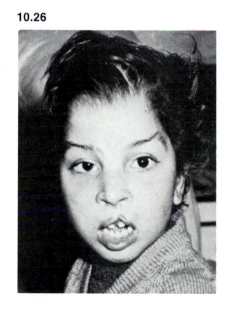

10.26

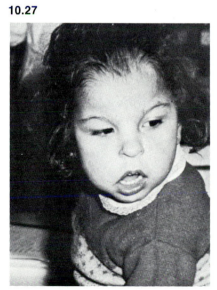

10.27

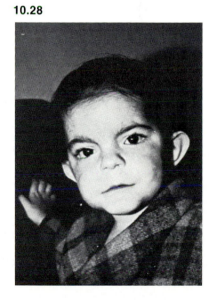

10.28

10p MONOSOMY

REFERENCES

BERGER R., LARROCHE J.C., TOUBAS P.L.: Deletion of the short arm of chromosome n° 10. *Acta paediatr. scand.*, **66**:659–662, 1977.

BOURROUILLOU G., COLOMBIES P., GALLEGOS D., MANELFE C., ROCHICCHIOLI P.: Monosomie partielle du bras court d'un chromosome 10: à propos d'une observation avec étude tomodensitométrique. *Ann. Génét.*, **24**:61–64, 1981.

FRANCKE U., KERNAHAN C., BRADSHAW C.: Del(10p) autosomal deletion syndrome: clinical, cytogenetic, and gene marker studies. *Humangenetik*, **26**:343–351, 1975.

PRIETO F., BADIA L., MORENO J.A., BARBERO P., ASENSI F.: 10p− syndrome associated with multiple chromosomal abnormalities. *Hum. Genet.*, **45**:229–235, 1978.

SHOKEIR M.H.K., RAY M., HAMERTON J.L., BAUDER F., O'BRIEN H.: Deletion of the short arm of chromosome n° 10. *J. med. Genet.*, **12**:99–113, 1975.

The first two observations were independently reported by Shokeir et al. (1975) and by Francke et al. (1975). Four cases are now known (Berger et al., 1977; Bourrouillou et al., 1981). One observation has not been considered due to the complexity of the associated chromosomal anomalies (Prieto et al., 1978).

GENERALIZATIONS

- Sex ratio: 2M/2F
- Mean birth weight: 2,660 g
- Mean parental ages:
     maternal: 28 years
     paternal: 31.75 years

PHENOTYPE

Craniofacial Dysmorphism

The skull is fairly normal in shape. The forehead is high and wide.

Palpebral fissures are horizontal or slightly slanted downward and outward. Epicanthus and ptosis may be observed and, in two cases, microphthalmy was present.

The tip of the nose is wide and bulbous.

A high arched palate is noted in two observations, and a harelip in another.

Microretrognathism is considerable.

The ears vary in appearance, but often are small and low-set.

Malformations

Various malformations have been observed, including congenital heart disease in three cases, cerebral malformations (atrophy of the olfactory bulbs, and an abnormality of the median line), kidney and digestive defects (umbilical hernia, pyloric stenosis), and malpositioning of the feet and hands.

Mental Retardation

Mental retardation appears to be quite severe in the two living children.

Prognosis

Two children died, at ages 2 days and 13 weeks.

CYTOGENETIC STUDY

The four cases considered here occurred *de novo*. The breakpoint is in 10p13 (3 cases) or in p14.

DERMATOGLYPHICS

These do not appear to be characteristic.

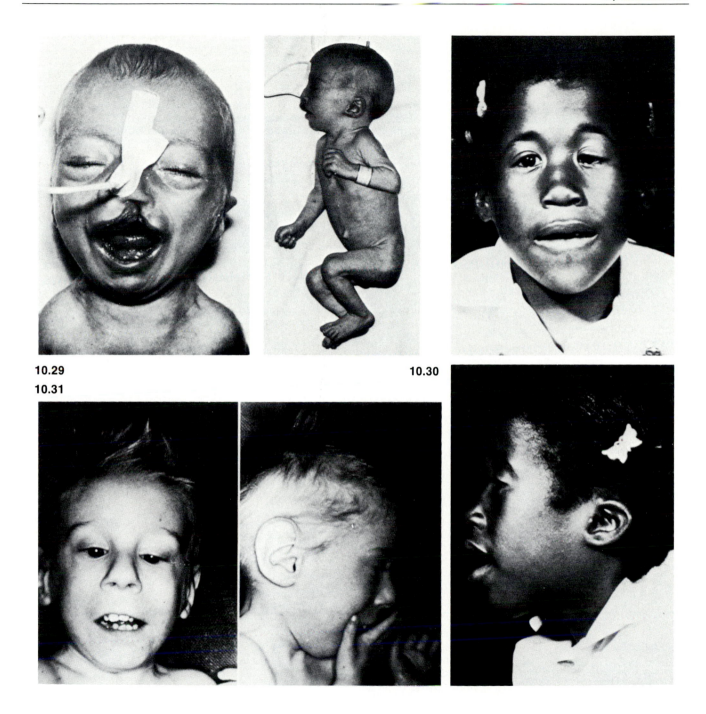

10.29

10.30

10.31

10q2 MONOSOMY

REFERENCES

LEWANDOWSKI R.C., KUKOLICH M.K., SEARS J.W., MANKINEN C.B.: Partial deletion 10q. *Hum. Genet.*, **42**:339–343, 1978.

TURLEAU C., GROUCHY J. de, PONSOT G., BOUYGUES D.: Monosomy 10qter. *Hum. Genet.*, **47**:233–237, 1979.

WEGNER R.D., KUNZE J., PAUST H.: Monosomy 10qter due to a balanced familial translocation: t(10;16) (q25.2;q24). *Clin. Genet.*, **19**:130–133, 1981.

Only three such cases are known (Lewandowski et al., 1978: Turleau et al., 1979; Wegner et al., 1981). The deletion is limited to the two last bands of 10q. Presently it is impossible to characterize a syndrome but the main features would seem to be the following:

- Prematurity, and a low birth weight
- Severe mental retardation
- Microcephaly
- A long face with effacement of the angles of the jaw
- Prominent nose bridge, with effacement of the fronto-nasal angle
- Anomalies of the external genitals, and aplasia of the perineum

Various other malformations have been observed, including congenital heart diseases, agenesis of the corpus callosum, and anomalies of the internal genital organs.

r(10)

REFERENCES

SIMONI G., ROSSELLA F., DALPRA L., VISCONTI G., PIRIASCHWARZ C.: Ring chromosome 10 associated with multiple congenital malformations. *Hum. Genet.*, **51**:117–121, 1979.

TSUKIDO R., TSUDA N., DEZAWA T., ISHII T., KOIKE M.: Ring chromosome 10:46,XX,r(10) (p15 q26). *J. med. Genet.*, **43**:148–151, 1980.

Five cases are known (see Simoni et al., 1979; and Tsukino et al., 1980). The phenotype is not very characteristic. In two adolescents, small stature and moderate mental retardation are noted.

Cardiac and renal anomalies appear to be frequent.

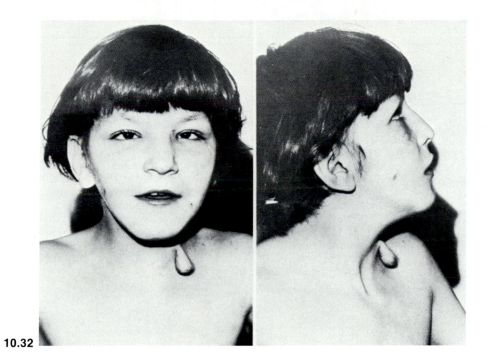

10.32

CHROMOSOME 11

Chromosome 11 is of importance both because of its gene assignments and also because of its genic and cytogenetic disorders.

Some important genes have been localized on the short arm: those for non-α globins, whose mutations are responsible for sickle cell anemia and β-thalassemia, among others; insulin; lactate dehydrogenase A; acid phosphatase-2; catalase. This latter enzyme may be the object of a gene mutation responsible for the recessive disease acatalasemia. This gene also is closely linked to the WAGR syndrome, which is due to a deletion of 11p13, and associates aniridia, ambiguous genitalia, mental and growth retardation, and nephroblastoma or gonadoblastoma.

In addition to the WAGR syndrome, cytogenetic diseases include partial trisomies for the long arm, most of which are due to the 3:1 malsegregation of a t(11;22); and partial monosomies for the long arm. Partial trisomies for the short arm are more rarely encountered.

A fragile site is situated in 11q13.

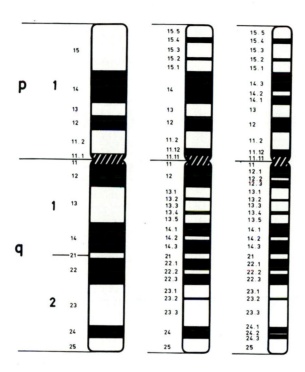

GENE MAP

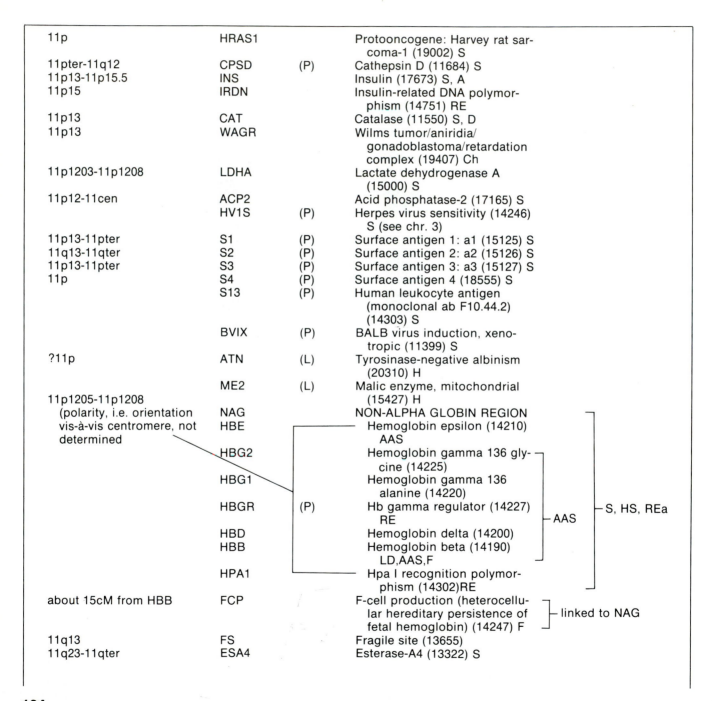

11p	HRAS1		Protooncogene: Harvey rat sarcoma-1 (19002) S
11pter-11q12	CPSD	(P)	Cathepsin D (11684) S
11p13-11p15.5	INS		Insulin (17673) S, A
11p15	IRDN		Insulin-related DNA polymorphism (14751) RE
11p13	CAT		Catalase (11550) S, D
11p13	WAGR		Wilms tumor/aniridia/gonadoblastoma/retardation complex (19407) Ch
11p1203-11p1208	LDHA		Lactate dehydrogenase A (15000) S
11p12-11cen	ACP2		Acid phosphatase-2 (17165) S
	HV1S	(P)	Herpes virus sensitivity (14246) S (see chr. 3)
11p13-11pter	S1	(P)	Surface antigen 1: a1 (15125) S
11q13-11qter	S2	(P)	Surface antigen 2: a2 (15126) S
11p13-11pter	S3	(P)	Surface antigen 3: a3 (15127) S
11p	S4	(P)	Surface antigen 4 (18555) S
	S13	(P)	Human leukocyte antigen (monoclonal ab F10.44.2) (14303) S
	BVIX	(P)	BALB virus induction, xenotropic (11399) S
?11p	ATN	(L)	Tyrosinase-negative albinism (20310) H
	ME2	(L)	Malic enzyme, mitochondrial (15427) H
11p1205-11p1208 (polarity, i.e. orientation vis-à-vis centromere, not determined	NAG		NON-ALPHA GLOBIN REGION
	HBE		Hemoglobin epsilon (14210) AAS
	HBG2		Hemoglobin gamma 136 glycine (14225)
	HBG1		Hemoglobin gamma 136 alanine (14220)
	HBGR	(P)	Hb gamma regulator (14227) RE
	HBD		Hemoglobin delta (14200)
	HBB		Hemoglobin beta (14190) LD,AAS,F
	HPA1		Hpa I recognition polymorphism (14302)RE
about 15cM from HBB	FCP		F-cell production (heterocellular hereditary persistence of fetal hemoglobin) (14247) F
11q13	FS		Fragile site (13655)
11q23-11qter	ESA4		Esterase-A4 (13322) S

S, HS, REa

AAS

linked to NAG

11q23-11qter	PBGD		Porphobilinogen (PBG) deaminase (uroporphyrinogen I synthase) (17600)S
11q	GST1	(P)	Glutathione S-transferase-1 (13835) S
Proximal to NAG	PTH		Parathyroid hormone (16345) REa
11q13-11qter	GANAB	(P)	Neutral alpha-glucosidase AB (10416) S
	FN	(L)	Fibronectin (?13560) S
	M4F2	(P)	Monoclonal antibody 4F2 (15807) S

MORPHOLOGY AND BANDING

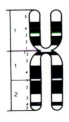

QFQ   GTG   RHG   RBA   THA   C

11q2 TRISOMY

Long philtrum
Retracted lower lip
Microretrognathia
Flexion contracture of the limbs

REFERENCES

FRANCKE U., WEBER F., SPARKES R.S., MATTSON D., MANN J.: Duplication 11 (q21 to q23→qter) syndrome. *Birth Defects, Orig. Art. Ser., XIII-3B*:167–186, 1977.

FRANCKE U.: Abnormalities of chromosomes 11 and 20. *In* J.J. YUNIS: *New Chromosomal Syndromes.* New York, Academic Press, 1977.

PIHKO H., THERMAN E., UCHIDA I.A.: Partial 11q trisomy syndrome. *Hum. Genet.,* **58**:129–134, 1981.

TUSQUES J., GRISLAIN J.R., ANDRÉ M., MAINARD R., RIVAL J.M., CADUDAL J.L.: Trisomie partielle 11q identifiée grâce à l'étude en dénaturation ménagée par la chaleur de la translocation équilibrée paternelle. *Ann. Génét.,* **15**:167–172, 1972.

The first observation of partial 11q trisomy was that of Tusques et al. (1972). Since then, a number of observations have been made, most of them resulting from a t(11;22) translocation, which would seem to be the most frequently observed translocation in humans. In another chapter, we will consider trisomies due to a t(11;22), which seem to make up a rather distinct entity. Here we will consider only those 11q trisomies that do not result from this t(11;22). Twelve such cases are known (ref. in Francke et al., 1977; and Pihko et al., 1981).

GENERALIZATIONS

- Sex ratio: 1M/1F
- Mean birth weight: 2,450 g
- Mean crown-heel length at birth: 48 cm

PHENOTYPE

In newborns, hypertonia of the limbs, leading to exaggerated flexion, contrasts with axial hypotonia. An aged appearance and waxy complexion or the absence of subcutaneous fatty tissue have been described.

Craniofacial Dysmorphism

The shape of the skull is usually normal. Nonetheless, microcephaly, brachycephaly, or cranial asymmetry have been reported in several instances.

The face is wide and may be moon-shaped, with chubby cheeks.

Palpebral fissures are horizontal or slightly slanting upward and outward.

The nose is short, snubbed and fleshy, with prominent alae, and a pronounced naso-frontal angulation.

The philtrum is long and prominent, and overlaps the lower lip, which is retracted. These children appear to be chewing on their lower lip.

Microretrognathia is observed in all cases, and seems to be the main feature of this syndrome. It is frequently accompanied by malformations of the palate and by glossoptosis, thus resembling Pierre Robin syndrome.

Neck, Thorax, and Abdomen

The neck is short, and the thorax narrow with very externally located nipples. The abdominal wall is often deficient and hernias are frequent.

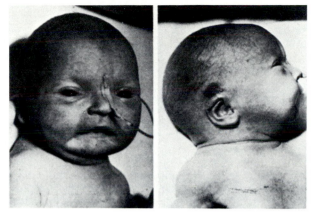

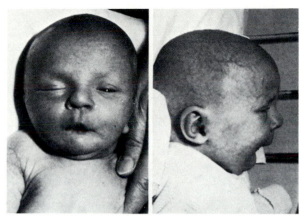

11.1

11.2

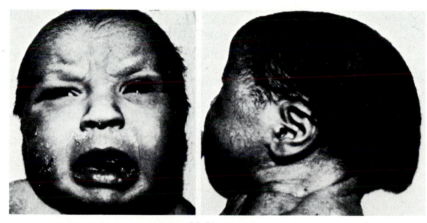

11.3

11.4

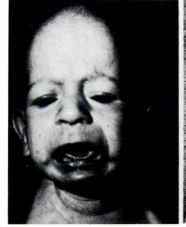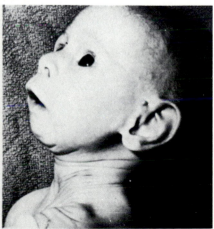

Limbs

The limbs have a highly characteristic appearance, with flexion in pronation of the forearms upon the arms giving the appearance of pleading posture, or of bird's wings, finger flexion upon the palm, and poor positioning of the feet, or rocker-bottom feet. The skin of the palms and soles of the feet are often wrinkled.

Genitalia

Hypoplasia of the penis is present nearly always in boys.

Malformations

Inner organ malformations often are very severe, and help determine the initial vital prognosis: they include congenital heart disease, which is almost always present, urogenital malformations, agenesis of the corpus callosum, agenesis of the gallbladder, ectopic spleen, and common mesentery. In addition to malpositioning of the limbs, such bone anomalies as agenesis of the collar bone, dysplasia of the acetabulum, and radiocubital synostosis may be observed.

Mental Retardation

This would seem to be fairly severe, but is difficult to evaluate due to the young age of the children. The case reported by Jacobsen et al. (1973) is unique in its prolonged survival (the child was examined at the age of 12), a phenotype with only slight dysmorphism, and an IQ of around 65.

Prognosis

Five out of eleven children died in the neonatal period or before the age of one year. In case of survival, growth retardation is likely to lessen with age.

CYTOGENETIC STUDY

In nearly all of the cases, the partial 11q trisomy results from the malsegregation of a parental translocation. The breakpoint is located in q23 in most cases, which suggests that a preferential breakpoint exists at this site. In the other cases, the breakpoint is more proximal, at the limit of the two regions, q1 and q2. The phenotype does not appear to be much different in these two situations. Mortality is higher in those trisomies which are more extended.

DERMATOGLYPHICS

Dermatoglyphics do not seem to be characteristic.

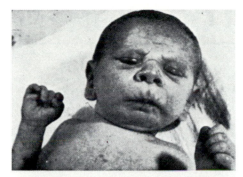

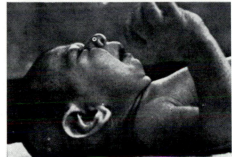

11.5

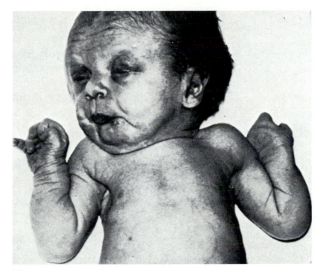

11.6

11.7

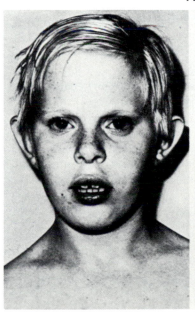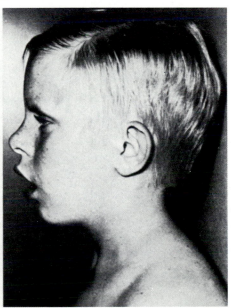

TRISOMY BY t(11;22) TRANSLOCATION

Long philtrum
Retracted lower lip
Microretrognathia
Preauricular pits and tags
Anal atresia or stenosis

REFERENCES

AURIAS A., LAURENT C.: Trisomie 11q. In-dividualisation d'un nouveau syndrome. *Ann. Génét.*, **18**:189–191, 1975.

AURIAS A., TURC C., MICHIELS Y., SINET P.M., GRAVELEAU D., LEJEUNE J.: Deux cas de trisomie 11q (q231→qter) par translocation t(11;22) (q231;q111) dans deux familles différentes. *Ann. Génét.*, **18**:185–188, 1975.

FELDMAN G.M., SPARKES R.S.: The problem of partial trisomy 22 reconsidered. *Hum. Genet.*, **45**:97–101, 1978.

FRACCARO M., LINDSTEN J., FORD C.E., ISELIUS L.: The 11q; 22q translocation: a European collaborative analysis of 43 cases. *Hum. Genet.*, **56**:21–51, 1980.

GIRAUD F., MATTEI J.F., MATTEI M.G., BERNARD R.: Trisomie partielle 11q et translocation familiale 11-22. *Humangenetik*, **28**:343–347, 1975.

KESSEL E., PFEIFFER R.A.: 47,XY,+der (11;22) (q23; q12) following balanced translocation t(11;22) (q23;q12) mat. *Hum. Genet.*, **37**:111–116, 1977.

SCHINZEL A., SCHMID W., AUF DER MAUR P., MOSER H., DEGENHARDT H.K., GEISLER M., GRUBISIC A.: Incomplete trisomy 22. I. Familial 11/22 translocation with 3:1 meiotic disjunction. Delineation of a common clinical picture and report of nine new cases from six families. *Hum. Genet.*, **56**:249–262, 1981.

ZACKAI E.H., EMANUEL M.S.: Site specific reciprocal translocation, t(11;22) (q23;q11), in several unrelated families with 3:1 meiotic disjunction. *Am. J. med. Genet.*, **7**:507–522, 1980.

The characterization of the trisomy resulting from meiotic 3:1 segregation of a t(11;22) parental translocation can be justified by the following considerations: on the one hand, this trisomy had first been described as a trisomy 22. The 3:1 segregation does, indeed, result in a 47-chromosome complement with a "non-21" G trisomy. This rearrangement was correctly described later as a partial 11q trisomy, when the t(11;22) was identified (Aurias et al., 1975; Aurias and Laurent, 1975; Giraud et al., 1975; Kessel and Pfeiffer, 1977; Feldman and Sparkes, 1978). The characterization of trisomy by t(11;22) is even more valid because of its frequency, since t(11;22) is probably the reciprocal translocation most often found in humans. More than 50 cases have been recently analyzed (Fraccaro et al., 1980; Zackai and Emmanuel, 1980; Schinzel et al., 1981).

GENERALIZATIONS

- Sex ratio: 4M/3F
- Mean birth weight: 2,730 g

PHENOTYPE

The phenotype is very similar to that of 11q2 trisomy. The following features are noted:

- Microcephaly
- A wide face with chubby cheeks
- A short, broad, flat nose
- A long, prominent philtrum
- Retraction of the lower lip
- Microretrognathia, high-arched palate or cleft palate, and glossoptosis, which may be suggestive of the Pierre Robin syndrome
- Low set, deformed ears, with blind pits and preauricular tags
- A micropenis with hypoplasia of the scrotum and/or cryptorchidism
- The most frequently encountered malformations are complex heart defects, renal anomalies, anal anomalies, and hip dislocation. Less frequently, malrotation of the intestine, uterine anomalies, diaphragm hypoplasia, atresia of the canal combined with aplasia of the auricle, and anomalies of the colar bone may be observed
- Growth retardation, hypotonia, and severe psychomotor retardation

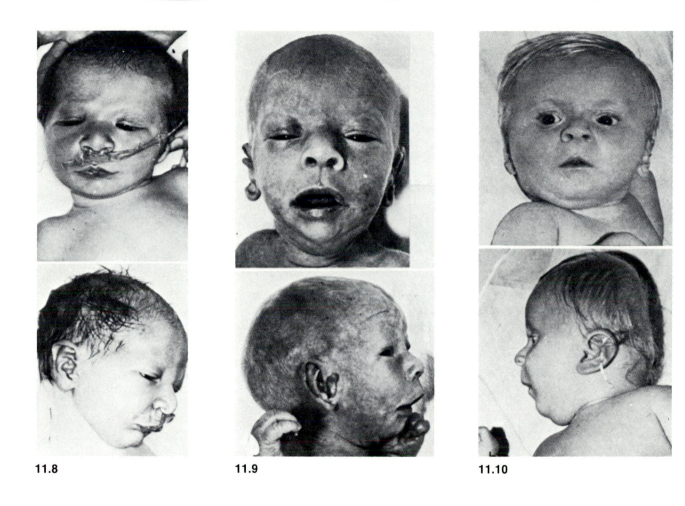

11.8   11.9   11.10

11.11

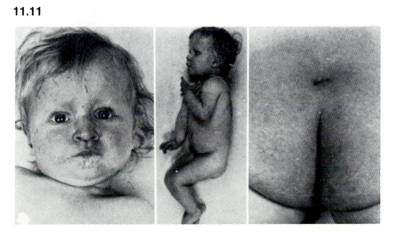

The prognosis is characterized by a high frequency of early deaths. In surviving children, dysmorphism seems to evolve toward midface hypoplasia, with prominence of the chin which contrasts with the severe micrognathia of early childhood.

The majority of the symptoms of trisomy by t(11;22) may be attributed to 11q2 trisomy. Features probably due to the associated 22 trisomy are as follows: anal anomalies (atresia, stenosis, or other malformations), preauricular tags (pits are observed in 11q trisomy), anomalies in development of the auricle, with atresia of the auditory canal, and orientation of palpebral fissures downward and outward. No ocular coloboma is observed, although this sign belongs to the "cat eye" syndrome, which is thought to be due to partial 22pter→q11 trisomy.

CYTOGENETIC STUDY

Trisomy is nearly always the consequence of the malsegregation of a familial t(11;22). The most frequently indicated breakpoints are 11q23 and 22q11. In several instances, other breakpoints have been indicated: 11q25 and 22q13. In fact, these probably involve the same t(11;22)(q23;q11). It seems unlikely that all carriers of this translocation are related. A cytogenetic property that enhances the recurrence of this same translocation (and whose identity can only be hypothesized) would seem to be a more probable assumption.

The sex ratio of balanced carriers is normal.

The carrier of the balanced translocation is nearly always the mother. This translocation in men would lead to a decrease in fertility. The frequency of spontaneous abortions is high no matter what the sex of the carrier of the balanced translocation.

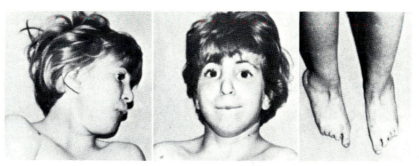

11.12

202

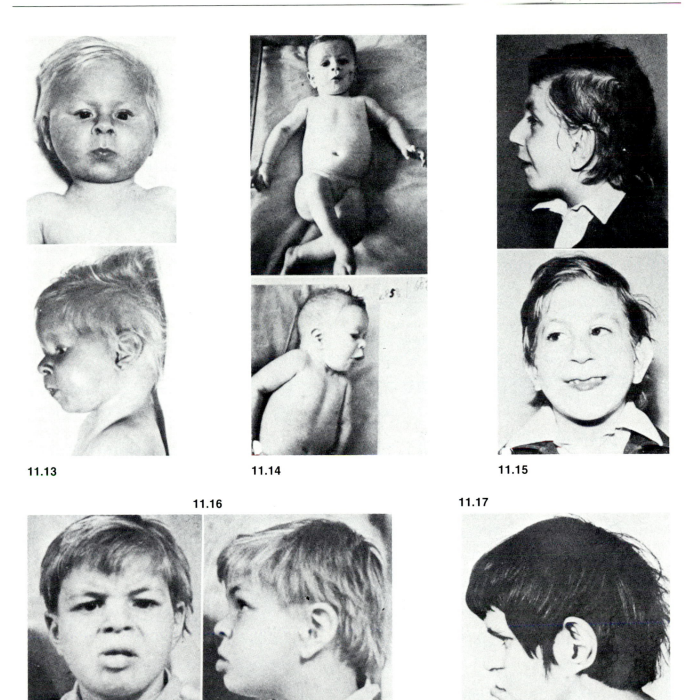

11.13

11.14

11.15

11.16

11.17

11q2 MONOSOMY

Trigonocephaly
Coarse features
Large mouth with downturned corners

REFERENCES

COCO R., PENCHASZADEH V.B.: Partial deficiency of long arm of chromosome n° 11. *J. Génét. hum.*, **25**:43–46, 1977.

JACOBSEN P., HAUGE M., HENNINGSEN K., HOBOLTH N., MIKKELSEN M., PHILIP J.: An (11;21) translocation in four generations with chromosome 11 abnormalities in the offspring. *Hum. Hered.*, **23**:568–585, 1973.

LAURENT C., BIEMONT M.C., VEYRON M., GUILHOT J., GUIBAUD P.: Deux nouveaux cas de monosomie 11q partielle avec point de cassure en 11q24. *Ann. Génét.*, **22**:239–241, 1979.

LEE M.L., SCIORRA L.J.: Partial monosomy of the long arm of chromosome 11 in a severely affected child. *Ann. Génét.*, **24**:51–53, 1981.

LEONARD C., COURPOTIN C., LABRUNE B., LEPERCQ G., KACHANER J., CAUT P.: Monosomie partielle par délétion du bras long du chromosome 11: del(11) (q23). *Ann. Génét.*, **22**:115–120, 1979.

LIPPE B.M., SPARKES R.S., FASS B., NEIDENGARD L.: Cranio-synostosis and syndactyly: expanding the 11q– chromosomal deletion phenotype. *J. med. Genet.*, **17**:480–482, 1980.

RIDLER M.A.C., McKEOWN J.A.: 11q aneuploidy: partial monosomy and trisomy in the children of a mother with a t(13;11) (p27;q23) translocation. *Hum. Genet.*, **52**:101–106, 1979.

The first observation of 11q2 monosomy was that of Jacobsen et al. (1973). Twenty-three observations are now known (ref. in Léonard et al., 1979; see also Coco and Penchazadeh, 1977; Ridler and McKeown, 1979; Laurent et al., 1980; Lippe et al., 1980; and Lee and Sciorra, 1981).

GENERALIZATIONS

- Sex ratio: 5F/1M
- Parental ages (in *de novo* cases):
    maternal: 26.7 years
    paternal: 28.4 years
- Mean birth weight: 2,750 g
- Mean crown-heel length at birth: 47.3 cm
- Mean head circumference at birth: 33.6 cm

PHENOTYPE

The most striking feature is trigonocephaly. Growth retardation, which is present at birth, persists with age.

Craniofacial Dysmorphism

The skull is deformed by trigonocephaly, which may be very marked, and occasionally by scaphocephaly. Macrocephaly or microcephaly have been observed. Facial features are rather coarse.

Palpebral fissures are normally oriented. Ptosis and coloboma, either uni- or bilateral, are frequently found, as well as hypertelorism. In rare instances, strabismus and epicanthus may be observed. The nose is bulbous with a wide, aplastic bridge. The philtrum is not well defined.

The mouth is large, with thin lips and downturned corners.

The chin recedes slightly.

The ears are low-set and slightly deformed, with a prominent anthelix and an underdeveloped lobe.

Neck, Thorax, and Abdomen

No characteristic anomalies are noted, but the neck may be short.

Limbs

Various malformations have been reported, including unusual flexion of the fingers on the thumb, clinodactyly, cutaneous syndactyly, shortness of the fingers, varus equine feet, hammertoe, and crowded toes.

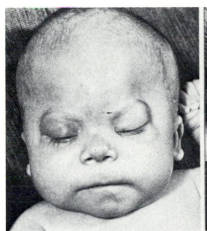 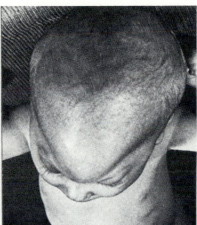 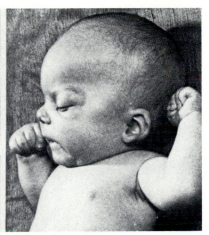

11.18

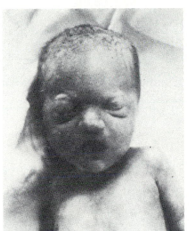 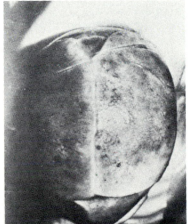

11.19

11.20

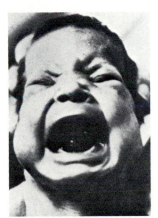

Genitalia

The very abnormal sex ratio of five girls to one boy was noted. Hypospadias and testicular ectopy were observed in boys.

Malformations

Cardiac and renal malformations may be observed. Other malformations have been reported on rare occasions. These include pyloric stenosis, spina bifida occulta, inguinal hernias, and deafness. Thrombopenia is frequently recorded.

Mental Retardation

Mental retardation is constantly present and generally quite severe, though it may, in rare cases, be moderate.

Prognosis

One-third of the children died soon after birth. Pulmonary and ear infections occur frequently. Considerable growth retardation is usual.

CYTOGENETIC STUDY

The deletion occurs *de novo* in the majority of cases. The breakpoint is nearly always in q23.

A familial translocation is found in two families in which 11q trisomic children are also found.

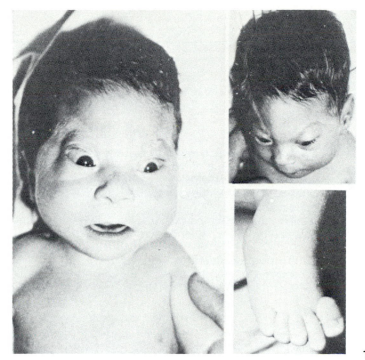

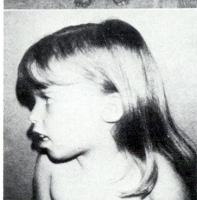

11.22

11.21

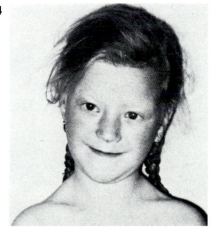

11.23

11.24

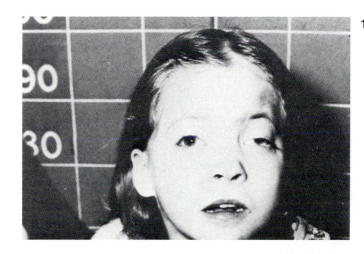

11p13 MONOSOMY

Aniridia
Ambiguous genitalia
Mental and growth retardation
Nephroblastoma or gonadoblastoma

REFERENCES

BRUSA P., TORRICELLI C.: Nefroblastoma di Wilms ed affezioni renali congenite nella casistica dell'IIPAI di Milano. *Minerva Paediat.*, **5**:457–463, 1953.

FRANCKE U., GEORGE D.L., BROWN M.G., RICCARDI V.: Gene dose effect: intraband mapping of the LDH A locus using cells from four individuals with different interstitial deletions of 11p. *Cytogenet. Cell Genet.*, **19**:197–207, 1977.

FRANCKE U., HOLMES L.B., RICCARDI V.M.: Aniridia-Wilms' tumor association: evidence for specific deletion of 11p13. *Cytogenet. Cell Genet.*, **24**:185–192, 1979.

FRANÇOIS J., COUCKE D., COPPIETERS R.: Aniridia-Wilms' tumour syndrome. *Ophthalmol.*, **174**:35–39, 1979.

JUNIEN C., TURLEAU C., GROUCHY J. de, SAID R., RETHORE M.O., BACCICHETTI C., DUFIER J.L.: Regional assignment of catalase (CAT) gene to band 11p13. Association with the aniridia-Wilms' tumor-gonadoblastoma (WAGR) complex. *Ann. Génét.*, **23**:165–168, 1980.

MILLER R.W., FRAUMENI J.F., MANNING M.D.: Association of Wilms' tumor with aniridia, hemihypertrophy and other congenital malformations. *New Engl. J. Med.*, **270**:922–927, 1964.

RICCARDI V.M., SUJANSKY E., SMITH A.C., FRANCKE U.: Chromosomal imbalance in the aniridia-Wilms' tumor association: 11p interstitial deletion. *Pediatrics*, **61**:604–610, 1978.

TURLEAU C., GROUCHY J. de, DUFIER J.L., LE HOANG PHUC, SCHMELCK P.H., RAPPAPORT R., NIHOUL-FEKETE C., DIEBOLD N.: Aniridia, male pseudohermaphroditism, gonadoblastoma, mental retardation, and del11p13. *Hum. Genet.*, **57**:300–306, 1981.

The association of an intercalary deletion, which may be very tiny, with an embryonic type tumor is now known in two instances. One of them is the association between a del 13q14 and retinoblastoma. The other is the WAGR association of a del 11p13, aniridia, mental and growth retardation, ambiguous genitalia, with either nephroblastoma (Wilms' tumor) or gonadoblastoma.

The association of aniridia and nephroblastoma was described for the first time by Brusa and Torricelli (1953) and confirmed by Miller et al. (1964) and Francois et al. (1979), among others.

The intercalary deletion of 11p was first identified by Francke et al. (1977) and Riccardi et al. (1978). Eleven cases have been reported in the literature. They were analyzed by Francke et al. (1979) and by Turleau et al. (1981).

GENERALIZATIONS

• Sex ratio: 7XY/4XX (see *Genitalia*)

PHENOTYPE

Growth retardation is frequently found, and is usually important. In two cases, development was situated at around the 50th percentile. Mental retardation is nearly constantly found. Only one patient had subnormal development.

Facial Dysmorphism

Various dysmorphisms have been observed, but none is characteristic.

Ocular Anomalies

Aniridia is the most constant element of this syndrome. It is bilateral and is accompanied by other ocular anomalies, including glaucoma (which may conceal aniridia), cataracts, nystagmus, ptosis, strabismus, and corneal opacities, which may lead to almost total blindness.

Genitalia

Anomalies of the genital organs make up some of the major signs of this syndrome. In XY subjects, varying degrees of sexual ambiguity may be observed, ranging from simple cryptorchidism to severe pseudohermaphroditism affecting the internal genital organs, with persistence of Müllerian structures and gonadal dysgenesis.

In XX subjects, the external genitals are usually female. In one case, an autopsy disclosed gonadal dysgenesis.

Tumors

A nephroblastoma was observed in four out of 11 children. The tumor was uni- or bilateral. The age at the time of the first diagnosis varied from 15 to 42 months. One observation concerned monozygotic twins, only one of whom developed the tumor.

A gonadoblastoma was observed in two cases of gonadal dysgenesis, a boy and a girl. Both cases involved a histological discovery at the time of either an autopsy or an exploratory laparotomy with castration.

Other Malformations

Renal malformations were reported in several cases, and were mainly malrotations.

CYTOGENETIC STUDY

The rearrangement usually appears *de novo* except in two cases. In one case, it involved an ins(2;11), and in the other an intrachromosomal translation.

The size of the monosomic segment is variable. Breakpoints are localized between p11.3 and p15.1. The smallest common region is the distal portion of the 11p13 band (Francke et al., 1979).

In three cases that occurred *de novo,* other structural rearrangements also were present.

LABORATORY FINDINGS

A number of known assignments are present on 11p, in the neighborhood of 11p13. An important enzyme is catalase, for which a gene dosage effect has been demonstrated, and whose gene appears to be very closely linked with the described syndrome (Junien et al., 1980). Since the chromosomal deletion may be at the limit of detection possibilities, the assay for catalase could prove to be extremely valuable as a detection tool in all cases of sporadic aniridia.

11p TRISOMY

REFERENCES

FALK R.E., CARREL R.E., VALENTE M., CRANDALL B.F., SPARKES R.S.: Partial trisomy of chromosome 11: a case report. *Am. J. ment. Def.*, **77**:383–388, 1973.

FRANCKE U.: Abnormalities of chromosomes 11 and 20. *In* J.J. YUNIS: *New Chromosomal Syndromes*. New York, Academic Press, 1977.

JUNIEN C., TURLEAU C., GROUCHY J. de, SAID R., RETHORÉ M.O., TENCONI R., DUFIER J.L.: Regional assignment of catalase (CAT) gene to band 11p13. Association with the aniridia-Wilms tumor-gonadoblastoma (WAGR) complex. *Ann. Génét.*, **23**:165–168, 1980.

PALMER C.G., POLAND C., REED T., KOJETIN J.: Partial trisomy 11, 46,XX,−3,−20, +der3,+der20,t(3;11;20) resulting from a complex maternal rearrangement of chromosomes 3,11,20. *Hum. Genet.*, **31**:219–225, 1976.

RETHORE M.O., JUNIEN C., AURIAS A., COUTURIER J., DUTRILLAUX B., KAPLAN J.C., LEJEUNE J.: Augmentation de la LDHA et trisomie 11 partielle. *Ann. Génét.*, **23**:35–39, 1980.

SANCHEZ O., YUNIS J.J., ESCOBAR J.L.: Partial trisomy 11 in a child resulting from a complex maternal rearrangement of chromosomes 11, 12 and 13. *Humangenetik*, **22**:59–65, 1974.

STROBEL R.J., RICCARDI V.M., LEDBETTER D.H., HITTNER H.M.: Duplication 11p11.3→14.1 due to meiotic crossing over. *Am. J. med. Genet.*, **7**:15–20, 1980.

Five cases of 11p trisomy having a common segment in 11p12 and p14, are known (Falk et al., 1973; Sanchez et al., 1974; Palmer et al., 1976; Rethoré et al., 1980; Strobel et al., 1980). The syndrome was characterized by Francke (1977) based on three observations. A more suggestive dysmorphic syndrome would seem to exist than for monosomy 11p. In addition, no characteristic ocular anomaly nor signs of malignancy are observed.

The principle features are as follows:

- A high, convex forehead, with frontal upsweep of hair; a prominent occiput
- A wide nose bridge
- Hypertelorism, accentuated even further by an epicanthus that may be very marked
- A short, wide, beaked nose
- Round, chubby cheeks
- A cleft lip, cleft palate, or bifid uvula
- Hypotonia, and severe mental retardation

Malformations

Inner organ malformations would not seem to be of importance.

Ocular malformations include nystagmus and strabismus. In one case (Strobel et al., 1980), a thorough ophthalmological examination was performed, since, in that family, one of the children was afflicted with del 11-aniridia. This examination disclosed normal irises and a normal macula.

CYTOGENETIC STUDY

Four out of five cases involved transmission of a parental rearrangement.

The unusual nature of these rearrangements should be pointed out: two cases consisted of a three-chromosome-rearrangement, and one case of intrachromosomal translation; the *de novo* case involved a trisomy 11p, but with intercalary deletion of the p13 band (Rethoré et al., 1981).

LABORATORY FINDINGS

In one case an augmentation of LDHA activity has been observed (Rethoré et al., 1980), confirming the localization of the gene in 11p12. The catalase activity was normal, confirming the localization of the corresponding gene in band 11p13 (Junien et al., 1980).

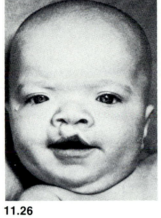

11.25

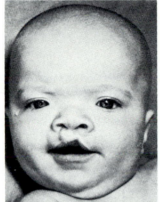

11.26

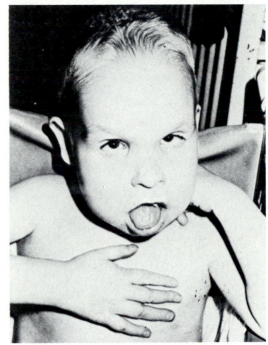

11.27

11.28

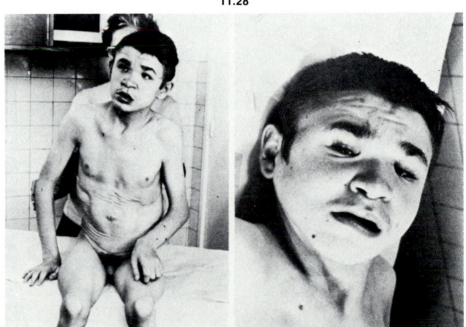

CHROMOSOME 12

The gene map of the short arm of chromosome 12 has been well established, mainly by the study of carriers of structural rearrangements. Lactate dehydrogenase B is one of the first regional assignments obtained through gene dosage effect. The one known hereditary disease caused by gene mutation is hemolytic anemia due to glyceraldehyde-3-phosphate dehydrogenase deficiency.

Chromosome disorders mainly include 12p trisomy. 12p monosomy and 12q trisomy are rare syndromes.

A fragile site, folic acid dependent, is present in 12q13.

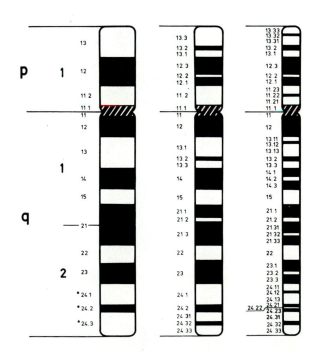

GENE MAP

12(?p)	KRAS2		Protooncogene ras 2: Kirsten rat sarcoma virus (19007) S
12p	KAR	(P)	Aromatic alpha-keto acid reductase (10792) S
	GPD	(P)	Alpha-glycerophosphate dehydrogenase (13846) S
12p13	GAPD		Glyceraldehyde-3-phosphate dehydrogenase (13840) S, D, R
12p121-12p122	LDHB		Lactate dehydrogenase B (15010) S, D
	LDHC	(L)	Lactate dehydrogenase C (15015) H
12p13	TPI1		Triosephosphate isomerase-1 (19045) S, D, R
	TPI2	(L)	Triosephosphate isomerase-2 (19046) S
12q21	PEPB		Peptidase B (16990) S
12p11-12p12	ENO2		Enolase-2 (13136) S
12p11-12qter	CS		Citrate synthase, mitochondrial (11895) S
12q12-12q14	SHMT		Serine hydroxymethyltransferase (13845) S, R
12q13	FS		Fragile site (13663)
	S8		Surface antigen 8 (18556) S
	BCT1		Branched chain amino acid transferase-1 (11352) S
	MIC3	(P)	Human leukocyte antigen
12q21-12q24.2	NF3	(L)	Familial intestinal neurofibromatosis (16224) Ch
12q24.1	IFI		Interferon, gamma or immune type (14757) S, A
	PAH	(L)	Phenylalanine hydroxylase (26160) REa

MORPHOLOGY AND BANDING

 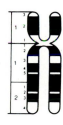      

QFQ　　　GTG　　　RHG　　　RBA　　　THA　　　C

12p TRISOMY

Corpulent infants
Turricephaly
Flat face

REFERENCES

BIJLSMA J.B., DE FRANCE H.F., BLEEKER-WAGEMAKERS L.M., DIJKSTRA P.F.: Double translocation t(7;12), t(2;6) heterozygosity in one family. *Hum. Genet.*, **40**:135–147, 1978.

HANSTEEN I.L., SCHIRMER L., HESTETUN S.: Trisomy 12p syndrome. Evaluation of a family with a t(12;21) (p12.1;p11) translocation with unbalanced offspring. *Clin. Genet.*, **13**:339–349, 1978.

KONDO I., HAMAGUCHI H., HANEDA T.: Trisomy 12p syndrome. *Hum. Genet.*, **46**:135–140, 1979.

PARSLOW M., CHAMBERS D., DRUMMOND M., HUNTER W.: Two cases of trisomy 12p due to rcpt (12;21) (p11;p11) inherited through three generations. *Hum. Genet.*, **47**:253–260, 1979.

RETHORE M.O.: Syndromes involving chromosomes 4, 9, and 12. *In* J.J. YUNIS, ed.: *New Chromosomal Syndromes.* New York, Academic Press, pp. 119–183, 1977.

SERVILLE F., JUNIEN C., KAPLAN J.C., GACHET M., CADOUX J., BROUSTET A.: Gene dosage effect for human triosephosphate isomerase and glyceraldehyde-3-phosphate dehydrogenase in partial trisomy 12p13 and trisomy 18p. *Hum. Genet.*, **45**:63–69, 1978.

SUERINCK E., SUERINCK A., KAPLAN J.C., MEYER J., JUNIEN C., NOËL B., RETHORE M.O.: Trisomie 12p par malsegregation d'une translocation paternelle t(12;22)(p11;p11). *Ann. Génét.*, **21**:243–246, 1978.

TENCONI R., PIOVAN E., PRETO A., MAGNABOSCO R., BACCICHETTI C.: Syndrome +12p. Case report and review. *Hum. Genet.*, **39**:97–101, 1977.

TENCONI R., GIORGI P.L., TARANTINO E., FORMICA A.: Trisomy 12p due to an adjacent 1 segregation of a maternal reciprocal translocation t(12;18)(p11;q23). *Ann. Génét.*, **21**:229–233, 1978.

UCHIDA I.A., LIN C.C.: Identification of partial 12 trisomy by quinacrine fluorescence. *J. Pediat.*, **82**:269–272, 1973.

The first observation of 12p trisomy was by Uchida and Lin (1973). Nineteen cases are known (ref. in Rethoré et al., 1977; Bijlsma et al., 1978; Hansteen et al., 1978; and also Tenconi et al., 1977, 1978; Serville et al., 1978; Suerinck et al., 1978; Parslow et al., 1979; Kondo et al., 1979).

GENERALIZATIONS

All known cases are due to malsegregation of a parental rearrangement.

- Sex ratio: 3M/2F
- Mean birth weight: 3,300 g
- Mean crown-heel length at birth: 48.25 cm

PHENOTYPE

The infants are plump, chubby, and hypotonic.

Craniofacial Dysmorphism

The cranium is deformed by turricephaly with a flat apex, giving it a square appearance. The forehead is high and bulging. The flat rectangular face gives the impression that the lower facial region projects forward.

Palpebral fissures are most often horizontal. A pronounced hypertelorism and bilateral epicanthus are present. The eyebrows are thick and implanted in an irregular manner.

The nose is very short, with a broad and poorly defined bridge.

The philtrum is broad.

The mouth is rather large and the lower lip everted.

The chin is small.

The ears are low or normally set. The helix is folded only along its upper portion; the anthelix is protuberant. The concha is deep set.

Neck, Thorax, and Abdomen

The neck is short and crowded with cutaneous folds.

The sternum is sometimes short or depressed.

Low-set or supernumerary nipples can be observed.

Limbs

The hands are stocky, the palms broad. The fingers may be permanently flexed, slender, and tapering.

The feet are flat. Dorsiflexion of the big toe and genu valgum have been reported. The first and second toes are abnormally wide-spaced.

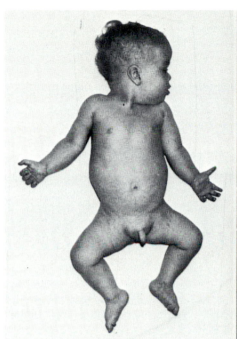

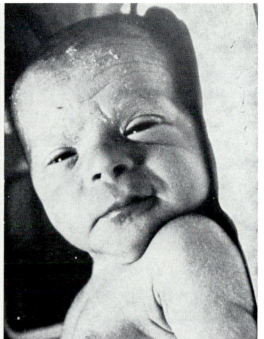

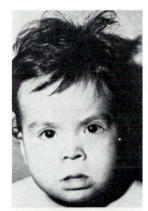

12.1

12.2

12.3

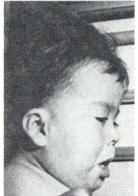

12.4

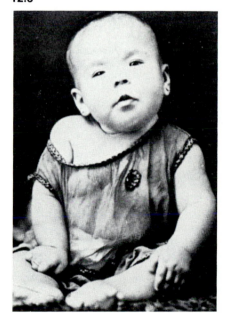

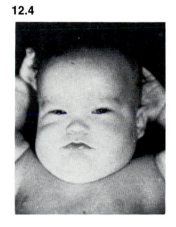

Genitalia

The testicles are sometimes ectopic.

Malformations

In one case, cardiomyopathy led to the child's death. A clinical heart murmur was reported in several cases.

Mental Retardation

Mental deficiency is severe. Some children never learn to walk or speak.

Prognosis

Three children died, one of whom was a girl afflicted with ovarian dysgerminoma. The babies are sometimes obese and hypotonic. The older children, on the contrary, are severely growth-retarded and hypertonic. In adolescents, the face shows signs of precocious aging.

CYTOGENETIC STUDY

The near totality of the cases are familial. The translocation most often occurs on the short arm of an acrocentric, especially a 21. The breakpoint on 12p is usually in p11. In several cases of 3:1 segregation, it is in p12.

DERMATOGLYPHICS

These do not seem characteristic. A single transverse palmar crease is present in three out of four patients.

LABORATORY FINDINGS

In the only case where LDH activity was quantitatively assayed, an increase of 1.5 was observed in red cells, whereas it was normal for fibroblasts and leukocytes (where subunit A predominates).

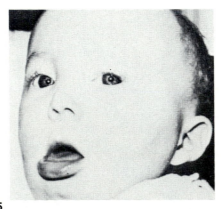 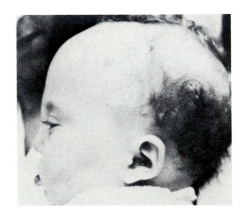

12.5

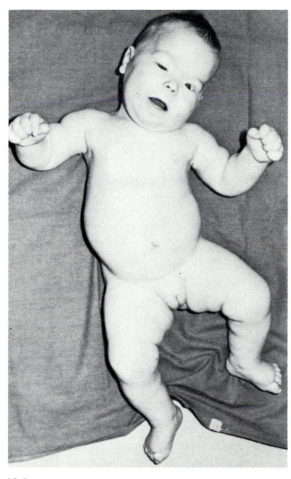

12.6

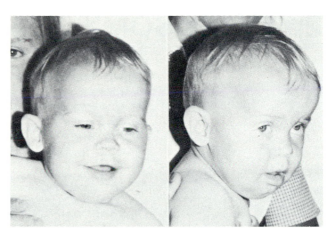

12.7

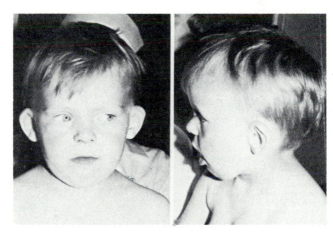

12.8

12.9

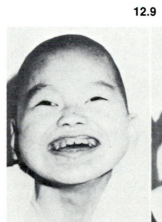

12p MONOSOMY

Growth retardation
Microcephaly
Narrow forehead
Pointed nose
Micrognathia

REFERENCES

MAGNELLI N.C., THERMAN E.: Partial 12p deletion: a cause for a mental retardation, multiple congenital abnormality syndrome. *J. med. Genet.*, **12**:105–108, 1975.

MALPUECH G., KAPLAN J.C., RETHORE M.O., JUNIEN C., GENEIX A.: Une observation de délétion partielle du bras court du chromosome 12. Localisation du gène de la lactico déshydrogénase-B. *Lyon méd.*, **233**:275–279, 1975.

MAYEDA K., WEISS L., LINDAHL R., DULLY M.: Localization of the human lactate dehydrogenase-B gene on the short arm of chromosome 12. *Amer. J. hum. Genet.*, **26**:59–64, 1974.

ORYE E., CRAEN M.: Short arm deletion of chromosome 12. Report of two new cases. *Humangenetik*, **28**:335–342, 1975.

TENCONI R., BACCICHETTI C., ANGLANI F., PELLEGRINO P.A., KAPLAN J.C., JUNIEN C.: Partial deletion of the short arm of chromosome 12 (p11;p13). Report of a case. *Ann.Génét.*, **18**:95–98, 1975.

The first observation of 12p monosomy was that of Mayeda et al. (1974), who simultaneously confirmed the localization of the LDH-B gene on 12p. Six observations are now known (Magnelli and Therman, 1975; Malpuech et al., 1975; Tenconi et al., 1975; Orye and Craen, 1975).

GENERALIZATIONS

All cases are due to a deletion occurring *de novo*.

- Sex ratio: 5M/1F
- Mean maternal age: 30.8 years
- Mean paternal age: 33.8 years
- Mean birth weight: 2,930 g

PHENOTYPE

The infants are microcephalic, of small size and corresponding weight, with relatively little dysmorphism.

Craniofacial Dysmorphism

Microcephaly is practically constant; the cranium is narrow, with a protruding occiput.

The palpebral fissures are horizontal or slanted downward and outward. Epicanthus and hypertelorism are rare.

The nose is long, with a protuberant ridge.

The upper lip is normal, except in one case where the philtrum is described as being long and poorly defined.

The chin is small and slightly receding.

The ears are long and very low-set. The margin of the helix can extend to the antitragus. The anthelix is hypoplastic in one case.

Neck, Thorax and Abdomen

The neck is normal, as are the thorax and abdomen.

Limbs

The hands are narrow, even "simian," with clinodactyly of the fifth digit. Camptodactyly and brachymetacarpy have been reported.

Brachymetatarsia was reported in one case.

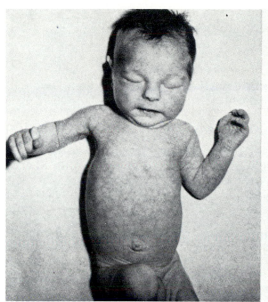
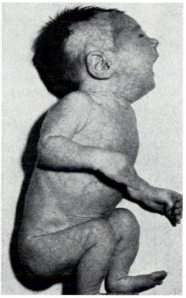

12.10

12.11

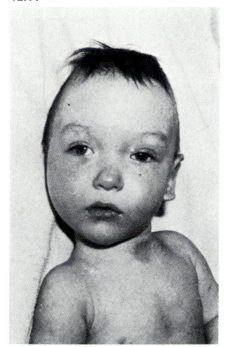

12.12

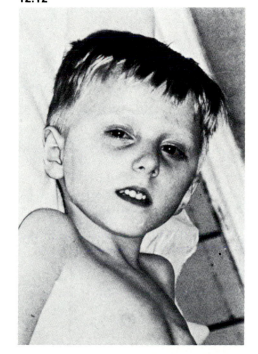

Genitalia

Cryptorchidism is frequently observed, and a micropenis is occasionally found.

Malformations

Visceral malformations are not generally present. Two patients out of six are afflicted with cardiopathy.

Mental Retardation

Mental retardation appears to be significant. The IQ in children is around 50. In the only known adult patient, deterioration is severe, with the IQ evaluated at 20. Seizures and epilepsy have been reported.

Prognosis

The oldest patient is 35 years old, measures 145 cm, weighs 45 kg, and has an IQ of 20. The other patients vary in age from several months to 10 years old. No fatal case in childhood is known.

CYTOGENETIC STUDY

In every known case, deletion occurs *de novo*. It usually involves an intercalary deletion of p12 or p11p12.

DERMATOGLYPHICS

Flexion creases are normal. Palmar and digital dermatoglyphics do not seem characteristic.

LABORATORY FINDINGS

In those cases where it was assayed, LDH activity showed a decrease of about 40 percent in red cells, where subunit B is predominant. Since monosomy always involves band p12, it can be deduced that the gene for LDH-B is precisely localized in this band.

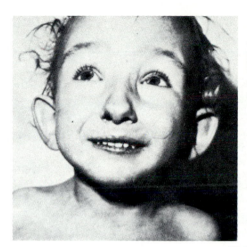
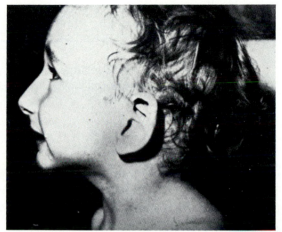

12.13

12.14

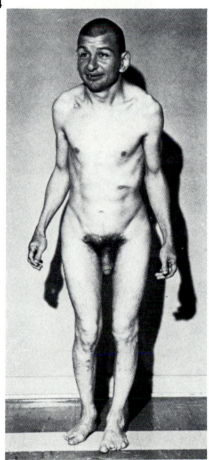

12q2 TRISOMY

Relatively large head
Short limbs

REFERENCES

HARROD M.J.E., BYRNE J.B., DEV V.G., FRANCKE U.: Duplication 12q mosaicism in two unrelated patients with a similar syndrome. *Am. J. med. Genet.,* **7**:123–129, 1980.

HEMMING L., BROWN R.: Partial trisomy 12q associated with a familial translocation. *Clin. Genet.,* **16**:25–28, 1979.

HOBOLTH N., JACOBSEN P., MIKKELSEN M.: Partial trisomy 12 in a mentally retarded boy and translocation (12;21) in his mother. *J. med. Genet.,* **11**:299–303, 1974.

MUELENAERE A. de, FRYNS J.P., Van den BERGHE H.: Partial distal 12q trisomy. *Ann. Génét.,* **23**:251–253, 1980.

Six cases of trisomy 12q24→qter are known (Hobolt et al., 1974; Hemming and Brown, 1979; Harrod et al., 1980; and Muelenaere et al., 1980).

GENERALIZATIONS

- Sex ratio: 2M/4F
- Mean birth weight: 2,940 g

PHENOTYPE

Craniofacial Dysmorphism

The skull is relatively large with frontal bossing. The head circumference, larger than normal, contrasts with growth retardation.

The face is rectangular with chubby cheeks.

The bridge of the nose is wide, with an undefined curve. The tip of the nose is snubbed.

The lips are thin, with downturned corners. The philtrum is short.

The chin is small.

Ears are low-set, round, and in posterior rotation, with a short lobe.

Neck, Thorax, and Abdomen

The neck is short with an excess of skin.

Minor anomalies can be observed, such as abnormal space between the nipples, prominence of the xiphoid appendix, pectus excavatum, sacral dimple, and hip dislocation.

Limbs

The limbs are short, especially in the proximal segment. Articular anomalies are also observed, including radiocubital synostosis, hyperextensibility, hyperflexion of the wrists, cubital deviation of the fingers, proximal thumbs, valgus flat feet, and hammer toe.

Genitalia

Cryptorchidism was reported in one case and atresic ovarian follicles in another.

Malformations

Inner organ malformations are uncommon. They may, however, include hydronephrosis with ureterocele, ectopic kidney, and incomplete rotation of the intestines with anal malpositioning.

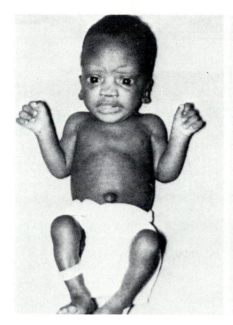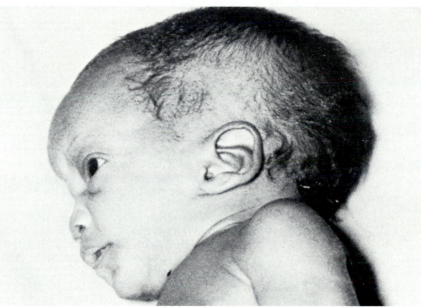

12.15

Mental Retardation

In newborns, hypertonia or hypotonia may be observed. In older children, mental retardation is severe.

Prognosis

One child died of pneumonia at the age of 6 days. Two patients are now about 20 years old.

Growth retardation is variable in extent.

CYTOGENETIC STUDY

Except for two cases of mosaicism occurring *de novo,* the rearrangement results from malsegregation of a parental translocation.

The breakpoint is in 12q24 in all of the observations.

DERMATOGLYPHICS

A single uni- or bilateral transverse palmar crease may be observed, as well as clinodactyly of the Vth with absence of a flexion crease.

CHROMOSOME 13

Chromosome 13 has few gene assignments but is highly important for its chromosomal disorders.

Only one enzymatic assignment has been confirmed, that of the esterase D, whose locus is closely linked with that of retinoblastoma. Measurement of its enzymatic activity is extremely important in all cases of retinoblastoma and especially in prenatal diagnosis of cases of familial chromosomal rearrangement.

Apart from the intercalary deletion responsible for retinoblastoma, cytogenetic diseases include complete 13 trisomy which is one of the major syndromes in cytogenetics, partial trisomies, and partial monosomies through deletions or by ring formation.

A polymorphism exists which is due to the variation in length of the short arm and the size of the satellite. Chromosomes with satellites in duplicate or triplicate also may be observed. This polymorphism is common to all the acrocentrics (groups D and G). In addition, a juxtacentromeric polymorphism visible in C − and Q − bands also exists for chromosome 13.

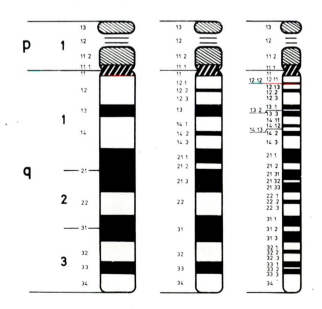

GENE MAP

13p12	RNR		Ribosomal RNA (18045) A
13q14	ESD		Esterase D (13328) S, F, D
13q14	RB1		Retinoblastoma-1 (18020) Ch
13q14	XRS	(L)	X-ray sensitivity (19443) Ch
13q14?	LP	(L)	Lipoprotein-Lp (15220) F
	CF7	(P)	Factor VII (22750) D
	CF10	(P)	Factor X (22760) D
	CBT	(L)	Carotid body tumor (16800) F

MORPHOLOGY AND BANDING

QFQ   GTG   RHG   RBA   THA   C

13 TRISOMY

Harelip
Microphthalmia
Hexadactyly
Early death

REFERENCES

BERGER R.: Trisomie 13. *Presse méd.*, **79**:1969–1972, 1971.

CONEN P.E., ERKMAN B.: Frequency and occurrence of chromosomal syndromes. I. D-trisomy. *Amer. J. hum. Genet.*, **18**:374–378, 1966.

HAMERTON J.L.: *Human Cytogenetics. General Cytogenetics, Vol. II*; New York, Academic Press, 1971.

HUEHNS E.R., LUTZNER M., HECHT F.: Nuclear abnormalities of the neutrophils in D_1 (13-15) trisomy syndrome. *Lancet*, **i**:589–590, 1964.

HUEHNS E.R., HECHT F., KEIL J.V., MOTULSKY A.G.: Developmental hemoglobin anomalies in a chromosomal triplication: D_1 trisomy syndrome. *Proc. nat. Acad. Sci.*, **51**:89–97, 1964.

MAGENIS R.E., HECHT F., MILHAM S.: Trisomy 13 (D_1) syndrome: studies on parental age, sex ratio, and survival. *J. Pediat.*, **73**:222–228, 1968.

PATAU K., SMITH D.W., THERMAN E., IN-HORN S.L., WAGNER H.P.: Multiple congenital anomaly caused by an extra autosome. *Lancet*, **i**:790–793, 1960.

REDHEENDRAN R., NEU R.L., BANNERMAN R.M.: Long survival in trisomy-13-syndrome: 21 cases including prolonged survival in two patients 11 and 19 years old. *Am. J. med. Genet.*, **8**:167–172, 1981.

SARAUX H., LAFOURCADE J., LEJEUNE J., DHERMY P., CRUVEILLER J., TURPIN R.: La trisomie 13 et son expression ophtalmologique. *Arch. Ophthal.*, **24**:581–602, 1964.

SMITH D.W., PATAU K., THERMAN E., IN-HORN S.L., DEMARS R.J.: The D_1 trisomy syndrome. *J. Pediat.*, **62**:326–341, 1963.

SPARKES R.S., SPARKES M.C., CRIST M.: Expression of esterase D and other gene markers in trisomy 13. *Hum. Genet.*, **52**:179–183, 1979.

TAYLOR A.I.: Autosomal trisomy syndromes: a detailed study of 27 cases of Edwards' syndrome and 27 cases of Patau's syndrome. *J. med. Genet.*, **5**:227–252, 1968.

UCHIDA I.A., PATAU K., SMITH D.W.: Dermal patterns and D1 trisomies. *Amer. J. hum. Genet.*, **14**:345–352, 1962.

The first observation of 13 trisomy was reported by Patau et al. in 1960. The numerous cases published since then have allowed a detailed description of the syndrome, which is one of the best known in human cytogenetics (Smith et al., 1963; Taylor, 1968; Berger, 1971; Hamerton, 1971). Isolated cases have long since ceased to be published. The designation ''Patau's syndrome'' is occasionally encountered, but is not accepted by the majority of authors.

GENERALIZATIONS

- Frequency: probably between 1 per 4,000 and 1 per 10,000 (Conen and Erkman, 1966).
- Sex ratio: there is a slightly greater number of affected females, in no way comparable in number with that observed in 18 trisomy (Magenis et al., 1968).
- Mean parental ages:

  maternal: 31.6 years
  paternal: 34.5 years

As in the case of 18 and 21 trisomies, the distribution of maternal ages is bimodal. It shows two peaks, one at 25 and the other at 38 years, the latter being due to maternal age effect.

- Mean birth weight: 2,600 g

PHENOTYPE

The major features immediately suggestive of chromosome 13 trisomy are harelip, microphthalmia, and hexadactyly.

Craniofacial Dysmorphism

The cranium is small and its shape abnormal. The forehead is receding, the temples may be narrow, the fontanels and suture lines are broad. An ulceration of the scalp often is observed at the vertex, occasionally subtended by a cranial lacuna.

Palpebral fissures are horizontal. Bilateral microphthalmia of variable intensity is present, but rarely attains anophthalmos. Cyclopia is observed in rare instances.

The nose is broad and flat.

In two-thirds of the cases, the upper lip is destroyed by a harelip, usually bilateral, and accompanied by a cleft palate. In the absence of harelip, either a normal appearance of the whole face or hypoplasia, which exceptionally may produce cebocephalia, are observed.

The ears are low-set. The helix is abnormal, flat, and poorly defined.

A frequent characteristic is the presence of one or several hemangiomas, usually on the face, forehead, or nape.

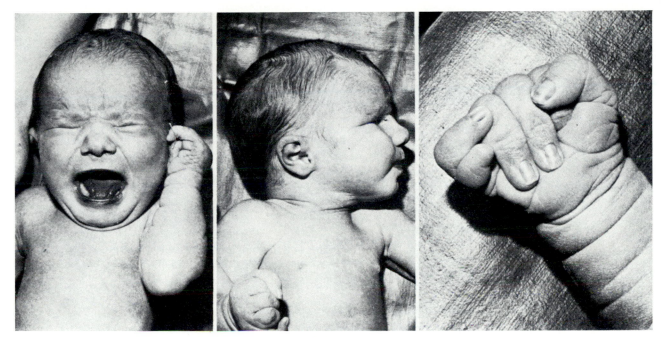

13.1

13.2

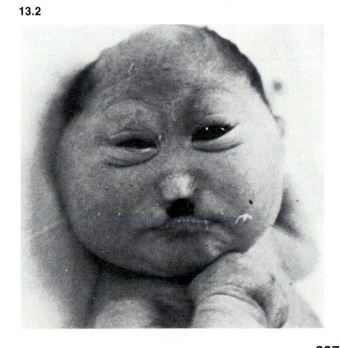

Neck, Thorax, and Abdomen

The skin on the nape may be loose.

The last ribs are often hypoplastic or absent. The pelvis is hypoplastic, with a flattened acetabular angle.

Limbs

The fingers are flexed, with or without overlapping, with or without camptodactyly. The most frequent feature is uni- or bilateral hexadactyly. The supernumerary finger is implanted on the ulnar margin of the hand. At times complete, it is usually reduced to a rudimentary appendage. The nails are narrow and strongly convex.

The feet are usually described as "rocker-bottom" feet, with protrusion of the heel and convexity of the sole. Uni- or bilateral hexadactyly is also noted here, although less frequently than for the hands.

Genitalia

In the male, cryptorchidism and anomalies of the scrotum are usually present.

In the female, bicornate uterus, hypertrophy of the clitoris, and double vagina are observed.

Malformations

Ocular and visceral malformations are constantly present in 13 trisomy.

Faulty development of the prechordal plate necessary to facial morphogenesis and induction of proencephalic development could be responsible for both midface defects and ocular and cerebral anomalies.

Ocular malformations mainly include microphthalmia, which may attain apparent anophthalmia; cataract; iridoschisis; persistence of the primitive vitreous body; and retinal dysplasia (Saraux et al., 1964).

Visceral malformations generally are revealed by autopsy. *Cerebral malformations* are the most typical, particularly arhinencephaly, which is often limited to the absence of the olfactory tractus and bulb. Also observed is holoproencephaly, or its extreme, cyclopia. Cerebral or cerebellar heteropies have been reported.

Cardiac malformations are seen in 80 percent of the cases; these include ventricular septal defect, ductus arteriosus defect, atrial septal defect, or rarer anomalies.

Urinary malformations are present in 30 to 60 percent of the cases; including polycystic kidneys, hydronephrosis, renal or ureteral duplication.

Digestive anomalies primarily involve the colon (malrotation). Heteropic islets of pancreatic or spleen tissue often are reported.

X-ray examination confirms anomalies of the cranium and ribs and occasionally shows vertebral anomalies (absence of one or two vertebrae), hyperplasia of the sacrum, and retarded bone age.

Deafness due to defects in the organ of Corti has been reported.

Mental Retardation

Because of early death of patients, mental retardation is difficult to determine but seems to be severe. Seizures, with hypsarrhythmia, hypotonia, and failure to thrive are the usual features in cases of more prolonged survival.

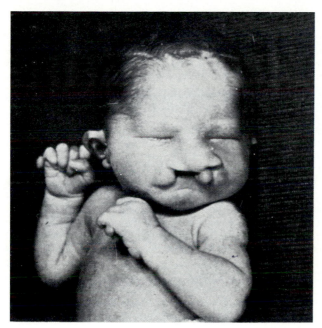

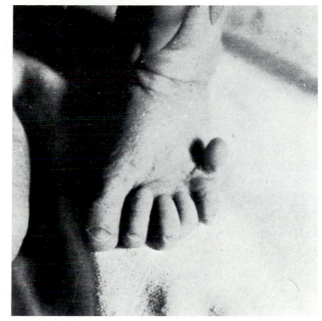

13.3

13.4

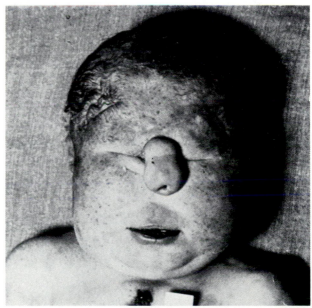

Prognosis

Mean life expectancy is 130 days and appears to be the same for both sexes. Nearly half the infants die within the first month and more than two-thirds before the sixth month. Survival beyond 3 years is exceptional. Cases of prolonged survival have been compiled by Redheendran et al. (1981). The oldest patient is 19.

CYTOGENETIC STUDY

In 80 percent of the cases, free 13 trisomy is involved.

In 20 percent of the cases, either a mosaic, such as 46,XX(Y)/47,XX(Y),+13, or trisomy due to a translocation is involved.

In cases of mosaicism, the severity of the clinical features may sometimes be attenuated, accompanied by a prolonged survival. A very unusual observation concerns a mosaic 46,XX/47,XX,+D woman, mother of two daughters having abnormal phenotypes and both mosaic like their mother.

A translocation, almost always a t(13qDq), and more precisely a t(13q14q), may occur *de novo* or can be transmitted by one of the parents. In the latter case, the risk of recurrence, or of free 21 trisomy by interchromosomal effect, is 5 percent regardless of the carrier parent's sex. It is accompanied by an increased risk of abortion (probably on the order of 20 percent). In the case of t(13q13q), the risk of recurrence or abortion is 100 percent.

Exceptionally, more complex rearrangements may be observed, such as pericentric inversions responsible for partial duplication-deficiency.

DERMATOGLYPHICS

The anomalies of flexion creases and dermatoglyphics are important features of the syndrome (Uchida et al., 1962).

A single transverse palmar crease is practically constant.

The axial triradius is very high, in t'' or t''', one of the most extreme positions observed in chromosomal anomalies.

The digital patterns show a high frequency of radial loops, particularly on the thumbs, and a high frequency of arches. Thenar and interdigital patterns are more frequent than in normal subjects.

LABORATORY FINDINGS

Blood Anomalies

Two features are particular to 13 trisomy:

- The persistence in the newborn of the embryonic Gower-2 hemoglobin, or $Hb\alpha_2^A\epsilon_2$ (Huehns et al., 1964).
- A high frequency of abnormal, sessile or pediculate, blotch-shaped projections, of the polynuclear neutrophil nuclei (Huehns et al., 1964).

Biochemical Dosage Effect

A gene dosage effect is observed for esterase D (ESD), whose gene is assigned to q14 (Sparkes et al., 1979).

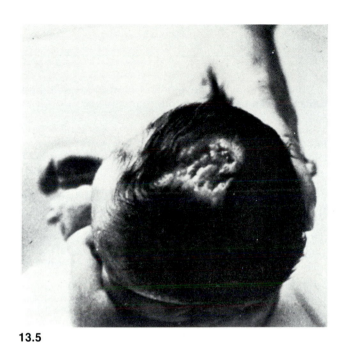

13.5

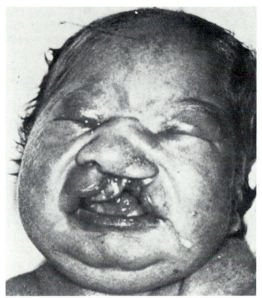

13.6

13.7

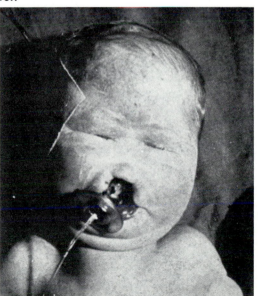

13q TRISOMIES

A number of observations of partial 13q trisomies have been reported involving segments of variable length, with breakpoints occurring at different sites on 13q. One of the most frequent sites is the interface between q14 and q21.

From a phenotypical standpoint, when the trisomy includes the q2 and q3 regions, it leads to a characteristic clinical syndrome.

Trisomies involving all or part of the q1 region (as well as the centromere and 13p) are markedly rare, and would not seem to be responsible for any clinical syndrome as recognizable as that of distal trisomy.

It should be emphasized that these distinctions are somewhat artificial in that there exists an overlap between some cases of proximal and distal trisomies. This would explain the existence of features common to both varieties of partial trisomy.

A small percentage of partial trisomies is due to *de novo* duplication. The majority are the result of the malsegregation of a parental rearrangement: a reciprocal translocation, or in a fair number of cases of distal trisomies, a pericentric inversion.

13q2 & 3 Trisomy

REFERENCES

BROHOLM K.A., EEG-OLOFSON O., HALL B.: An inherited chromosome aberration in a girl with signs of de Lange syndrome. *Acta paed. Scand.*, **57**:547–552, 1968.

GIRAUD F., MATTEI J.F., MATTEI M.G.: Trisomie 13 partielle par translocation t(2;13) maternelle. *Ann. Génét.*, **20**:203–208, 1977.

GROUCHY J. de, TURLEAU C., DANIS F., KOHOUT G., BRIARD M.L.: Triscmie 13qter par duplication en tandem 46,XX, dir dup 13 (q21→qter), 9qh+. *Ann. Génét.*, **21**:247–251, 1978.

HADEBANK M.: Partial trisomy 13q21→qter de novo due to a recombinant chromosome rec (13)dup q. *Hum. Genet.*, **52**:91–99, 1979.

HORNSTEIN L., SOUKUP S.: A recognizable phenotype in a child with partial duplication 13q in a family with t(10q;13q). *Clin. Genet.*, **19**:81–86, 1981.

McCORQUODALE M., ERICKSON R.P., ROBINSON M., ROSZCZIPKA K.: Kleeblattschädel anomaly and partial trisomy for chromosome 13 (47,XY,+der(13),t(3;13) (q24;q14). *Clin. Genet.*, **17**:409–414, 1980.

NIEBUHR E.: Partial trisomies and deletions of chromosome 13. In J.J. YUNIS: *New Chromosomal Syndromes*, New York, Academic Press, pp. 273–299, 1977.

SCHINZEL A., HAYASHI K., SCHMID W.: Further delineation of the clinical picture of trisomy for the distal segment of chromosome 13. *Hum. Genet.* **32**:1–12, 1976.

SCHUTTEN H.J., SCHUTTEN B.T., MIKKELSEN M.: Partial trisomy of chromosome 13. Case report and review of literature. *Ann. Génét.*, **21**: 95–99, 1978.

WENGER S.L., STEELE M.W.: Meiotic consequences of pericentric inversions of chromosome 13. *Am. J. med. Genet.*, **9**:275–283, 1981.

About 30 cases of partial trisomies of the 13q2 & 3 regions are known (Schinzel et al., 1976; Niebuhr, 1977; Giraud et al., 1977; Schutten et al., 1978; Grouchy et al., 1978; Hadebank, 1979; McCorquodale et al., 1980; Hornstein and Soukup, 1981; and Wenger and Steele, 1981).

Facial dysmorphism very rapidly led to a comparison with Cornelia de Lange syndrome (Broholm et al., 1968). This resemblance is also characteristic of 3q2 trisomy.

Birth weight is usually normal. Respiratory distress and neonatal feeding difficulties are common.

Features most often observed include:

- Microcephaly with trigonocephaly and temporal retraction
- Hemangiomata on the face, forehead, and neck
- Thick, converging eyebrows, long, curved lashes
- Hypotelorism
- A short, bulbous nose
- A long philtrum
- Thin lips, with the lower lip possibly slightly everted
- Poorly implanted, decayed teeth, and high-arched palate
- A normal or slightly receding chin
- Malformed ears with poorly developed lobe and a prominence of the anterior branch of the anthelix
- A narrow thorax, with too widely spaced nipples
- Umbilical and inguinal hernias
- Cryptorchidism (or even pseudohermaphroditism)
- Anomalies of the hands and feet, including hexadactylies, long spatular fingers and toes, convex nails, and thumb anomalies

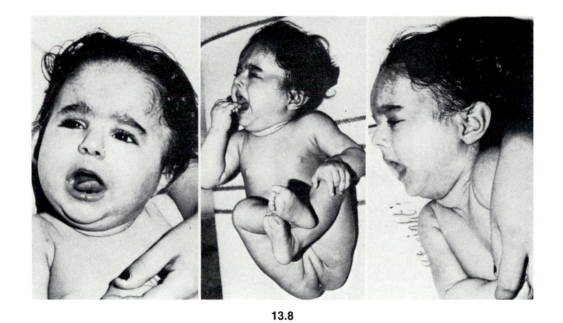

13.8

13.9

13.10

13.11

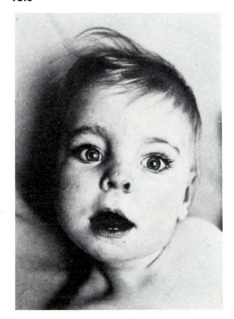

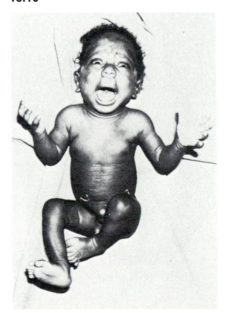

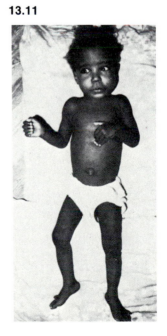

More rarely, the following may be observed:

- Microphthalmia and colobomata
- A cleft lip or palate
- Inner organ, cardiovascular, digestive, and kidney malformations
- Clubfoot, costal and vertebral anomalies

Among biological features specific to trisomy 13, nuclear projections are observed when the trisomy involves q21, q14, and q13.

The prognosis is marked by severe mental retardation, seizures, and sometimes growth retardation. Death in the first year is rare.

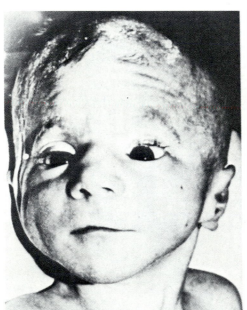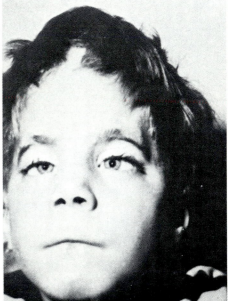

13.12

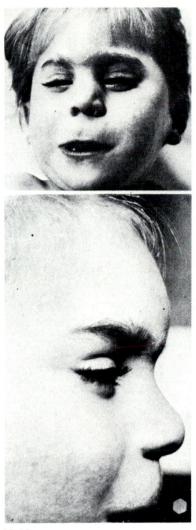

13.13

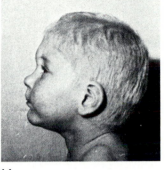
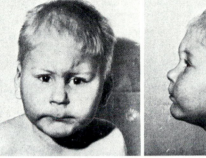

13.14

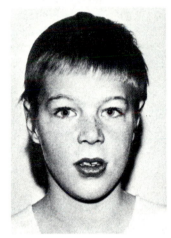
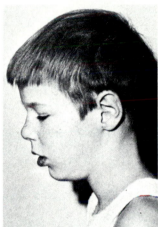

13.15

13.16

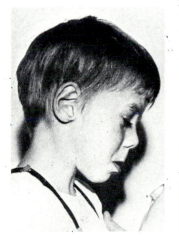
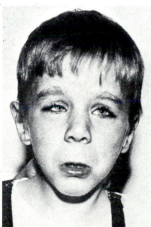

13q1 Trisomy

REFERENCES

GIRAUD F., MATTEI J.F., MATTEI M.G.: Trisomie 13 partielle par translocation t(2;13) maternelle. *Ann. Génét.*, **20**:203–208, 1977.

MOEDJONO S.J., SPARKES R.S.: Partial trisomy of 13 (pter→q12) due to 47,XY, +der(13), t(13;22) (q12;q13)mat. *Hum. Genet.*, **50**:241–246, 1979.

Ten cases of 13q1 trisomy are known. (See Giraud et al., 1977; Moedjono and Sparkes, 1979.)

Specific features are:

- Low birth weight
- Mental retardation
- Micrognathia
- Clinodactyly of the Vth

A large number of nuclear projections and a persistence of embryonic hemoglobin (hematological features that may help in the cytogenetic diagnosis) are present.

In the more extensive cases of proximal trisomy, the following may be observed:

- A receding forehead
- Cerebral anomalies
- Eye anomalies (microphthalmia, colobomata)
- Inner organ malformations

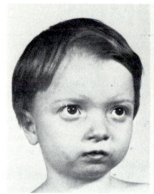
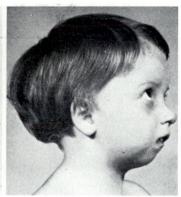

13.17

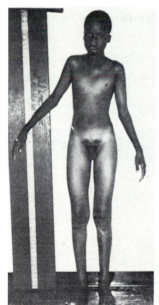
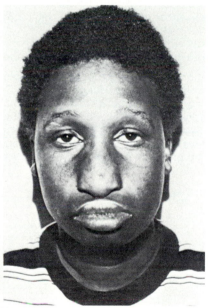

13.18

13.19

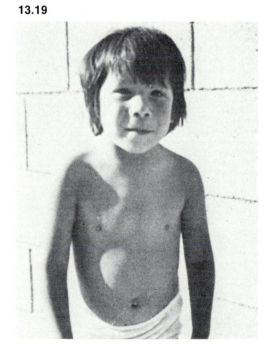

13q MONOSOMIES

REFERENCES

LEJEUNE J., LAFOURCADE J., BERGER R., CRUVEILLER J., RETHORE M.O., DUTRIL-LAUX B., ABONYI D., JEROME H.: Le phénotype (Dr). Etude de trois cas de chromosome D en anneau. *Ann. Génét.*, **11**:79–87, 1968.

LELE K.P., PENROSE L.S., STALLARD H.B.: Chromosome deletion in a case of retinoblastoma. *Amer. J. hum. Genet.*, **27**:171–174, 1963.

NICHOLS W.W., MILLER R.C., HOFFMAN E., ALBERT D., WEICHSELBAUM R.R., NOVE J., LITTLE J.B.: Interstitial deletion of chromosome 13 and associated congenital anomalies. *Hum. Genet.*, **52**:169–173, 1970.

NIEBUHR E., OTTOSEN J.: Ring chromosome D(13) associated with multiple congenital malformations. *Ann. Génét.*, **16**:157–166, 1973.

NIEBUHR E.: Partial trisomies and deletions of chromosome 13. *In* J.J. YUNIS: *New Chromosomal Syndromes*. New York, Academic Press, pp. 273–295, 1977.

WANG H.C., MELNYCK J., McDONALD L.T., UCHIDA I.A., CARR D.H., GOLDBERG B.: Ring chromosomes in human beings. *Nature*, **195**:733–734, 1962.

The first observation of a ring chromosome in humans was made by Wang et al. in 1962. It probably involved a D. The first clinical description of the r(13) syndrome was that of Lejeune et al. in 1968.

The first observation of partial deletion of a D chromosome, probably a 13, associated with a retinoblastoma, was reported in 1963 by Lele et al.

Since these first observations, a number of cases of partial deletions of chromosome 13 have become known. They include terminal deletions, rings, and intercalary deletions.

Niebuhr and Ottosen in 1973, and Niebuhr in 1977, attempted to characterize four groups of patients as a function of their clinical features, based on 72 observations from the literature. These four groups were as follows:

Group 1   Carriers of an r(13), with normal thumbs
Group 2   Carriers either of an r(13) or of a del(13q) with absence or hypoplasia of the thumbs
Group 3   Patients with retinoblastoma
Group 4   Patients with 13q−, normal thumbs, and no ocular tumors.

Such a classification, and notably the distinction between carriers and noncarriers of a thumb anomaly, is not a true reflection of reality. This is due to both the variation in expressivity of a number of clinical signs, and to the great diversity of breakpoints observed in each individual case (Nichols et al., 1979).

We have chosen to describe the phenotype for monosomy 13q3, either total or partial, which includes rings and terminal deletions, and intercalary deletions which include band q14 and are accompanied by a retinoblastoma.

13q3 Monosomy

Microcephaly
Greek profile
Rabbitlike incisors
Agenesis of the radial axis

GENERALIZATIONS

- Sex ratio: 1M/1F
- Mean parental ages:
  maternal: 26 years
  paternal: 28.9 years
- Mean birth weight: 2,250 g

PHENOTYPE

Craniofacial Dysmorphism

The most distinctive sign is the absence of a defined nasal bridge, producing a Greek profile.

Microcephaly is almost constantly present and is often severe. It may be accompanied by trigonocephaly.

The face may be clearly asymmetric.

The bridge of the nose is wide and protuberant; its ridge is continuous with the slope of the forehead, thus giving the appearance of a Greek profile.

Hypertelorism is often visible and may be confirmed by X-ray examination. Palpebral fissures are normally oriented. Epicanthus and palpebral ptosis are frequently encountered. Eye anomalies are usually present, the most common being microphthalmia, iris or retina coloboma, and cataracts.

The nose has no characteristic features apart from the effaced bridge.

The philtrum is short and exposes the upper incisors that are set in a ''rabbitlike'' forward slant, which is highly characteristic. The palate is flat.

The chin is usually small.

The ears are large, with a deep sulcus helix butting up against a well-developed lobe.

Neck, Thorax, and Abdomen

The neck may be short, and pterygium colli has been observed in several patients. Bone anomalies are frequently found, including hip dislocations, lumbar and sacral agenesis, supernumerary ribs, etc.

Genitalia and Perineum

Genitals are abnormal, with cryptorchidism, hypospadias, epispadias, bifid scrotum, and bifid uterus. Anal atresia was observed in one-fourth of the cases and, more rarely, a perineal fistula.

239

Limbs

The most significant anomalies are hypoplasia or absence of the thumb, agenesis of the first metacarpal, a fusion of the fourth and fifth metacarpals, and syndactyly. These features are a countertype to the polydactyly of 13 trisomy. In rare cases of r(13), supernumerary fingers and bifid thumbs have been observed, and have been interpreted as being the consequence of ring mechanics leading to duplication-deficiencies.

The feet are sometimes malformed, with disorderly implantation of the toes and absence of the fifth toe.

Malformations

Inner organ malformations include the following:

- Cerebral anomalies which are similar to those observed in trisomy 13: arhinencephaly, aplasia of the falx cerebri, agenesis of the corpus callosum, anencephaly, and holoproencephaly
- Congenital heart disease, including atrial and ventricular septal defect, etc.
- Aplasia of the gall bladder, and hypoplastic kidneys

Mental Retardation

The IQ is much lower than 50 in the majority of cases.

Prognosis

Considerable growth and mental retardation will determine the severity of the prognosis. Life expectancy is difficult to evaluate from data in the literature. Most observations concern very young children and, in a few cases, adolescents. One child out of five died before the age of 6 months.

CYTOGENETIC STUDY

The 72 cases of partial 13q deletions studied by Niebuhr (1977) included 61% of r(13) and 28% of terminal or interstitial del(13), with the remaining cases being mosaicism or complex aberrations.

DERMATOGLYPHICS

Dermatoglyphic anomalies are not often encountered, and have no distinguishing features.

Anomalies of the thumb may lead to an absence of the axial triradius and anomalies in palmar flexion creases.

FINE SEMIOLOGY

The most distal monosomies, which only concern 13q33 or 34, are characterized by the absence of metacarpal and thumb anomalies, as well as a lesser frequency of genital anomalies and anal atresia.

In r(13), certain features of partial or complete 13 trisomy may be seen, due to the partial duplication of the rings. These include supernumerary fingers and trigonocephaly.

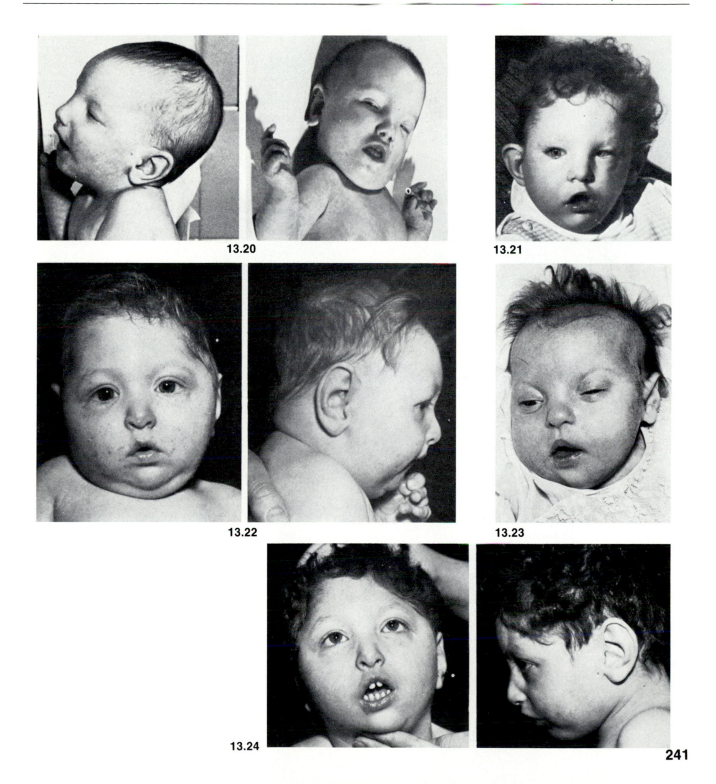

13.20

13.21

13.22

13.23

13.24

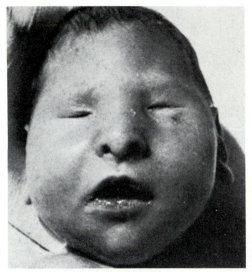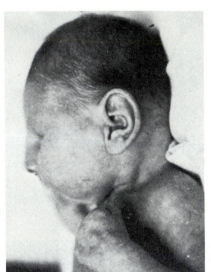

13.25

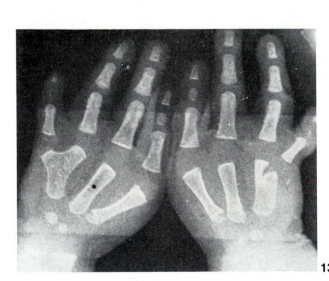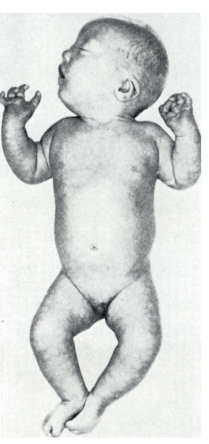

13.26

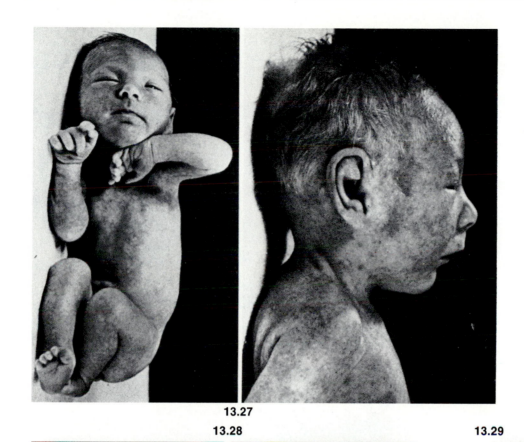

13.27

13.28

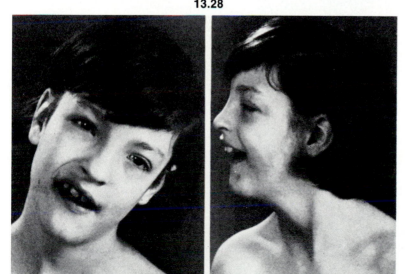

13.29

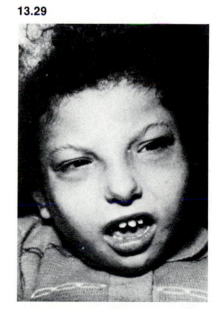

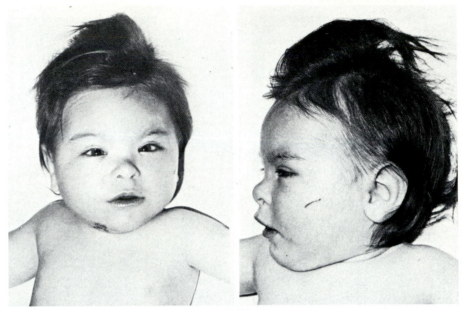

13.30

13.31

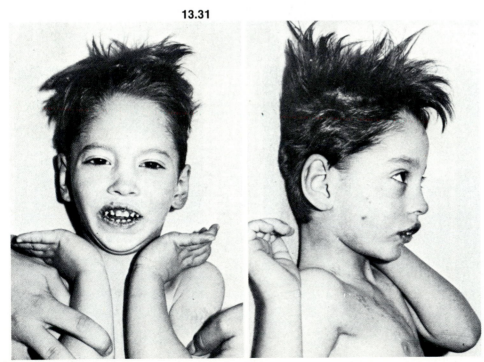

13q14 Monosomy and Retinoblastoma

REFERENCES

DESPOISSE S., JUNIEN C., TURLEAU C., RIVERA H., RETHORE M.O., GROUCHY J. de: Gene dosage studies for esterase D. In: *VIth International Workshop on Human Gene Mapping.* Oslo, June 29th-July 3rd, 1981.

GROUCHY J. de, TURLEAU C., CABANIS M.O., RICHARDET J.M.: Rétinoblastome et délétion intercalaire du chromosome 13. *Arch. fr. Pédiat.,* **37**:531–535, 1980.

RICCARDI V.M., MINTZ-HITTNER H., FRANCKE U., PIPPIN S., HOLMQUIST G.P., KRETZER R.L., FERRELL R.: Partial triplication and deletion 13q: study of a family presenting with bilateral retinoblastoma. *Clin. Genet.,* **15**:332–345, 1979.

RIVERA H. TURLEAU C., GROUCHY J. de, JUNIEN C., DESPOISSE S., ZUCKER J.M.: Retinoblastoma-del(13q14): report of two patients, one with a trisomic sib due to maternal insertion. Gene-dosage effect for esterase D. *Hum. Genet.,* **59**:211–214, 1981.

SPARKES R.S., SPARKES M.C., WILSON M.G., TOWNER J.W., BENEDICT W., MURPHREE A.L., YUNIS J.J.: Regional assignment of genes for human esterase D and retinoblastoma to chromosome band 13q14. *Science,* **208**:1042–1044, 1980.

STRONG L.C., RICCARDI V.M., FERRELL R.E. SPARKES R.S.: Familial retinoblastoma and chromosome 13 deletion transmitted via an insertional translocation. *Science,* **213**:1501–1503, 1981.

The characterization of this group of patients is justified by the existence of retinoblastoma. First of all, this is generally the major clinical feature that brings the patient in for consultation. Secondly, it is the sign that largely determines the treatment and prognosis.

From a cytogenetic standpoint, this is a very heterogeneous group, with the deletion capable of extending on both sides of q14, from q11 to q22.

More than 20 observations have been studied by banding techniques. The smallest segment common to all of these observations is band q14 or a portion of it (see Grouchy et al., 1980).

GENERALIZATIONS

- Sex ratio: 1M/2F
- Mean birth weight: usually normal

PHENOTYPE

The phenotype usually has been described in a rather brief and succinct manner by the authors. It is unusual to observe both a retinoblastoma and signs of 13q3 monosomy simultaneously. Facial dysmorphism, described in several cases, is not pathognomonic. Eye anomalies other than retinoblastoma have been reported, including microphthalmy and coloboma.

Mental retardation is usually found, but varies in importance. Growth retardation is also variable.

Associated malformations are a function of the extent of the deletion and are rarely observed.

CYTOGENETIC STUDY

The deletion habitually occurs *de novo.* It may also result from a parental insertion. The corresponding trisomy has been observed with apparently normal or only minor phenotypical expression (Riccardi et al., 1979; Rivera et al., 1981; Strong et al., 1981).

LABORATORY FINDINGS

The study of patients afflicted with del(13)-retinoblastoma has allowed the precise assignment of the gene for esterase D (ESD) to 13q14 (Sparkes et al., 1980; Junien et al., 1981). Dosage of enzymatic activity should be envisaged in all cases of retinoblastoma.

CHROMOSOME 14

Chromosome 14 carries the family of genes responsible for the heavy chain of immunoglobulins.

Chromosome disorders mainly comprise 14q proximal trisomy. We also have considered 14 trisomy mosaicism and ring chromosomes.

A polymorphism exists due to the variation in length of the short arm and the size of the satellite. Chromosomes with satellites in duplicate or triplicate may also be observed.

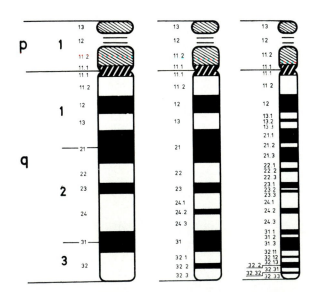

GENE MAP

14p12	RNR		Ribosomal RNA (18045) A
	LCH		Lentil agglutinin binding (15103) S
14q12-14q20	NP		Nucleoside phosphorylase (16405) S, D
14q21-14qter	WARS		Tryptophanyl-tRNA synthetase (19105) S
	EMP195	(P)	External membrane protein-195 (13374) S
	CKBB		Creatine kinase, brain type (12328) S
	PGFT	(P)	Phosphoribosylglycineamide formyltransferase (17246) S
	PFGS	(P)	Phosphoribosyl formylglycin-amide synthetase (10258) S
	ESAT	(P)	Esterase activator (13325) S
	SPH1	(L)	Spherocytosis, Denver type (18290) F
	PI		Protease inhibitor (Pi; alpha-1-antitrypsin) (10740) F, S
14q32	IGH		IMMUNOGLOBULIN HEAVY CHAIN GENE FAMILY REa
	IGHV		Variable region genes (about 250) (14707)
	IGHD		D (for diversity) region genes (many) (14708)
	IGHJ		J (for joining) region genes (more than 4) (14701)
	IGHM1		Constant region of heavy chain of IgM1 (14702)
	IGHM2		Constant region of heavy chain of IgM2 (14703)
	IGHD		Constant region of heavy chain of IgD (14717)
	IGHG4		Constant region of heavy chain of IgG4 (14713)
	IGHG2		Constant region of heavy chain of IgG2 (14711)
	IGHG3		Constant region of heavy chain of IgG3 (14712)
	IGHG1		Constant region of heavy chain of IgG1 (14710)
	IGHE		Constant region of heavy chain of IgE (14718)
	IGHA1		Constant region of heavy chain of IgA1 (14690)
	IGHA2		Constant region of heavy chain of IgA2 (14700)
	TRC	(L)	T-cell idiotypic receptor for MHC antigen (18688) H

MORPHOLOGY AND BANDING

QFQ   GTG   RHG   RBA   THA   C

14q1 TRISOMY

Prominent nose
Characteristic mouth

REFERENCES

ALLDERDICE P., MILLER O.J., MILLER D.A., BREG W.R., GENDEL E., ZELSON C.: Familial translocation involving chromosomes 6, 14 and 20 identified by quinacrine fluorescence. *Humangenetik,* **13**:205–209, 1971.

JUNIEN C., KAPLAN J.C., RAOUL O., RETHORE M.O., TURLEAU C., GROUCHY J. de: Effet de dosage sesquialtère de la nucléoside phosphorylase érythrocytaire et leucocytaire dans deux cas de trisomie partielle 14q. *Ann. Génét.,* **23**:86–88, 1980.

LOPEZ-PAJARES I., DELICADO A., COBOS P.V., LLEDO G., PERALTA A.: Partial trisomy 14q. *Hum. Genet.,* **46**:243–247, 1979.

RAOUL O., RETHORE M.O., DUTRILLAUX B., MICHON L., LEJEUNE J.: La trisomie partielle 14q. I. Trisomie partielle 14q due à une translocation maternelle t(10;14) (p15.2;q22). *Ann. Génét.,* **18**:35–39, 1975.

RAY M., HUNTER A.G.W., SACHDEVA R.K., CHRISTIE N.: Partial trisomy 14 with a 46,XY, − 13, + der(14),t(13;14) (q12;q22)mat karyotype. *Ann. Génét.,* **22**:47–49, 1979.

SMITH A., DEN DULK G., ELLIOTT G.: A severely retarded 18-year-old boy with tertiary partial trisomy 14. *J. med. Genet.,* **17**:230–232, 1980.

TURLEAU C., GROUCHY J. de, BOCQUENTIN F., ROUBIN M., CHAVIN-COLIN F.: La trisomie 14q partielle. II. Trisomie 14q partielle par translocation maternelle t(12;14)-(q24.4;q21). *Ann. Génét.,* **18**:41–44, 1975.

The first case of proximal 14 trisomy was described by Allderdice et al. in 1971. The syndrome was delineated in 1975 by Raoul et al. and Turleau et al., based on eight cases. Nearly 25 cases are known (Lopez-Pajares et al., 1979; Ray et al., 1979; Smith et al., 1979).

GENERALIZATIONS

- Sex ratio: 2F/1M
- Mean birth weight: 2,550 g (full-term babies)

PHENOTYPE

All children with proximal 14 trisomy have very characteristic features, especially in the shape of the nose and mouth. Growth retardation is severe, as is mental impairment.

Craniofacial Dysmorphism

The skull may be normal in shape; in several observations, however, microcephaly has been observed, and may or may not be associated with various deformities including asymmetry, sagittal craniostenosis, oxycephaly, and trigonocephaly.

Palpebral fissures are normally oriented. They may be small, and microphthalmia of variable intensity may be observed. Palpebral ptosis and blepharophimosis were reported.

The nose is prominent. The nasal bridge may be broad; hypotelorism rather than hypertelorism is present.

The upper lip is rather long and the philtrum poorly defined.

The mouth is remarkable and characterizes the syndrome. It resembles the arc of a circle, concave side down. Cupid's bow is replaced by a medial notch.

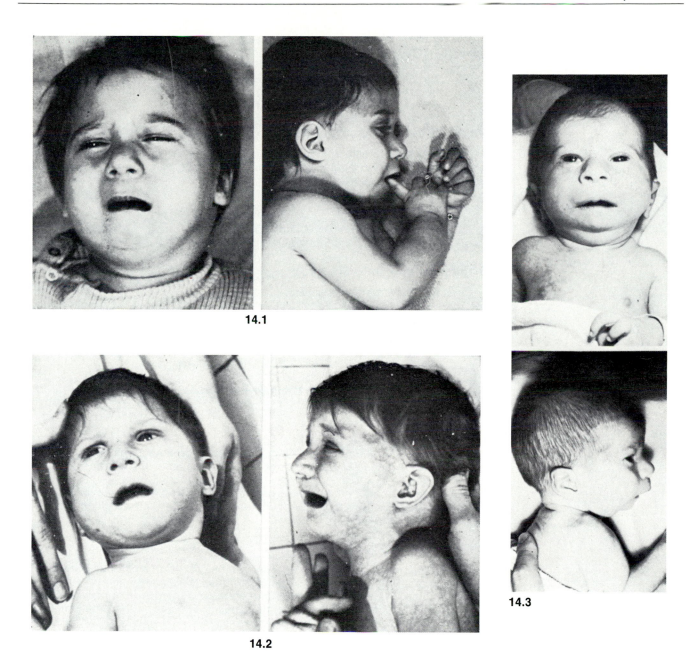

14.1

14.2

14.3

The vermilion of the lips is thin. Above all, the commissures do not form an acute angle, but are "ovalized" by the sagging of the external portions of the lower lip. Of normal size at rest, the mouth opens widely with the onset of crying.

Cleft palate or a high-arched palate are almost constantly present.

The chin is normal or slightly receding.

The ears are usually normally set and folded.

Neck, Thorax, and Abdomen

The neck is short, with an occasional excess of skin or a low-set hairline.

A broad thorax, horizontal ribs, cyphosis, lordosis, trunk obesity, hernias, and dimples have been reported.

Limbs

Anomalies of the limbs are frequent: abnormal positioning of the fingers and toes, clubfeet, radial aplasia, and bilateral hip dislocation.

Genitalia

Cryptorchidism may be observed.

Malformations

Visceral malformations are rare, apart from several cardiopathies.

Mental Retardation

Mental deficiency is severe in the majority of cases.

Prognosis

The vital prognosis is generally not impaired. The first months of life are often marked by feeding difficulties and repeated respiratory infections. In certain cases, these may result in the death of the child. Muscle tone incoordination and seizures are also observed.

Growth retardation is nearly always considerable, falling below the third percentile. The oldest patient is 18, stands 114 cm tall, the size of a 5- to 7-year-old child, and has had no puberty development.

CYTOGENETIC STUDY

In almost all cases, trisomy is due to the 3:1 disjunction of a parental translocation. It is characterized by a 47-chromosome karyotype, the supernumerary chromosome corresponding to the short arm, the centromere, and the proximal portion of 14q.

The length of the trisomic segment is variable. The site of the breakpoint varies from q12 to q24.

DERMATOGLYPHICS

The features of the palmar and digital dermatoglyphics do not seem significant.

LABORATORY FINDINGS

A gene dosage effect is observed for nucleoside phosphorylase (NP) (see Junien et al., 1980).

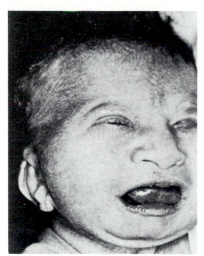
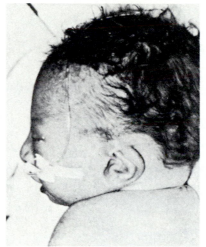

14.4

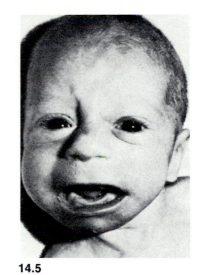

14.5

14.6

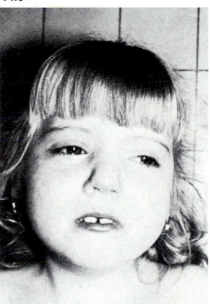

14.7

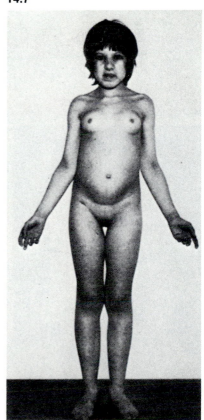

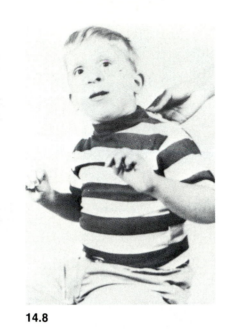

14.8

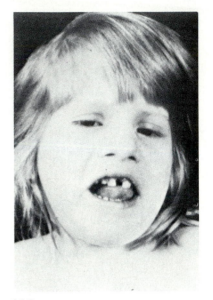

14.9

14.10

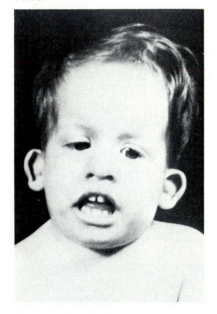

14.11

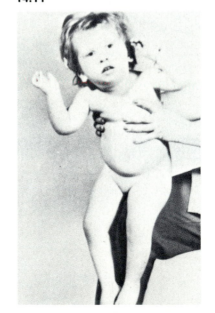

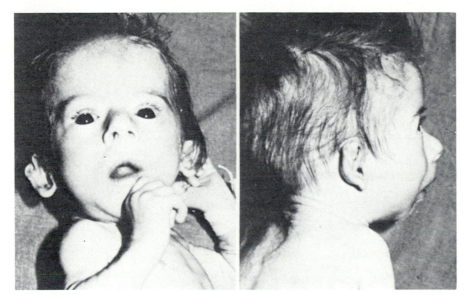

14.12

14.13

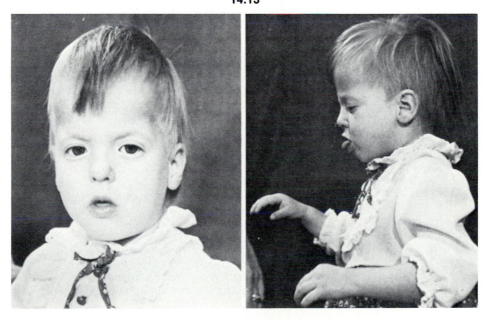

14 TRISOMY

REFERENCES

JENKINS M.B., KRIEL R., BOYD L.: Trisomy 14 mosaicism in a translocation 14q15q carrier: probable dissociation and isochromosome formation. *J. med. Genet.*, **18**:68–71, 1981.

RETHORE M.O., COUTURIER J., CARPENTIER S., FERRAND J., LEJEUNE J.: Trisomie 14 en mosaïque chez une enfant multimalformée. *Ann. Génét.*, **18**:71–74, 1975.

TURLEAU C., GROUCHY J. de, CORNU A., TURQUET M., MILLET G.: Trisomie 14 en mosaïque par isochromosome dicentrique. *Ann. Génét.*, **23**:238–240, 1980.

Since the first observation reported by Rethoré et al. (1975), five observations of trisomy 14 mosaicism, authenticated by banding, have been found (see Turleau et al., 1980; Jenkins et al., 1981).

GENERALIZATIONS

- Sex ratio: 4F/1M
- Mean birth weight: 3,240 g (4 cases)
- Mean crown-heel length at birth: 45 cm (3 cases)
- Mean maternal age: 26 years
- Mean paternal age: 30.3 years

PHENOTYPE

Despite the few known cases, 14 trisomy mosaicism appears to have its own distinct phenotype. Newborns are normal in weight but small in stature. Developmental retardation persists, and is considerable.

Craniofacial Dysmorphism

Given the length of the infant, true microcephaly cannot be considered to exist. The forehead is high and protruding.

Palpebral fissures are horizontal, narrow, deep-set and often asymmetric, with a uni- or bilateral ptosis. An opalescent cornea was reported in two observations.

The nose is short and bulbous.

The upper lip may be long and protruding.

The palate is high arched, with cleft palate in one case.

Clear-cut microretrognathia is present.

In infants, the ears are low-set and posteriorly rotated with a small auricle, a helix that is very much folded in its upper portion, and a lobe that is perpendicular to the skull. In two older children, the ears are detached.

Neck, Thorax, and Abdomen

The neck is short. The thorax is often narrow and deformed.

Genitalia

A small penis, underdeveloped testes, and unilateral cryptorchidism were noted in the only boy.

Malformations

Heart disease is present in three cases. Various other malformations have been reported, including facial asymmetry, with stenosis of the auditory canal, corporal asymmetry, hip dislocation, renal insufficiency, asthma, and dermatosis.

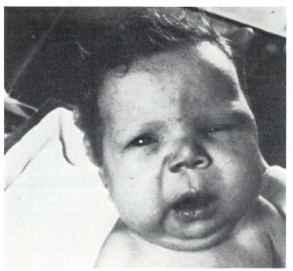

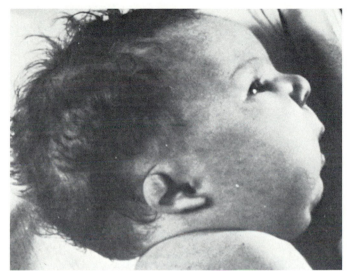

14.14

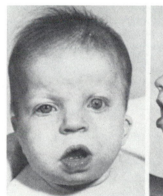
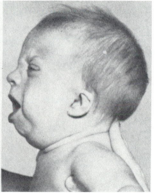

14.15

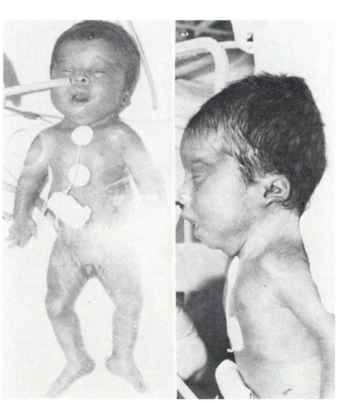

14.16

Mental Retardation

Mental retardation is severe.

Prognosis

Two out of five children died before the age of 3 months. The oldest child is 13½.

CYTOGENETIC STUDY

All cases occurred *de novo* and are due to mosaicism. In three cases out of five, free trisomy is involved. Two cases involve a t(14q14q) or an isochromosome.
In three cases, the proportion of trisomic cells is low, less than or equal to 10%.

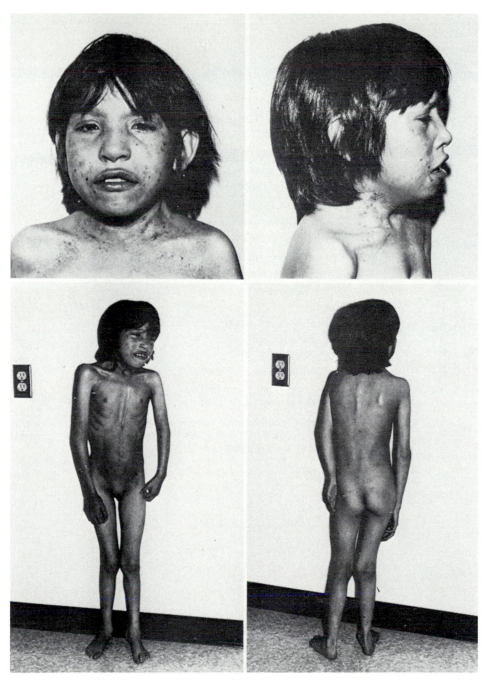

14.17

r(14)

REFERENCES

GILGENKRANTZ S., CABROL C., LAU-SECKER C., HARTLEYB M. E., BOHE B.: Le syndrome Dr. Etude d'un nouveau cas (46,XX,14r). *Ann. Génét.*, **14**:23–31, 1971.

LIPPE B.M., SPARKES R.S.: Ring 14 chromosome: association with seizures. *Am. J. med. Genet.*, **9**:301–306, 1981.

SCHMIDT R., EVIATAR L., NITOWSKY H.M., WONG M., MIRANDA S.: Ring chromosome 14: a distinct clinical entity. *J. med. Genet.*, **18**:304–307, 1981.

Since the first publication by Gilgenkrantz et al. (1971), eight cases of r(14) have been reported (see Lippe and Sparkes, 1981; Schmidt et al., 1981). One of these concerns a set of monozygotic twins.

GENERALIZATIONS

- Sex ratio: 7F/1M
- Mean birth weight: 2,700 g (6 cases)
- Mean parental ages:
    maternal: 26.5 years
    paternal: 28.4 years

PHENOTYPE

The principal features are as follows:

- Dolichocephaly
- A high forehead
- Epicanthus
- Palpebral fissures slanting downward and outward
- Large, low-set ears
- A short neck
- Mental retardation
- Seizures
- Repeated respiratory infections

Some of these patients had congenital heart disease, others manifested pigmentation troubles, and still others had neurological anomalies. Some of these features are suggestive of tuberous sclerosis.

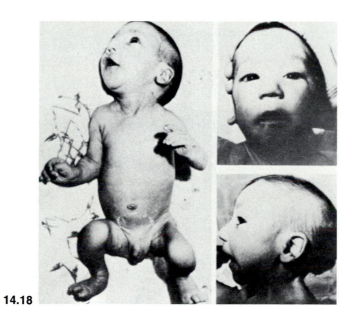

14.18

CHROMOSOME 15

REFERENCES

SCHRECK R.R., BREG W.R., ERLANGER B.F., MILLER O.J.: Preferential derivation of abnormal human G-group-like chromosomes from chromosome 15. *Hum. Genet.*, **36**:1–12, 1977.

WISNIEWSKI L., HASSOLD T., HEFFELFINGER J., HIGGINS J.V.: Cytogenetic and clinical studies in five cases of inv dup (15). *Hum. Genet.*, **50**:259–270, 1979.

Among gene assignments, that of the gene coding for hexosaminidase-A, whose mutation causes Tay-Sachs disease, should be considered.

In certain cases, the Prader-Willi syndrome would seem to be due to a short intercalary deletion, probably in 15q1.

Chromosomal syndromes include two partial trisomies, proximal and distal, and rings.

A polymorphism exists that is due to the variation in length of the short arm and the size of the satellite. Chromosomes with satellites in duplicate or triplicate may be observed.

Specific stainings exist for the juxtacentromeric region of 15, which are the basis for identification of most cases of proximal 15 trisomy (Schreck et al., 1977; Wisniewski et al., 1979).

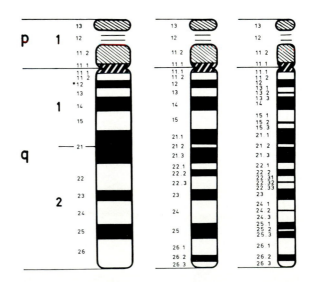

GENE MAP

15p12	RNR		Ribosomal RNA (18045) A
15q11–15qter	MANA	(P)	Alpha-mannosidase-A, cyto-plasmic (15458) S
15q11	PWS		Prader-Willi syndrome (17627) Ch
15q11–15q12	CVS	(P)	Coronavirus 229E sensitivity (12246) S
15q21–15q22	B2M		Beta-2-microglobulin (10970) S, D, H
15q22-15qter	MPI		Mannosephosphate isomerase (15455) S
15q22-15qter	PKM2		Pyruvate kinase-3 (17905) S, D
15q22-15q25.1	HEXA		Hexosaminidase-A (27280) S
15q21-15qter	IDH2		Isocitrate dehydrogenase, mito-chondrial (14765) S
	BVIN	(P)	BALB virus induction, N-tropic (11398) S
	GLUC	(P)	Neutral alpha-glucosidase C (10418) S
15p21-15q22	SORD		Sorbitol dehydrogenase (18250) S, H
15q25-15q26	FES		Onc gene: feline sarcoma virus (19003) S, A

MORPHOLOGY AND BANDING

QFQ GTG RHG RBA THA C

261

15q TRISOMIES

15q1 Trisomy

Oval face
High cheekbones
Deep orbits
Seizures

REFERENCES

CASTEL Y., RIVIERE D., BOUCLY J.Y., TOUDIC L.: Trisomie 15q partielle par translocation maternelle t(7;15)(q35;q14). *Ann. Génét.*, **19**:75–79, 1976.

GENEIX A., JAFFRAY J.Y., MALET P., FOULON E., JALBERT P., CROST P.: A new case of partial trisomy 15q –. *Hum. Genet.*, **51**:335–338, 1979.

SCHRECK R.R., BREG W.R., ERLANGER B.F., MILLER O.J.: Preferential derivation of abnormal human G-group-like chromosomes from chromosome 15. *Hum. Genet.*, **36**:1–12, 1977.

TAYSI K., DEVIVO D.C., SEKHON G.S.: Partial trisomy 15 and intractable seizures. *Acta paediatr. Scand.*, **68**:445–447, 1979.

WANG H., HUNTER A.G.W.: A supernumerary «G» like chromosome originating from a maternal 13;15 translocation in a nondysmorphic, retarded girl. *Clin. Genet.*, **15**:273–277, 1979.

WISNIEWSKI L., HASSOLD T., HEFFELFINGER J., HIGGINS J.V.: Cytogenetic and clinical studies in five cases of inv dup(15). *Hum. Genet.*, **50**:259–270, 1979.

The syndrome was described by Castel et al. in 1976, based on 14 cases. These partial D trisomies may explain a certain number of "false" G trisomies observed before the discovery of banding techniques. More than 30 cases of proximal 15 trisomy are currently known (Geneix et al., 1979; Taysi et al., 1979; Wang and Hunter, 1979; Wisniewski et al., 1979).

GENERALIZATIONS

- Sex ratio: 2F/1M
- Parental ages: in *de novo* cases, the mean maternal age seems high: 32.6 years (5 cases). Paternal ages are not known.
- Mean birth weight: 3,050 g

PHENOTYPE

Facial dysmorphism is moderate, and clinical diagnosis does not seem possible. Cytogenetic diagnosis has often been performed relatively late, between 4 and 21 years.

Craniofacial Dysmorphism

Slight microcephaly is present in less than half of the cases.

Features common to the patients are an oval face with high cheekbones, "full" cheeks, and eyes set deeply in their orbits. Microphthalmia has occasionally been reported. The nose is somewhat full.

Other anomalies reported are strabismus, high-arched or fissured palate, and low-set ears.

Limbs

Short, thick fingers, malpositioned fingers and toes, and clubfoot have been reported.

Genitalia

No particular anomaly is noted.

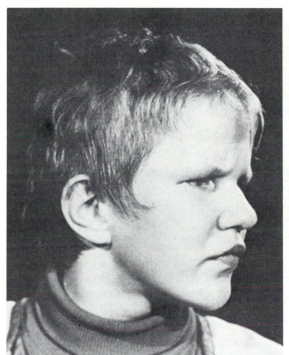

15.1

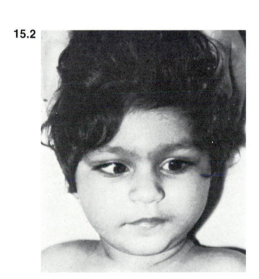

15.2

Malformations

No severe visceral malformations seem to be present.

Mental Retardation

Mental impairment is severe. The IQ is about 20. It reaches 50 in less severely afflicted patients. In several cases, patients are described as "hyperactive" or aggressive, or show autistic behavior. Patients are frequently subject to seizures.

Prognosis

Growth retardation is irregular.
Prognosis for life expectancy is not impaired.
Several cytogenetic forms may be distinguished

- Trisomy 15q1 may result from the 3:1 disjunction of a balanced parental translocation
- It may occur *de novo,* with the presence of a supernumerary acrocentric having the size of a G
- It may also arise *de novo,* with the presence of a supernumerary chromosome carrier of satellites at both extremities, which has been interpreted as resulting from an inverted duplication of a 15 (mirror duplication of the proximal part of 15q). In fact, this rearrangement causes partial 15 tetrasomy (Schreck et al., 1977; Wisniewski et al., 1979).

The trisomy concerns the q1 region, but also may include the q21 segment.

DERMATOGLYPHICS

Palmar and digital prints are not characteristic.

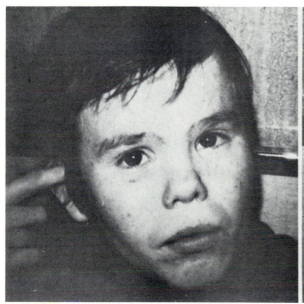
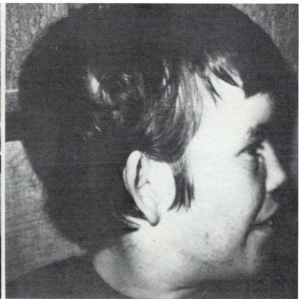

15.3

15.4

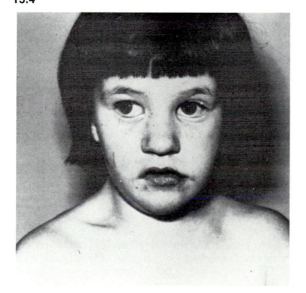

15.5

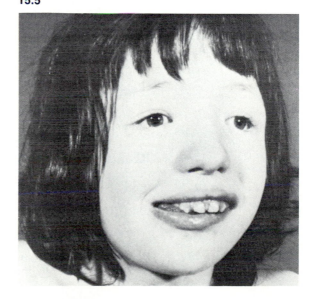

15q2 Trisomy

Microdolichocephaly
Narrow palpebral fissures
Protuberant philtral borders
Micrognathia

REFERENCES

FUJIMOTO A., TOWNER J.W., EBBIN A.J., KAHLSTROM E.J., WILSON M.G.: Inherited partial duplication of chromosome n° 15. *J. med. Genet.*, **11**:287–291, 1974.

TURLEAU C., GROUCHY J. de, CHAVIN-COLIN F., ROUBIN M.: Trisomie 15q distale. *Ann. Génét.*, **20**:214–216, 1977.

TZANCHEVA M., KRACHOUNOVA M., DAMJANOVA Z.: Two familial cases with trisomy 15q distal due to a rcp(5;15) (p14;q21). *Hum. Genet.*, **56**:275–277, 1981.

ZABEL B., BAUMANN W.: Trisomie partielle pour la partie distale du bras long du chromosome 15 par translocation X/15 maternelle. *Ann. Génét.*, **20**:285–289, 1977.

Five cases of distal 15q2 trisomy are known (Fujimoto et al., 1974; Turleau et al., 1977; Zabel and Baumann, 1977; Tzancheva et al., 1981).

GENERALIZATIONS

• Sex ratio: 3M/2F

PHENOTYPE

Craniofacial Dysmorphism

This is highly characteristic. Microdolichocephaly, with a very prominent occiput, is present. The face is rather long, with round, flaccid cheeks.

Palpebral fissures are slightly slanted downward and outward. They are barely opened, and give a rather peculiar appearance. Strabismus is often present.

The nose is snubbed.

The philtrum is long, with very prominent margins that clearly emphasize Cupid's bow.

Micrognathia is very severe.

The ears are low-set, rather large, and more or less well-folded, with prominence of the anthelix.

Neck, Thorax, and Abdomen

The neck is short. In one case, the thorax was deformed, with severe scoliosis and bilateral dislocation of the hips.

Limbs

Various osteoarticular anomalies, such as clubfoot, articular blocking, and syndactylies, are observed.

Genitalia

In boys, cryptorchidism with or without hypospadias may be noted.

Malformations

Heart disease, including atrial and ventricular septal defects and pulmonary stenosis, was noted in nearly all cases. Auditory troubles have been reported in two observations. Cranial dysmorphism is accompanied by various radiological anomalies. Cataracts and a diaphragmatic hernia were reported, in one case.

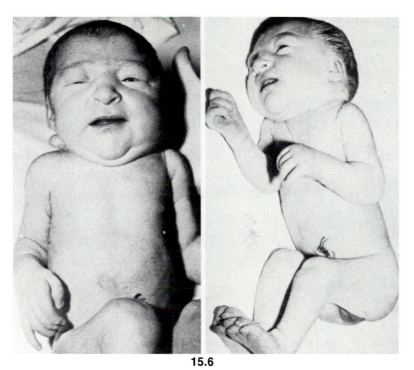

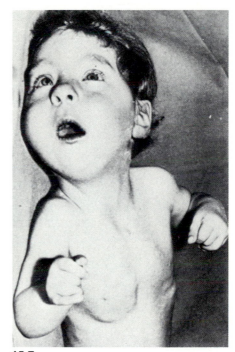

15.6

15.7

15.8

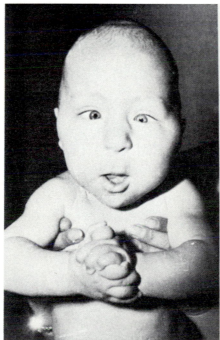

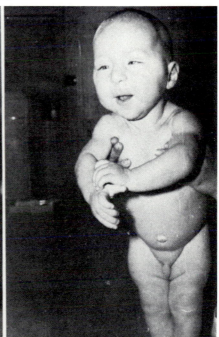

Mental Retardation

Mental retardation is very severe in all cases. Muscular hypertonia was reported on several occasions.

Prognosis

These children were observed at a very young age. Two of them died (one at 3 days, one at 5 months).

CYTOGENETIC STUDY

In all cases, the trisomy results from malsegregation of a reciprocal parental translocation. The breakpoint varies between q15 and q22.

DERMATOGLYPHICS

No systematic anomaly has been noted.

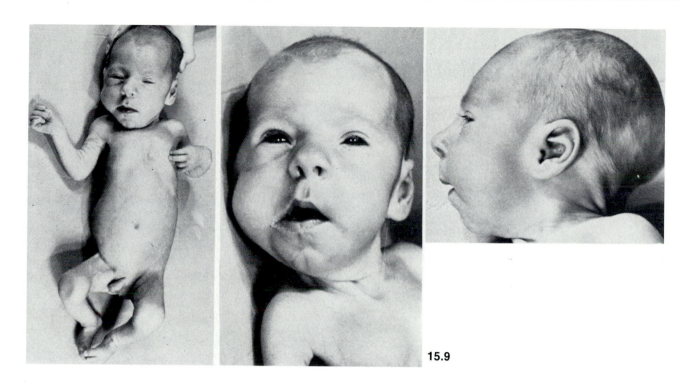

15.9

15.10

15.11

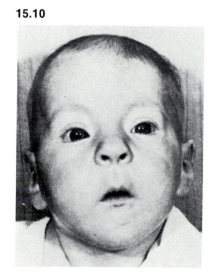

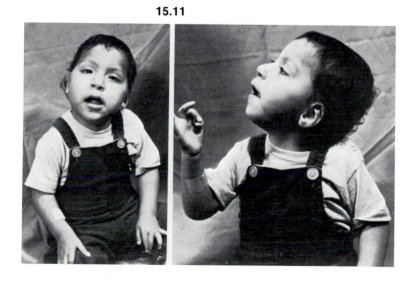

THE PRADER-WILLI SYNDROME AND 15q1 MONOSOMY

REFERENCES

EMBERGER J.M., RODIERE M., ASTRUC J., BRUNEL D.: Syndrome de Prader-Willi et translocation 15-15. *Ann. Génét.*, **20**:297–300, 1977.

FRACCARO M., ZUFFARDI O., BÜHLER E.M., JURIK L.P.: 15/15 translocation in Prader-Willi syndrome. *J. med. Genet.*, **14**:275–276, 1977.

FUJITA H., SAKAMOTO Y., HAMAMOTO Y.: An extra dic(15p) (q11) chromosome in Prader-Willi syndrome. *Hum. Genet.*, **55**:409–411, 1980.

GUANTI G.: A new case of rearrangement of chromosome 15 associated with Prader Willi syndrome. *Clin. Genet.*, **17**:423–427, 1980.

HAWKEY C.J., SMITHIES A.: The Prader-Willi syndrome with a 15/15 translocation: case report and review of the literature. *J. med. Genet.*, **13**:152–156, 1976.

LEDBETTER D.H., RICCARDI V., AIRHART S.D., STROBEL R.J., KEENAN B.S., CRAWFORD J.D.: Deletions of chromosome 15 as a cause of Prader-Willi syndrome. *New Engl. J. Med.*, **304**:325–329, 1981.

PRADER A., LABHART A., WILLI H.: Ein Syndrom von Adipositas, Kleinwuchs, kryptorchismus und oligophrenic nach myatonicartigen Zustand in Neugeborenenalter. *Schweiz. med. Wochenschr.*, **86**:1260–1261, 1956.

WISNIEWSKI L.P., WITT M.E., GINSBERG-FELLNER F., WILNER J., DESNICK R.J.: Prader-Willi syndrome and a bisatellited derivative of chromosome 15. *Clin. Genet.*, **18**:42–47, 1980.

Hawkey and Smithies (1976), Emberger et al. (1977), and Fraccaro et al. (1977) were the first to describe an association between a t(15q15q) and Prader-Willi syndrome. About 20 observations are now known of patients with Prader-Willi syndrome who are carriers of a rearrangement of the juxtacentromeric region which is compatible with a short intercalary deletion of 15q1 (see Fujita et al., 1980; Wisniewski et al., 1980; Guanti, 1980; Ledbetter et al., 1981).

The existence of patients with Prader-Willi syndrome and a normal karyotype leads to the following two hypotheses (Fraccaro et al., 1978):

1. The Prader-Willi syndrome is heterogeneous and only certain cases are due to a del(15).
2. Some patients are carriers of a deletion that cannot be detected through currently existing techniques.

PHENOTYPE

The syndrome described by Prader, Labhart, and Willi (1956) is fairly frequent. More than 170 cases have been found in the literature by Hawkey and Smithies (1976). Major features are neonatal hypotonia with feeding difficulties, poor spontaneous movement, spells of cyanosis, a weak or absent cry, hypogonadism with cryptorchidism, mental retardation, obesity, hyperphagia, and shortness of stature. Less frequently encountered features are diminished intrauterine movement, breech delivery, behavioral troubles, craniofacial dysmorphism (narrow bifrontal diameter, almond-shaped eyes, strabismus, fish-shaped mouth), and acromicria.

CYTOGENETIC STUDY

Chromosomal aberrations that have been observed include:

- t(15;15)
- Translocations involving a 15 and a nonacrocentric
- Interstitial del(15q11q13) (Ledbetter et al., 1981)
- More complex rearrangements such as an isodicentric or a chromosome carrier of two satellites resulting from an inverted duplication (Fujita et al., 1980; Wisniewski et al., 1980).

In all cases, these rearrangements are compatible with a short intercalary deletion in the 15q1 region. The smallest area common to all these observations is the 15q11 band.

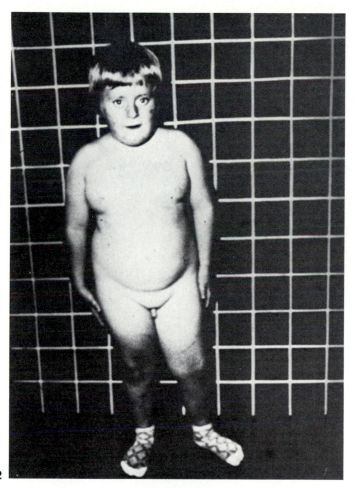

15.12

r(15)

REFERENCES

YUNIS E., LEIBOVICI M., QUINTERO L.: Ring(15) chromosome. *Hum. Genet.,* **57**: 207–209, 1981.

A dozen cases of r(15) have been reported (see Yunis et al., 1981).

No truly characteristic facial dysmorphism exists. Principal symptoms include:

- Intrauterine growth retardation, with a mean birthweight of 2,020 g
- Severe, persistent growth retardation
- Microcephaly
- Variable psychomotor retardation

Anomalies of the radial axis and a birdlike profile have also been reported.

15.13

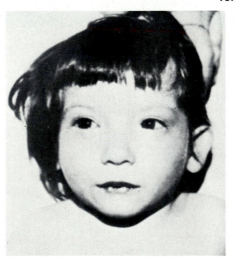
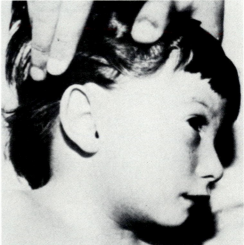

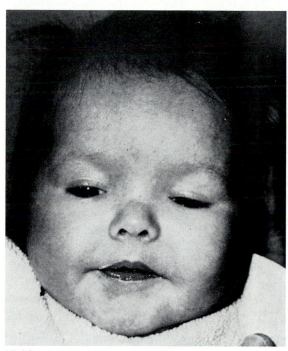

15.14

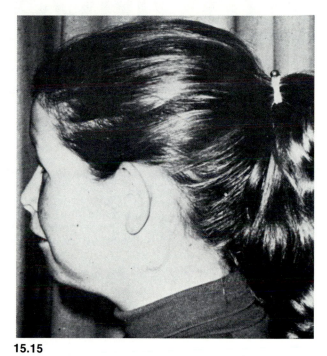

15.15

15.16

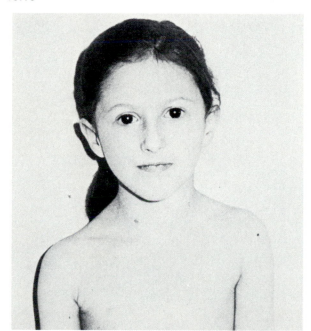

CHROMOSOME 16

Chromosome 16 would appear to be a highly important chromosome. Indeed, although very few aneusomies of chromosome 16 are observed in live newborns, 16 trisomies are found very frequently in early spontaneous abortions (Boué and Boué, 1975).

Gene assignments are numerous and of great interest. They include the α-globin genes, whose mutations cause thalassemias; and the gene for haptoglobin, one of the first to have been localized by linkage with a chromosome polymorphism, the secondary juxtacentromeric constriction in the present case.

Chromosomal pathology remains rare. Nonetheless, the two partial trisomies of the long and short arm seem to be highly characteristic.

The secondary juxtacentromeric constriction is the object of a polymorphism. There are two fragile sites, one of them in p12 and the other in q22.

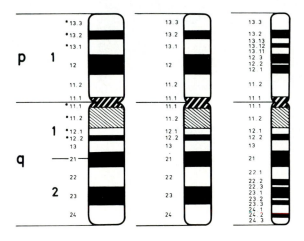

GENE MAP

	CTH	(P)	Cystathionase (21950) S
	AG		ALPHA GLOBIN GENE CLUSTER
16p12-16pter	HBZ		Hemoglobin zeta (2 loci) (14230, 14231) RE
16p12-16pter	HBA		Hemoglobin alpha (1, 2, or 3 loci) (14180) S-HS
16p12-16p13	PGP		Phosphoglycolate phosphatase (17228) S

16p12-16q21	DIA4	(P)	Diaphorase-4 (12586) S
16p12	FS		Fragile site (13656)
16q12-16q22	APRT		Adenine phosphoribosyltrans-ferase (10260) S, D
16q22	HP		Haptoglobin (10410) Fc
16q22	LCAT		Lecithin-cholesterol acyltrans-ferase (24590) F, LD
16q22	FS		Fragile site (13657)
	CTRB	(P)	Chymotrypsinogen B (11889) REa
	TK2	(P)	Thymidine kinase, mitochon-drial (18829) S
16p11-16pter	GOT2	(P)	Glutamate oxalacetic transami-nase, mitochondrial (13815) S
	LIPB	(P)	Lysosomal acid lipase-B (24798) S
	GCF2	(P)	Growth rate controlling factor-2 (13923) S
	ESB3	(P)	Esterase-B3 (13329) S
16p11-16pter	GPT1	(P)	Glutamate pyruvate transami-nase, soluble red cell (13822) S
	EBS1	(P)	Epidermolysis bullosa, Ogna type (13195) F
	HAGH	(P)	Glyoxalase II (hydroxyacyl glutathione hydrolase) (13876) S
	NHCP2	(P)	Nonhistone chromosomal pro-tein-2 (11888) S
	AVR	(P)	Antiviral state regulator (10747) D
	DIPI	(P)	DI (defective interfering) parti-cle induction, control of (12602) S
	VMD1		Macular dystrophy, atypical vitelliform (15384) F

MORPHOLOGY AND BANDING

QFQ   GTG   RHG   RBA   THA   C

16p TRISOMY

Small, round skull
Prominent maxillae
Micrognathia
Thumb anomalies

REFERENCES

DALLAPICCOLA B., CURATOLO P., BALESTRAZZI P.: De novo trisomy 16q11→pter. *Hum. Genet.*, **49**:1–6, 1979.

LESCHOT N.J., DE NEF J.J., GERAEDTS J.P.M., BECKER-BLOEMKOK M.J., TALMA A., BIJLSMA J.B., VERJAAL M.: Five familial cases with a trisomy 16p syndrome due to translocation. *Clin. Genet.*, **16**:205–214, 1979.

ROBERTS S.H., DUCKETT D.P.: Trisomy 16p in a liveborn infant and a review of partial and full trisomy 16. *J. med. Genet.*, **15**:375–381, 1978.

YUNIS E., GONZALEZ J.T., TORRES DE CABALLERO O.M.: Partial trisomy 16q−. *Hum. Genet.*, **38**:347–350, 1977.

The first two observations of trisomy 16p were reported by Yunis et al. (1977) and by Roberts and Duchett (1978). Leschot et al. (1979) described a family with five cases of 16p trisomy, two of which were the object of an induced abortion. An eighth case was observed by Dallapiccola et al. (1979).

GENERALIZATIONS

- Sex ratio: 6F/2M (the males were the result of the pregnancies that ended in induced abortions)
- Mean birth weight: 2,120 g
- Mean crown-heel length at birth: 42.4 cm

PHENOTYPE

Craniofacial dysmorphism would seem to be quite specific, despite the small number of observations. Intrauterine and postnatal growth retardation is considerable.

Craniofacial Dysmorphism

The skull is small and round. The face is round and flat.

Lashes and eyebrows are scant. Palpebral fissures vary in their orientation. Hypertelorism is present. The bridge of the nose is wide and shallow. The nose is short with anteverted nares and a bulbous tip.

A prominence of the maxillae and micrognathia are present. The upper lip overlaps the lower one to a great extent. The palate may be cleft or high-arched.

The ears appear to be quite characteristic: they are round and low-set; the tragus and the lower part of the helix are hypoplastic; the anthelix is protuberant, but its posterior branch is absent.

Neck, Thorax, and Abdomen

The umbilical artery is often single. Diastasis recti, an umbilical hernia, a narrow pelvis, and dysplasia of the hip also have been reported.

Limbs

The most striking symptom is aplasia or hypoplasia of the thumb, or its proximal implantation. The fingers frequently overlap.

A hammer big toe was noted in two cases, and a clubfoot in one case.

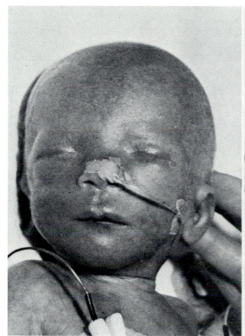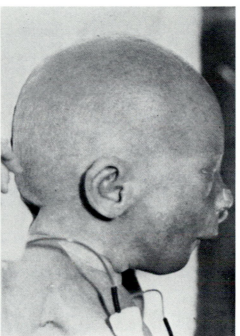

16.1

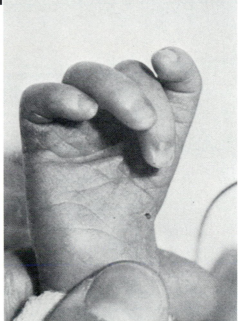

Malformations

These tend to vary, and are not truly suggestive of this syndrome: congenital glaucoma in one case; three cases of heart disease in the family described by Leschot et al. (1979) (two Fallot tetralogies and one atrial septal defect); one bicornuate uterus; one case of ovarian agenesis; and atrophy of the thymus.

Mental Retardation

Mental impairment is very severe. The neonatal period is often marked by feeding difficulties, seizures, hypertonia, and poor reflexes.

Prognosis

Half of these children died during the first year. The oldest child seen in consultation was 11½.

CYTOGENETIC STUDY

Only one case occurred *de novo*. Among the three parental translocations, two are t(16p;21q). The breakpoint is either in q11 or p11.

DERMATOGLYPHICS

A single transverse palmar crease has been noted on several occasions.

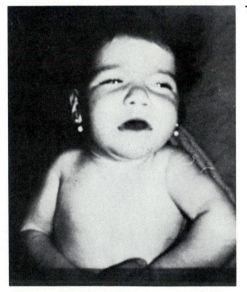

16.2

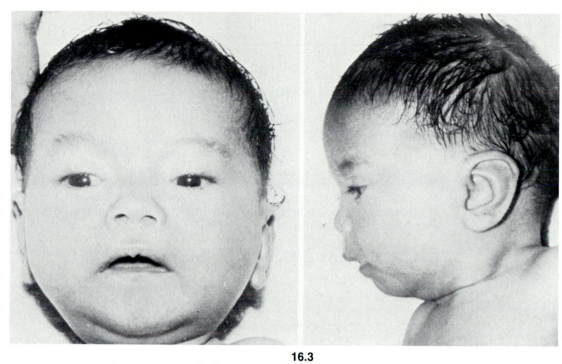

16.3

16.4

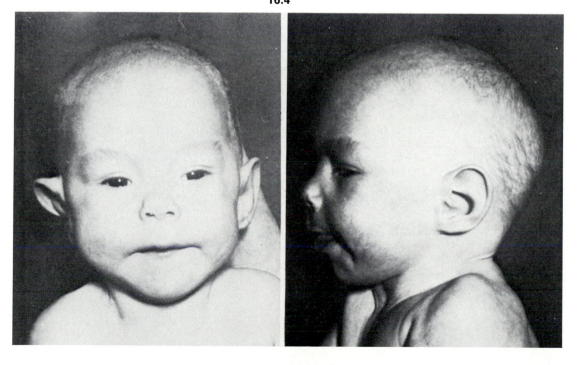

16q TRISOMY

Enophthalmia
Short, prominent nose
Thin upper lip
Osteoarticular anomalies
Perineal anomalies

REFERENCES

BALESTRAZZI P., GIOVANNELLI G., LAN-DUCCI RUBINI L., DALLAPICCOLA B.: Partial trisomy 16q resulting from maternal translocation. *Hum. Genet.* 49:229–235, 1979.

GARAU A., CRISPONI G., PERETTI D., VANNI R., ZUFFARDI O.: Trisomy 16q21→qter. *Hum. Genet.,* 53:165–167, 1980.

RETHORE M.O., LAFOURCADE J., COUTURIER J., HARPEY J.P., HAMET M., ENGLER R., ALCINDOR L.G., LEJEUNE J.: Augmentation de l'activité de l'adénine phosphorybosyl transférase chez un enfant trisomique 16q22.2→16qter par translocation t(16;21)(q22.2;q22.2)pat. *Ann. Génét.,* 25:36–42, 1982.

RIDLER M.A.C., McKEOWN J.A.: Trisomy 16q arising from a maternal 15p; 16q translocation. *J. med. Genet.,* 16:317–319, 1979.

SCHMICKEL R., POZNANSKI A., HIMEBAUGH J.: 16q trisomy in a family with a balanced 15–16 translocation. *Birth Defects, Orig. Art. Ser.,* XI-5:229–236, 1975.

The first well-documented observation was reported by Schmickel et al. (1975). This patient, as well as that reported by Ridler and McKeown (1979), are trisomic for the whole of 16q. Two other children are q21→qter trisomic (Balestrazzi et al., 1979; Garau et al., 1980). A fifth patient, described by Rethoré et al. (1981), is trisomic for a shorter segment, q22→qter.

GENERALIZATIONS

- Sex ratio: 4M/1F
- Mean birth weight: 2,130 g
- Mean crown-heel length at birth: 42.5 cm
- Mean head circumference at birth: 30 cm

These measurements do not take into account the patient of Rethoré et al. (1981), who had normal measurements at birth.

PHENOTYPE

The phenotype is very similar in all five children.

Craniofacial Dysmorphism

The head is often asymmetrical, with a high forehead and temporal retraction.
Palpebral fissures slant downward and outward, with an abnormal crease of the eyelids.
The bridge of the nose is flat. Hypertelorism is present.
The nose is short and prominent.
The upper lip is thin and not visible. The lower lip recedes.
The chin is small.
Ears are low-set, often poorly folded, with prominence of the middle portion of the anthelix and antitragus.

Neck, Thorax, and Abdomen

The thorax is often abnormal and may be long and narrow, with kyphoscoliosis or gibbosity due to hemivertebrae.

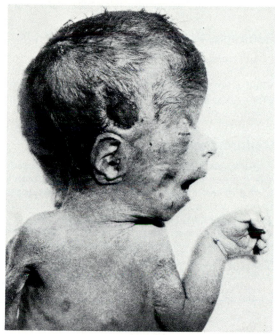

16.5

Limbs

The most suggestive anomalies are flexion contractures of the upper limbs, malpositioning of the hands and feet, digit-like or too long a thumb, absence of a toe, hammertoe, and osteosclerosis of the distal phalanges, which is visible on X-ray films.

Genitalia, Perineum

Anomalies of the genitals and of the perineum are encountered constantly. In boys, a small penis, with or without sexual ambiguity, is noted; one boy was designated a female. Cryptorchidism is sometimes found. In both sexes, anomalies of the anal orifice are always present, including imperforation, anterior displacement, and formation of a cloaca.

Malformations

Bone anomalies have already been discussed and mainly concern the vertebrae, ribs, phalanges, and skull. Heart defect involving atrial septal defect has been reported in two cases. Cutaneous anomalies appear to be frequent: these may include dry skin, wrinkled skin with a decrease in subcutaneous tissue, excessive fragility with recurring lesions, and temporal scalp.

Cleft palate and malrotation of the colon have been sporadically reported.

Mental Retardation

One child had an IQ of 43 at the age of 5 months. Mental retardation is severe in the oldest child, age $3\frac{1}{2}$.

Prognosis

The neonatal period is marked by nutritional difficulties and digestive troubles. Growth retardation persists and remains severe. Most children (4 out of 5) died during the first few months.

CYTOGENETIC STUDY

The breakpoint is variable, from p11 to q22. The phenotype and prognosis do not appear to be modified by the size of the trisomy.

LABORATORY FINDINGS

In the only case in which gene dosages were performed, the rate of haptoglobin and adenine phosphoribosyltransferase showed a gene dosage effect (Rethoré et al., 1981).

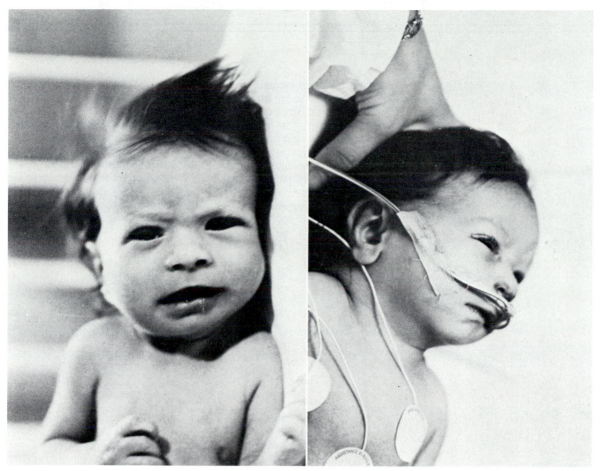

16.6

CHROMOSOME 17

Like chromosome 16, chromosome 17 may be one of those chromosomes whose gene balance is indispensable for development of the embryo. Indeed, very few imbalances are known for infants born alive.

Among known genes, the following should be considered: that of thymidine kinase, the first to have been localized in humans by interspecific cellular hybridization; the gene families responsible for synthesis of growth hormones and of collagene 1; the gene coding for galactokinase, and that of acid-alpha-glucosidase.

Genetic diseases include type II galactokinase deficiency, glycogenosis, and a form of pituitary dwarfism.

The only cytogenetic syndrome that can be described is trisomy 17qter.

A fragile site is present in 17p12.

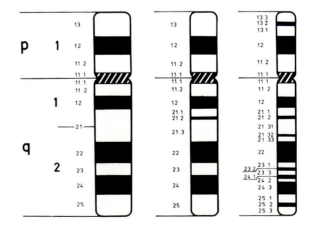

GENE MAP

?17p13	MDLS	(L)	Miller-Dieker lissencephaly syndrome (24720) Ch
17p11-17pter	MYH	(P)	Myosin, skeletal, heavy chain (16073) REa
17p12	FS		Fragile site (13665) Ch
17q21-17q22	TK1		Thymidine kinase-1 (18830) S, Ch, R
17q21-17q22	GALK		Galactokinase (23020) S, Ch, R
17q21-17q22	A12M4		Adenovirus-12 chromosome modification site-17 (10297) V
17q21-17qter	GAA		Acid alpha-glucosidase (23230) S
	S9	(P)	Surface antigen 9 (18557) S
17q22-17q24			GROWTH HORMONE GENE FAMILY S, REb, A
	GHN		Growth hormone, normal (13925) S, REb, A
	CSAL		Chorionic somatomammotropin-like (15020) S, REb, A
	CSA		Chorionic somatomammotropin A (15020) S, REb, A
	GHV		Growth hormone variant (13925) S, REb, A
	CSB		Chorionic somatomammotropin B (15020) S, REb, A
	CKBB	(L)	Creatine kinase, brain type (12328) S
17q21-17q22	COL1A1		Collagen I alpha-1 polypeptide (12015) S, M, A
	COL4A1	(L)	Collagen IV alpha-1 polypeptide (12013) REa

MORPHOLOGY AND BANDING

QFQ   GTG   RHG   RBA   THA   C

17q2 TRISOMY

Squinty eyes
Large mouth
Brachyrhizomelia
Hexadactyly

REFERENCES

BERBERICH M.S., CAREY J.C., LAWCE H.J., HALL B.D.: Duplication (partial trisomy) of the distal long arm of chromosome 17: a new clinically recognizable chromosome disorder. *Birth Defects, Orig. Art. Ser., XIV-6C*:287–295, 1978.

GALLIEN J.U., NEU R.L., WYNN R.J., STEINBERG WARREN N.. BANNERMAN R.M.: Brief clinical report: an infant with duplication of 17q21→17qter. *Am. J. med. Genet.,* **8**:111–115, 1981.

GROUCHY J. de, TURLEAU C., SACHS Ch., POMAREDE R., RAPPAPORT R.: Growth hormone and distal trisomy 17qter. *Clin. Genet.,* **20**:322 (Letter to the Editors), 1981.

TURLEAU C., GROUCHY J. de, BOUVERET J.P.: Distal trisomy 17q. *Clin. Genet.,* **16**:54–57, 1979.

Five patients are known to be trisomic for 17q2. Three cases are familial (Berberich et al., 1978) and two are isolated (Turleau et al., 1979; Gallien et al., 1981).

GENERALIZATIONS

- Sex ratio: 3F/2M
- Mean birth weight: 2,875 g
- The crown-heel length at birth was given in three observations (47.5, 47.5, and 33 cm).

PHENOTYPE

A shortness of stature involving mainly the proximal segments of the limbs is present (although this brachyrhizomelia was not specifically emphasized by Berberich et al.). Craniofacial dysmorphism is characteristic.

Craniofacial Dysmorphism

Microcephaly with a high forehead, frontal bossing, and temporal retraction are all present. The hairline is set high, with a widow's peak.
Palpebral fissures are narrow and creased, giving a squinty appearance.
The nose is short and flattened, with wide nares.
The philtrum is long.
The mouth is large with downturned corners. The upper lip, which is thin, overlaps the lower one, which is also thin. Cleft palate is nearly always present.
The chin is small and receding.
Ears are round and appear to be low-set. The upper part of the helix is underdeveloped and tends to tilt backward.

Neck, Thorax, and Abdomen

The neck is short and thick, with an excess of skin, resembling in one case a buffalo hump.

Limbs

The most noteworthy symptoms are brachyrhizomelia, already mentioned, hexadactyly in two isolated cases, and ligament hyperlaxity.

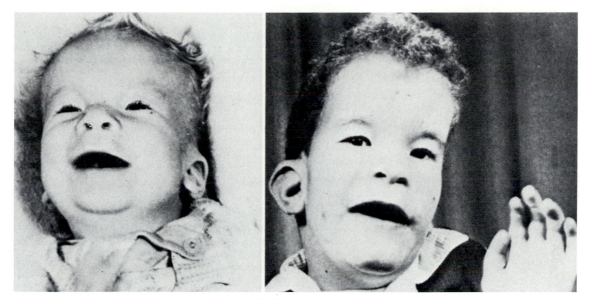

17.1

17.2

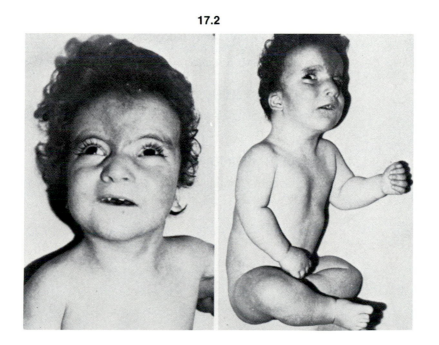

Genitalia

One of the females had labia majora which resembled scrotal folds, as well as a rudimentary uterus and the absence of the vagina. Another had hypoplastic, cystic ovaries. The third had irregular menstrual periods after the age of 13.

Both males had cryptorchidism.

Malformations

The following malformations have been described for the four patients with trisomy q23→qter:

- Cerebral anomalies
- Urinary tract malformations
- Cardiac malformations, including atrial septal defect, alone or accompanied by subaortic stenosis.

The patient described by Gallien et al. (1981), who was trisomic for 17q21→qter, was afflicted with much more severe malformations, which caused death 2 hours after birth. The autopsy revealed complex heart disease, gastrointestinal anomalies, urogenital abnormalities, and cerebral malformations. X-ray films showed a number of bone anomalies.

The severity of these malformations clearly explains the rareness of this syndrome, and of anomalies of chromosome 17, in general.

Mental Retardation

Mental retardation is very severe.

Prognosis

Of the three children who died, one was stillborn, the second died at 3 months, and the third at 10 years. The other two are still alive at ages 6 and 17.

When death does not occur early, growth retardation is very severe.

CYTOGENETIC STUDY

The five known cases are due to a parental translocation. The breakpoint is in q23 or q21.

LABORATORY FINDINGS

In one patient (Grouchy et al., 1981), assay of the growth hormone showed a normal level.

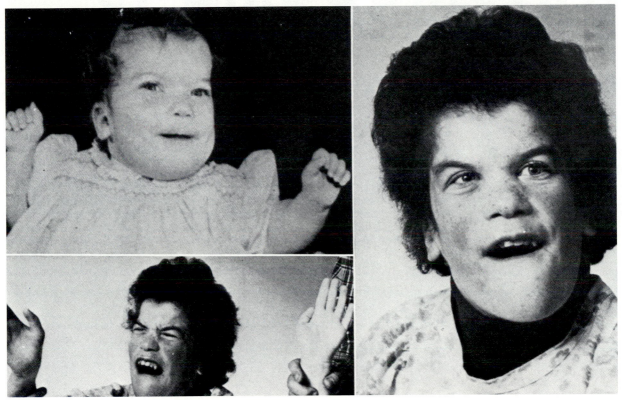

17.3

CHROMOSOME 18

REFERENCES

NIELSEN K.B., DYGGVE H., FRIEDRICH U., HOBOLTH N., LYNGBYE T., MIKKELSEN M.: Small metacentric nonsatellited extra chromosome. *Hum. Genet.*, **44**:59–69, 1978.

WILSON M.G., TOWNER J.W., FORSMAN I., SIRIS E.: Syndromes associated with deletion of the long arm of chromosome 18 [del(18q)]. *Am. J. med. Genet.*, **3**:155–174, 1979.

Chromosome 18 has few gene assignments; only the gene for peptidase A has been localized in 18q23. This enzyme shows a gene dosage effect in the various aneusomies of chromosome 18.

Chromosomal diseases, however, are important and have been described very early. Trisomy 18 is one of the "classical" cytogenetic syndromes. This is also true of 18p monosomy, which was the first deletion observed in man, of 18qter monosomy, and of r(18). We have not considered interstitial monosomies of 18q which were suggested by Wilson et al. (1979).

Partial trisomies are also known. Only 18qter trisomies are sufficiently homogeneous to allow characterization of a syndrome. Proximal 18q trisomies constitute a slightly different syndrome from that of complete 18 trisomy. A number of observations are available for 18p trisomy, especially those of i(18p), but cytogenetic identification of these small metacentrics is very difficult to establish (Nielsen et al., 1978).

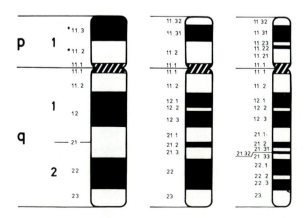

GENE MAP

18q23	PEPA		Peptidase A (16980) S, D
	CG	(I)	Human chorionic gonadotropin (11885) S (see chr. 10)
	NLP2	(L)	Neoplastic lymteration-2 (16185) Ch
	ASNRS	(P)	Asparaginyl-tRNA synthetase (10841) REa

MORPHOLOGY AND BANDING

QFQ GTG RHG RBA THA C

Type and Countertype

	18 Trisomy	*18q Monosomy*
Nose	Thin and well developed	Midface retraction
Chin	Retromicrognathia	Prominent
Ears	Aplastic anthelix	Prominent anthelix
	Faunesque ears	Deep concha
	Hypoplastic helix	Helix well folded
Pelvis	Narrow	Froglike attitude
Hands	Short, overlapping fingers	Tapered fingers
Dermatoglyphics	Arches	Whorls

18 TRISOMY

Faunlike ears
Micrognathia
Occipital protuberance
Overlapping fingers
High frequency of arches
Narrow pelvis
Rocker-bottom feet
Considerable growth retardation
Premature death

REFERENCES

BERGER R.: Trisomie 18. *Nouv. Presse méd.*, 1:745–748, 1972.

BUTLER J.L., SNODGRASS G.J., FRANCE N.E., SINCLAIR L., RUSSEL L.A.: E (16-18) trisomy syndrome: analysis of 13 cases. *Arch. Dis. Child.*, **40**:600–611, 1965.

CONEN P.E., ERKMAN B.: Frequency and occurrence of chromosomal syndromes. II. E-trisomy. *Amer. J. hum. Genet.*, **18**:387–398, 1966.

EATON A.P., KONTRAS S.B., SOMMER A., WEHE R.A.: Long-term survival in trisomy 18. *Birth Defects, Orig. Art. Ser.*, XI-5:327–328, 1975.

EDWARDS J.H., HARNDEN D.G., CAMERON A.H., CROSSE V.M., WOLFF O.H.: A new trisomic syndrome. *Lancet*, i:787–790, 1960.

LAFOURCADE J., LEJEUNE J., BERGER R., RETHORE M.O., ARCHAMBAULT L.: La trisomie 18. Cinq observations nouvelles. Revue de la littérature. *Sem. Hôp. Paris*, **41**:24–35, 1965.

NIELSEN J., HOLM V., HAAHR J.: Prevalence of Edwards' syndrome. Clustering and seasonal variation? *Humangenetik*, **26**:113–116, 1975.

SMITH D.W., PATAU K., THERMAN E., IN-HORN S.L.: A new autosomal trisomy syndrome. *J. Pediat.*, **57**:338–345, 1960.

TAYLOR A.I.: Autosomal trisomy syndromes: a detailed study of 27 cases of Edwards' syndrome and 27 cases of Patau's syndrome. *J. med. Genet.*, **5**:227–252, 1968.

UCHIDA I., BOWMAN J.M., WANG H.C.: The 18-trisomy syndrome. *New Engl. J. Med.*, **266**:1198–1201, 1962.

WEBER W.W.: Survival and the sex ratio in trisomy 17–18. *Amer. J. hum. Genet.*, **19**:369–377, 1967.

The first patient with chromosome 18 trisomy was described in 1960 by Edwards et al. The authors believed that the supernumerary chromosome was a 17. Very rapidly, other authors reported similar observations, and it became evident that the clinical syndrome was correlated with 18 trisomy (Smith et al., 1960; Uchida et al., 1962; Lafourcade et al., 1965; Butler et al., 1965; Taylor, 1968; Berger, 1972). The phenotype is characterized by a severe malformative syndrome, almost always leading to death in the early weeks after birth.

GENERALIZATIONS

- Frequency: 1 in 8,000 births. The possibility of a concentration of births at certain periods of the year or in certain regions has been suggested (Conen and Erkman, 1966; Nielsen et al., 1975).
- Sex ratio: 4F/1M
- Mean maternal age: 32.5 years (bimodal distribution with two peaks, at 25–30 years and 40–45 years) (Weber, 1967)
- Mean paternal age: 34.9 years
- Pregnancy: postmaturity in almost every case. Mean duration: 42 weeks. Low fetal activity, hydramnios, small placenta, and single umbilical artery are frequent.
- Mean birth weight: 2,240 g (paradoxically low).

PHENOTYPE

Growth retardation is severe, with hypoplasia of the skeletal muscles and panniculus adiposus. Sucking is poor. The baby is at first hypotonic and then hypertonic. He or she responds weakly to noise and seems afflicted with severe mental retardation.

Craniofacial Dysmorphism

Dolichocephaly is considerable, the occiput protuberant, and the bitemporal diameter small. Microcephaly and gaping fontanels are sometimes present.

The bridge of the nose is usually slender or even protuberant. It is rarely flattened and broad. The nose is often upturned.

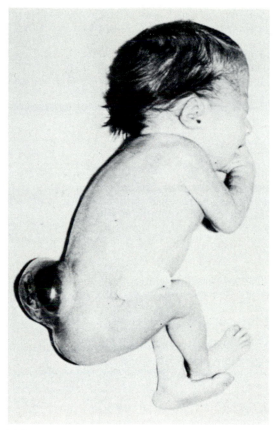

18.1

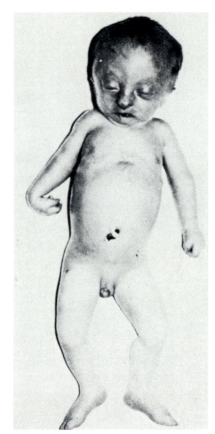

18.2

18.3

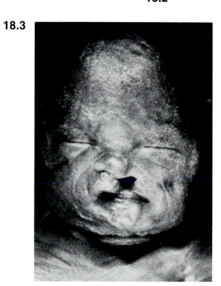

Palpebral fissures are horizontal and may be short, with bilateral epicanthal folds. Hypoplasia of the orbital ridges, corneal opacities, and microphthalmia may be observed.

The mouth is small and the palate narrow.

Considerable micrognathia is present.

The ears are very characteristic: they are low-set and "faunlike." The pinnae are flat, their upper portions pointed.

Protuberant occiput, micrognathia, and faunlike ears are the major features of the dysmorphism.

Neck, Thorax, and Abdomen

The neck is often short, with an excess of skin.

The sternum is almost always short, with a reduced number of ossification points. The nipples are small. The abdominal muscle girdle is defective, as evidenced by umbilical and inguinal hernias or diastasis of the recti abdominis.

The narrowness of the pelvis is a basic feature of 18 trisomy.

Limbs

The positioning of the hands is also very characteristic: the fists are closed and cannot be extended, the index overlaps the third digit, and the fifth overlaps the fourth.

The nails are hypoplastic. The infant often holds its arms extended up alongside its head in a "pleading" posture.

Malformations of the lower limbs include a limitation of thigh abduction or congenital dislocation of the hips, and especially "rocker-bottom" feet, with a protrusion of the calcaneum. The big toe is often short and in dorsiflexion. Syndactyly of the second and third digits is frequent.

Genitalia

Cryptorchidism is constant in the male. In the female, hypotrophy of the clitoris with hypoplasia of the labia majora is often noted.

Perineal anomalies and an imperforate anus may also be observed.

Malformations

Visceral malformations mainly include (in more than 95 percent of the cases) cardiac malformations, which are almost always the cause of death. Most frequent are ventricular septal defects and patent ductus arteriosus. Less frequent are atrial septal defects and bicuspid aortic or pulmonary valvules.

The lungs may be subject to abnormal segmentation.

Gastrointestinal malformations frequently include Meckel's diverticulum; and less frequently, heteropic islets of pancreatic or spleen tissue, pyloric stenosis, omphalocele, malrotation of the intestine, and diaphragmatic hernia.

Renal malformations are frequent: ectopic or horseshoe kidney; hydronephrosis, megalo-ureter or double ureter.

Other malformations are much rarer: meningocele; harelip or cleft palate; atresia of the choana; tracheo-esophageal fistula; bifid uterus; ovarian hypoplasia; phocomelia; lobster-claw deformity; atresia of the external auditory canal.

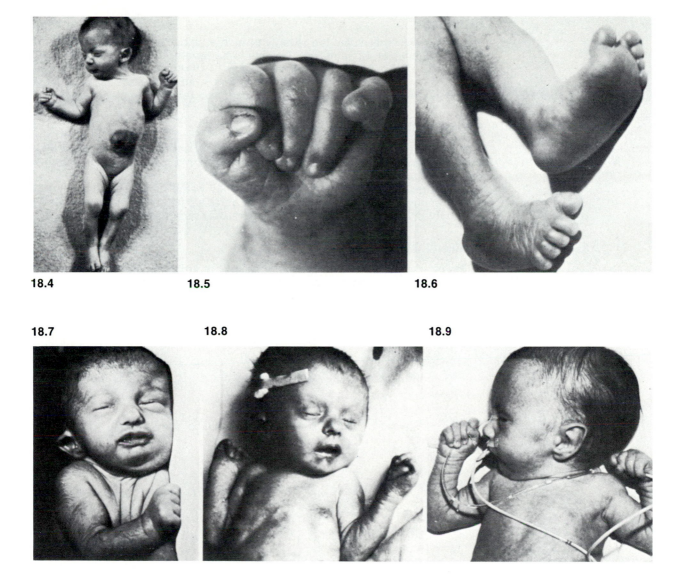

18.4   18.5   18.6

18.7   18.8   18.9

Prognosis

It is difficult to evaluate the severity of mental impairment because of the young age of the patients. Nonetheless, it seems to be severe.

Mean survival is 2–3 months for males and 10 months for females (Weber, 1967).

Few patients have survived childhood; these lived to the age of 15 or 19 years. Patients with mosaicism have a prolonged survival rate (Eaton et al., 1975).

CYTOGENETIC STUDY

In 80 percent of the cases, a free homogeneous trisomy is involved: 47,XY(X),+18.

In 10 percent of the cases there is mosaicism with a normal cell population and a trisomic population: 46,XY(X)/47,XY(X),+18.

The situation is more complex in the other 10 percent of the cases, where either double aneuploidy such as 48,XXY,+18, or a translocation transmitted or arising *de novo,* with more or less complete 18 trisomy, may be present.

DERMATOGLYPHICS

The most significant feature is the frequency of arches on the fingertips, 10 times higher than in the general population. These can be seen on 7 to 10 fingers.

More rarely noted is an absence of the distal flexion crease of the fingers.

A single transverse palmar crease is present in 30 percent of the cases.

The axial triradius is in t' or t".

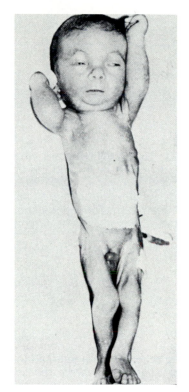

18.10

18.11

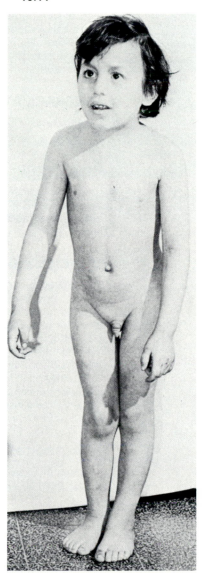

18q2 TRISOMY

Oval-shaped face
Prominent nose bridge
Microretrognathia
Excess of arches

REFERENCES

KUKOLICH M.K., ALTHAUS B.W., SEARS J.W., MANKINEN C.B., LEWANDOWSKI R.C.: Abnormalities resulting from a familial pericentric inversion of chromosome 18. *Clin. Genet.*, **14**:98–104, 1978.

MATSUOKA R., MATSUYAMA S., YAMA-MOTO Y., KUROKI Y., MATSUI I.: Trisomy 18q. *Hum. Genet.*, **57**:78–82, 1981.

NIAZI M., COLEMAN D.V., SALDANA GAR-CIA P.: Partial trisomy 18 in a family with a translocation (18;21) (q21;q22). *J. med. Genet.*, **15**:148–151, 1978.

TURLEAU C., GROUCHY J. de: Trisomy 18qter and trisomy mapping of chromosome 18. *Clin. Genet.*, **12**:361–371, 1977.

The phenotype for 18q2 trisomy was described by Turleau and Grouchy (1977) on the basis of four personal observations and five cases taken from the literature. Two more recent observations have since been made (Niazy et al., 1978; Kukolich et al., 1978).

The clinical picture of this partial trisomy differs from that of complete 18 trisomy, mainly due to the lesser frequency of internal malformations and to the low rate of early mortality.

GENERALIZATIONS

- Sex ratio: 7F/4M
- Mean birth weight: 2,700 g

PHENOTYPE

Morphological anomalies are moderate at birth, and generally do not lead to a diagnosis of 18 trisomy. In particular, no severe growth retardation, such as that seen in complete trisomy, is observed. Nor does one observe the characteristic finger flexion, prominence of the calcaneum, narrow pelvis, or shortness of the sternum.

Craniofacial Dysmorphism

In newborns, the face is round, the nose is snubbed with anteverted nares, the chin is small, and the ears are low-set with a poorly folded helix.

In older children, the forehead is high with dolichocephaly. The hairline is set high upon the forehead. The face is elongated with a very weak jawline.

Orientation of the palpebral fissures varies. The curve of the eyebrow is effaced. The most striking feature is the prominence of the bridge of the nose and absence of the nasofrontal angle. The tip of the nose is round and bulbous.

The mandible is small and receding, the palate high-arched. Teeth are implanted in an abnormal manner and there is frequent tooth decay.

Ears have a normal implantation, but are often in posterior rotation with a poorly folded helix.

Neck, Thorax, and Abdomen

The neck is short with an excess of skin. Occasionally, slight pectus excavatum as well as hernias may be noted.

Limbs

Limbs are usually normal. The hands may be large, with short fingers, and the toes may overlap.

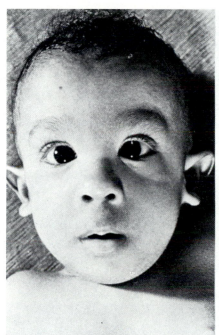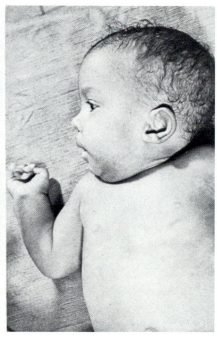

18.12

18.13

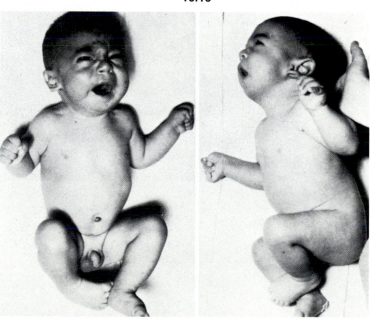

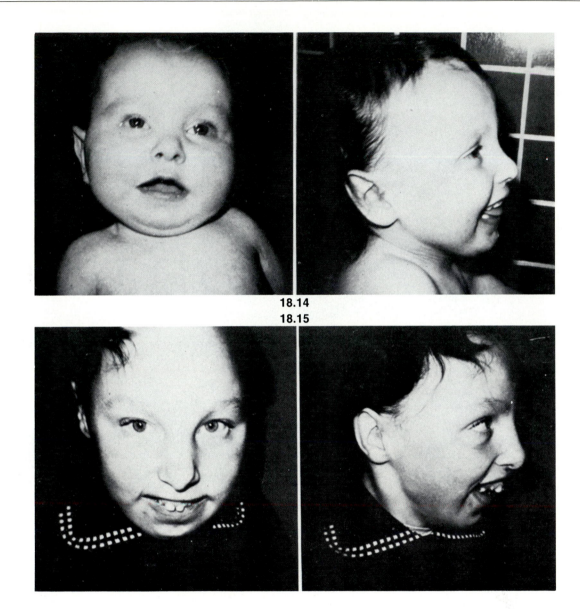

18.14
18.15

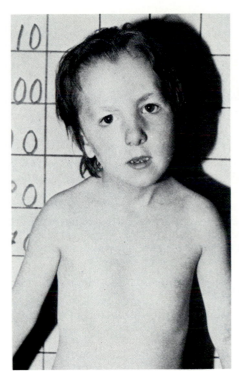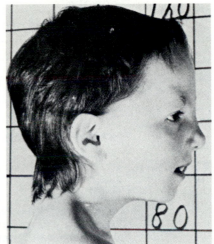

18.16

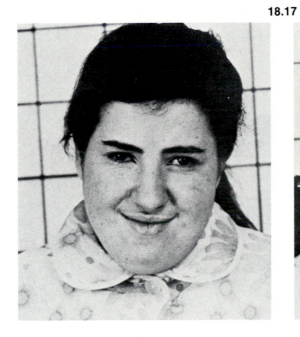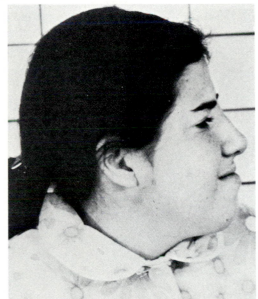

18.17

Genitalia

Cryptorchidism was observed in one boy. Hypoplasia of the labia majora was observed in a girl.

Malformations

These are rare. Benign congenital heart disease was seen in one of the cases, and a renal anomaly in another case.

Mental Retardation

The extent of mental retardation varies from an IQ of 80 to an IQ of less than 10.

Prognosis

Growth retardation varies from -1 to -3 DS, and seems to affect height more than weight. The vital prognosis is fairly good, and no early deaths have occurred. The oldest patient was 18 when she was seen in consultation for the last time. Her IQ was 80, and puberty and sexual development were normal.

CYTOGENETIC STUDY

Two mechanisms are responsible for formation of a partial 18q trisomy: either a parental translocation, or a parental pericentric inversion.

The translocations often occur with the short arm of an acrocentric. In the other cases, the difficulty in correctly identifying the breakpoints may explain why certain observations are contradictory (such as that of Matsuoka et al., 1981).

In the cases of pericentric inversions, partial 18q trisomy is accompanied by 18p monosomy, which may result in ptosis of the eyelids.

These two mechanisms are responsible for the birth of 18q trisomic and monosomic children in the same family.

DERMATOGLYPHICS

The excess of arches found in 18 trisomy are also found here.

LABORATORY FINDINGS

Peptidase A (PEPA) shows gene dosage effect in 18qter trisomy.

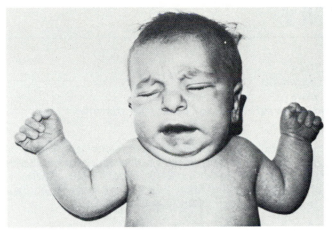

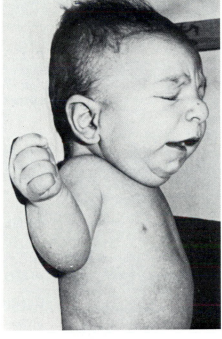

18.18

18.19

18.20

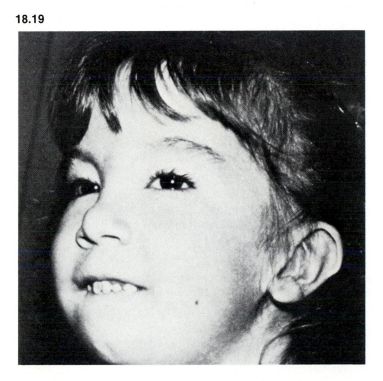

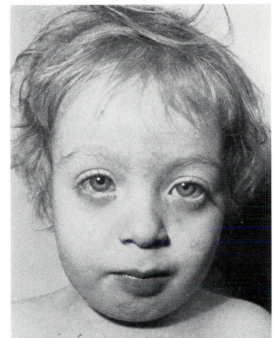

18p & q1 TRISOMY

Birdlike head
Microretrognathia
Faunlike ears
Lower mortality rate than for complete trisomy

REFERENCES

FRIED K., BAR-YOCHAI A., ROSENBLATT M., MUNDEL G.: Partial 18 trisomy (with 47 chromosomes) resulting from a familial maternal translocation. *J. med. Genet.*, 15:76–78, 1978.

HERNANDEZ A., CORONA-RIVERA E., PLASCENCIA L., NAZARA Z., IBARRA B., CANTU J.M.: De novo partial trisomy of chromosome 18 (pter→q11:). Some observations on the phenotype mapping of chromosome 18 imbalances. *Ann. Génét.*, 22:165–167, 1979.

STERN L.M., MURCH A.R.: Pseudohermaphrodism with clinical features of trisomy 18 in an infant trisomic for parts of chromosomes 16 and 18:47,XY, der(18), t(16;18)(p12;q11)mat. *J. med. Genet.*, 12:305–307, 1975.

TURLEAU C., CHAVIN-COLIN F., NARBOUTON R., ASENSI D., GROUCHY J. de: Trisomy 18q−. Trisomy mapping of chromosome 18 revisited. *Clin. Genet.*, 18:20–26, 1980.

The description of trisomy involving the short arm, and the proximal portion of the long arm of chromosome 18 is based on five observations (Stern and Murch, 1975; Fried et al., 1978; Hernandez et al., 1979; Turleau et al., 1980).

These observations suggest that this trisomy has a phenotype similar to that of complete 18 trisomy, though slight nuances may exist.

GENERALIZATIONS

Birth weight is generally low—about 2,000 g—except in one case.

PHENOTYPE

Craniofacial Dysmorphism

The occiput is prominent.
Palpebral fissures are small, slightly slanting downward and outward.
The nose is pointed with a prominent bridge and small nares. The mouth is small. Very pronounced microretrognathia is present.
Ears are low-set, small and somewhat faunlike, with hypoplastic lobe and tragus.
The main variations between this phenotype and that of 18 trisomy are the birdlike appearance of the head, the severity of the micrognathia, and somewhat higher-set ears.

Neck, Thorax, and Abdomen

Shortness of the sternum, hypoplasia of the first ribs, minor vertebral anomalies, and congenital hip dislocation may be observed.

Limbs

Finger flexion is less pronounced than with 18 trisomy. The calcaneum is protuberant, and the big toe is short and in dorsal flexion. Nails are hypoplastic.

Genitalia

In one case, pseudohermaphroditism was noted, but one cannot discount the possibility that it was due to the associated partial 16 trisomy.

Malformations

Internal malformations are occasionally present, but seem to be less severe than for complete trisomy. Coarctation of the aorta, patent ductus arteriosus, and ectopic or polycystic kidneys have been observed.

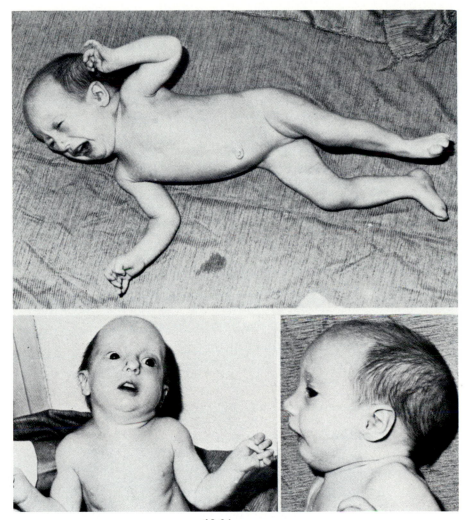

18.21

Prognosis

Two infants died, at 3 weeks and 5 months. The others are still alive at 4 years, 3 years, and 4 months. Survival thus appears to be less affected than with 18 trisomy.

CYTOGENETIC STUDY

Three of the five observations are due to a familial rearrangement, with breakpoints varying between 18q11 to 18q21.2. Two cases occurred *de novo*.

DERMATOGLYPHICS

The excess of arches found in 18 trisomy is not found here. A single palmar transverse crease may be present, as well as immaturity of the dermal ridges.

18.22

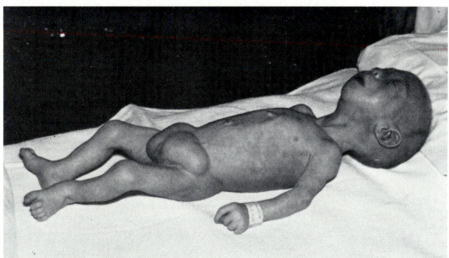

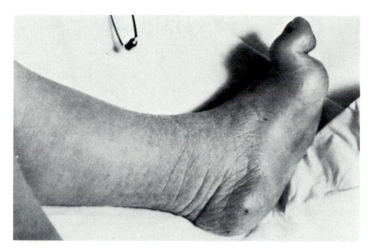

18.23

18.24

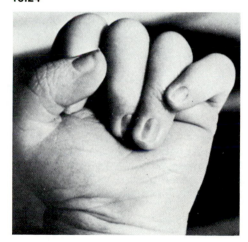

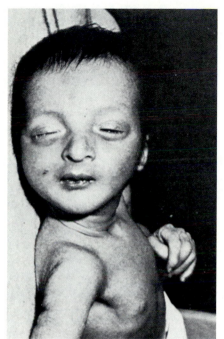

18.25

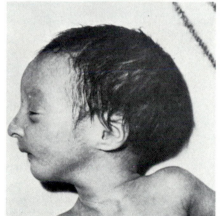

18p MONOSOMY OR 18p – SYNDROME

Small size
Round face
Broad, detached ears
Wide mouth
Dental anomalies

REFERENCES

GROUCHY J. de, LAMY M., THIEFFRY S., ARTHUIS M., SALMON C.: Dysmorphie complexe avec oligophrénie: délétion des bras courts d'un chromosome 17–18. *C.R. Acad. Sci. (Paris).* **256**:1028–1029, 1963.

GROUCHY J. de: The 18p–, 18q– and 18r syndromes. *Birth Defects: Orig. Art. Ser., V-*5:74–87, 1969.

RETHORE M.O.: Chromosome deletions and ring chromosome syndromes. *In* N.C. MYRIANTHOPOULOS, ed.: *Handbook of Clinical Neurology. Vol. 31, part II, Congenital malformations of the brain and skull.* Amsterdam, North Holland publ., pp. 549–620, 1977.

SCHINZEL A., SCHMID W., LUSCHER U., NATER M., BROOK C., STEINMANN B.: Structural aberrations of chromosome 18. I. The 18p– syndrome. *Arch. Genetik,* **47**:1–15, 1974.

UCHIDA I.A., McRAE K.N., WANG H.C., RAY M.: Familial short arm deletion of chromosome 18 concomitant with arrhinencephaly and alopecia congenita. *Amer. J. hum. Genet.,* **17**:410–419, 1965.

The deletion of the short arm of chromosome 18 was described in 1963 by Grouchy et al. This was the first deletion observed in the human species.

More than 70 cases have since been reported in the literature, and many more are no longer subject to publication (Grouchy, 1969; Schinzel et al., 1974; Rethoré, 1977).

In the majority of observations (approximately 84 percent) dysmorphism is not very characteristic, and psychomotor retardation is more or less severe. In 16 percent of the observations there is severe malformation of the cranium and face.

GENERALIZATIONS

Deletion most often occurs *de novo.*

- Sex ratio: 3F/2M
- Mean parental ages in *de novo* cases:
    maternal: 32 years
    paternal: 38 years
- Mean birth weight: 2,800 g

PHENOTYPE

In patients without severe associated cephalic malformation, there is a relatively significant diversity of clinical features. Nonetheless, a resemblance is found in a number of the patients; they are always small, and their posture is often characteristic: they stand with widespread legs, leaning slightly forward.

Craniofacial Dysmorphism

The cranial circumference is often at the lower limit of normality, but not to the extent of true microcephaly.

The face is round.

Palpebral fissures are horizontal. The nasal bridge is often flattened and may give the impression of hypertelorism, which in fact is only rarely present. Bilateral epicanthus is frequent, as well as bilateral ptosis of the eyelid and strabismus.

The upper lip is often short and juts forward. The philtrum is broad. Cupid's bow is often obliterated, and the margin of the lip is flat and broad. The lower lip is everted.

The chin, small and receding in the infant, becomes normal in the older child.

The teeth are of poor quality with frequent caries; the lateral incisors may be absent.

The ears are often low-set and posteriorly rotated. The pinnae are large, soft, floppy, and detached; the anthelix is aplastic.

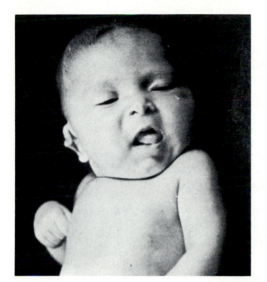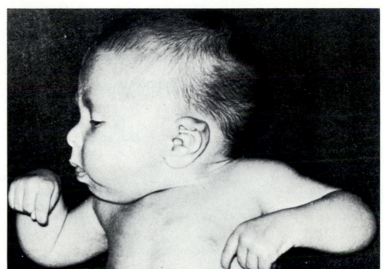

18.26

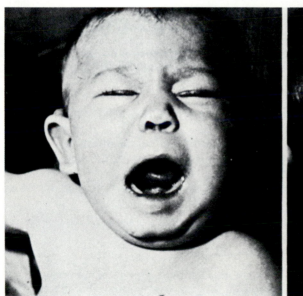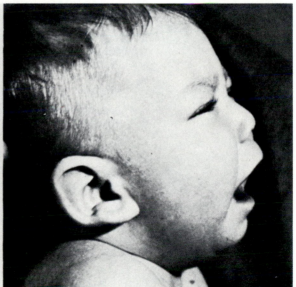

18.27

Neck, Thorax, and Abdomen

The neck is short and, in the female, often webbed. The hairline may be low.
The thorax is wide and depressed, and the nipples are too wide-spaced.
These combined features often suggest Turner's syndrome.
Diastasis of the recti abdominis or umbilical or inguinal hernias attest to the muscular hypotonia.

Limbs

The hands are broad and short. The three phalanges of the fingers are of decreasing width and are clearly separated by the flexion creases, as if they could telescope one into the other (Rethoré, 1977). Clinodactyly of the fifth digit is often present.

Edema on the backs of the hands or cubitus valgus may further contribute to the resemblance to Turner's syndrome.

Palmature of the toes, flat feet, concave feet, or rocker-bottom feet have, in rare instances, been reported.

Genitalia

Testicular ectopy, hypospadias, and small penis have, occasionally, been reported in the male.

Malformations

Very severe cephalic malformations are found in 16 percent of the cases. Most frequent is cebocephalia, with a single nasal orifice, or cyclocephalus with proboscis, accompanied, in both cases, by arhinencephaly, fusion of the frontal lobes or hemispheres, agenesia of the pituitary gland, and microphthalmia. Much rarer is lycostoma or occipital meningocele.

In the other patients, associated malformations are infrequent, but include:

- Cerebral malformations: agenesia of the corpus callosum, hydrocephaly, ventricular distention
- Ocular malformations: nystagmus, tapetoretinal degeneration, cataract, etc.
- Cardiac malformations: patent ductus arteriosus, aortic stenosis, ventricular hypertrophy, heart failure
- Diverse skeletal malformations: congenital hip dislocation, vertebral anomalies, etc.

Thyroid or growth-hormone insufficiencies and diabetes have been reported.

Mental Retardation

The intensity of mental retardation is variable. The IQ ranges between 25 and 75, with a majority of cases around 50. Verbal and manual abilities are often highly dissociated. Behavioral disorders, agitation, excessive emotivity, and fear of strangers may be present. Cases of autism, schizophrenia, hemiparesis, and deafness have been reported.

Seizures or EEG disorders are more common.

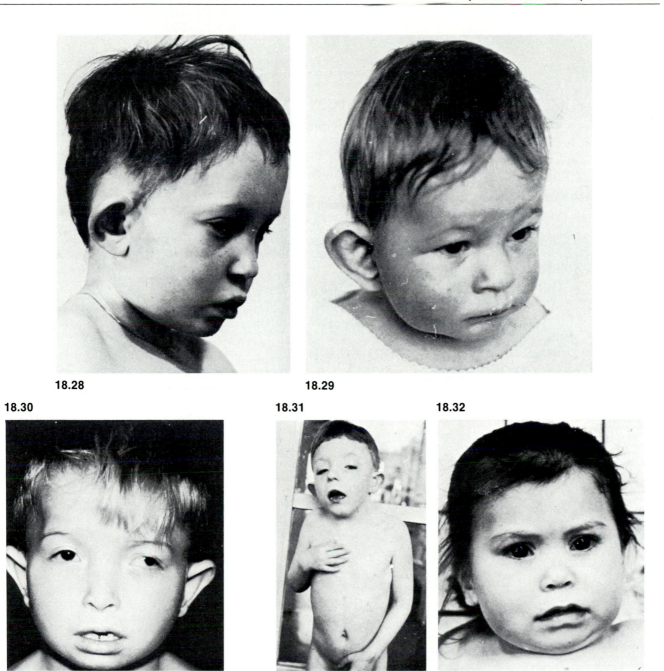

18.28

18.29

18.30

18.31

18.32

Prognosis

Apart from cases of associated severe malformations, the life expectancy of the patients does not seem diminished. One patient is 61 years old.

CYTOGENETIC STUDY

In two-thirds of the observations there is a clearcut deletion arising *de novo*. In one case, there is mosaicism.

In one-third of the observations the following are possible:

- A *de novo* translocation of chromosome 18 with loss of 18 p
- Malsegregation of a parental translocation
- A maternal mosaicism 46,XX/46,XX,18p –
- An associated sexual aneuploidy

In exceptional cases, 46,XX,18p – women, with a subnormal IQ, bore children who were also 18p – (Uchida et al., 1965).

DERMATOGLYPHICS

Palmar and digital dermatoglyphics are of no particular interest. The axial triradius can be in t′ more often than is normal.

LABORATORY FINDINGS

In half of the cases a decreased level or absence of IgA has been observed.

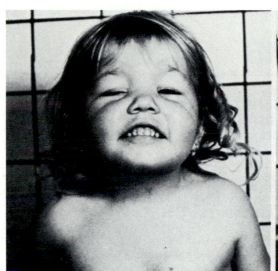

18.33

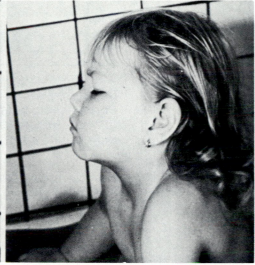

18.34

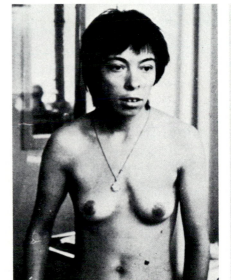

18.35

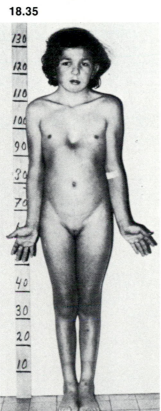

18q2 MONOSOMY OR 18q− SYNDROME

Depressed midface
Carp-shaped mouth
Strongly folded ears
High frequency of whorls

REFERENCES

GROUCHY J. de, ROYER P., SALMON C., LAMY M.: Délétion partielle des bras longs du chromosome 18. *Pathol. Biol.*, **12**:579–582, 1964.

GROUCHY J. de: The 18p−, 18q− and 18r syndromes. *Birth Defects, Orig. Art. Ser., V*-5:74–87, 1969.

LEJEUNE J., BERGER R., LAFOURCADE J., RETHORE M.O.: La délétion du bras long du chromosome 18. Individualisation d'un nouvel état morbide. *Ann. Génét.*, **9**:32–38, 1966.

RETHORE M.O.: Chromosome deletions and ring chromosome syndromes. *In* N.C. MYRIANTHOPOULOS, ed.: *Handbook of Clinical Neurology. Vol. 31, part II, Congenital malformations of the brain and skull.* Amsterdam, North Holland publ., pp. 549–620, 1977.

SCHINZEL A., HAYASHI K., SCHMID W.: Structural aberrations of chromosome 18. II. The 18q− syndrome. Report of three cases. *Humangenetik*, **26**:123–132, 1975.

SUBRT I., POKORNY J.: Familial occurrence of 18q−. *Humangenetik*, **10**:181–187, 1971.

The first observation of partial deletion of the long arm of chromosome 18 (18q−) was reported by Grouchy et al. in 1964. The clinical syndrome was delineated in 1966 by Lejeune et al. with additional data provided by two new observations. Since then, more than 50 observations have been reported in the literature (Grouchy, 1969; Schinzel et al., 1975; Rethoré, 1977). As the syndrome appears to be frequent, the publication of isolated cases is no longer justifiable.

GENERALIZATIONS

In the majority of cases the deletion occurs *de novo*.

- Sex ratio: 2M/3F
- Mean parental ages in *de novo* cases:
    maternal: 27 years
    paternal: 29.2 years
- Mean birth weight: 2,940 g
- Mean crown-heel length at birth: 48.3 cm

PHENOTYPE

In the infant, hypotonia is constant and intense. The infant lies on his back in a "froglike" position, the lower limbs flexed, externally rotated, and in hyperabduction.

Staturoponderal retardation increases with age. The crown-heel length of the patients rarely exceeds 1.50 m.

Craniofacial Dysmorphism

Craniofacial dysmorphism is notable and permits diagnosis by simple clinical examination.

The cranium is round, and microcephaly is moderate.

Observation of the profile reveals the characteristic feature of the syndrome: a depressed midface between a normal forehead and an apparent protrusion of the mandible.

Palpebral fissures are horizontal and the eyes deeply imbedded under the orbital rims. When the infant cries, the contraction of the subcutaneous muscles causes the appearance of a triangular protuberance at the inner tip of the eyebrows. Minor anomalies, such as nystagmus, epicanthus, strabismus, and ptosis, or much more severe anomalies, may be noted.

The shape of the nose is normal. The major axis of the nostrils is slanted upward

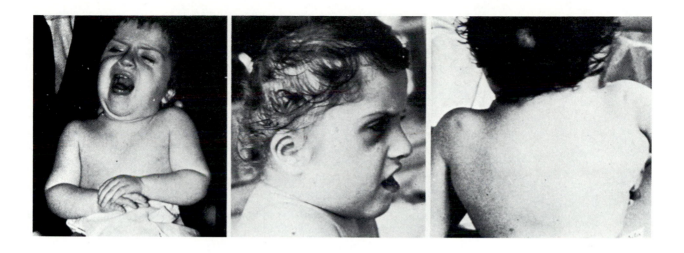

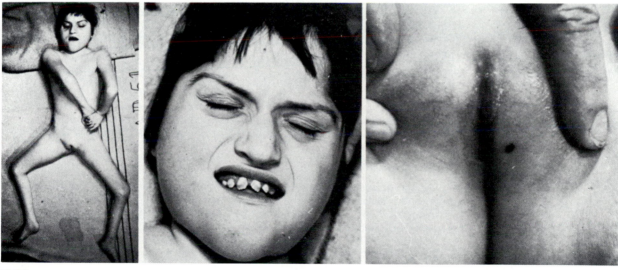

18.36

and outward. The wings of the nose are implanted so as to form, on each side, a small triangle with its base to the outside.

The upper lip is short. The philtral margins are not defined. Cupid's bow has little or no definition, giving the upper lip the shape of an arc of a circle, the vermilion being oriented toward the front. The lower lip is everted and projects beyond the upper lip; this produces the aspect of a "carplike" mouth.

A subcutaneous nodule may be present on the cheeks at the usual site of the dimples.

The chin may be slightly protuberant.

The ears are normally set and oriented. Their appearance is characteristic. The helix and anthelix are strongly developed and delimit a deep sulcus. The antitragus is protuberant. The concha is deep. The external auditory canal may be atresic.

Neck, Thorax, and Abdomen

The thorax is normal, except for too widespaced nipples and the presence of unilateral or bilateral subacromial dimples.

Limbs

Dimples are also present in the epitrochleal regions, on the back of the hands, and on the lateral faces of the knees.

The characteristic features of the hands are long, thin, tapered fingers and protuberant fingertips, shaped like "teardrops."

Implantation of the toes is disorderly. Clinodactyly, palmatures, and clubfoot has been noted.

Genitalia

The genital organs are almost always abnormal. Unilateral or bilateral testicular ectopy, hypoplastic scrotum, hypospadias, and small penis are noted in the male. In females there is more or less complete atrophy of the labia minora.

Malformations

Ocular malformations are practically constant: amblyopia, with or without nystagmus; optic atrophy; coloboma; corneal anomalies.

Osteoarticular malformations are present in half of the cases: supernumerary or hypoplastic ribs, costal synostoses, spina bifida occulta, coxa valga, etc.

Cardiac or renal malformations are not very characteristic.

Cerebral malformations have been revealed at autopsy, but they are very scarce.

Mental Retardation

The severity of mental retardation is variable. One-fourth of the patients have an IQ below 30. They are probably the most severely afflicted among carriers of chromosomal anomalies. They maintain the froglike position observed in infants and are reduced to an entirely vegetative, bedridden life.

One-fourth of the patients have an IQ between 30 and 50, another fourth between 50 and 70, and a final fourth above 70.

Psychotic behavior is often observed. Language difficulties are frequent. The patient may never speak; the possibility of associated deafness must be taken into

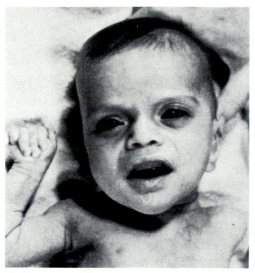
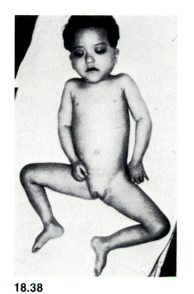
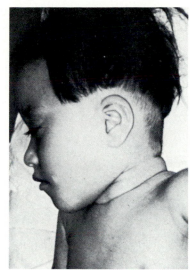

18.37

18.38

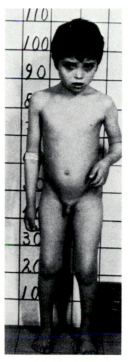
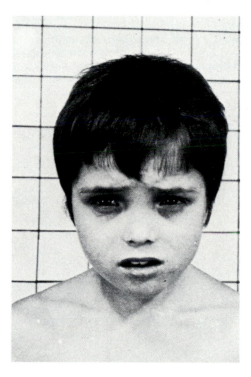
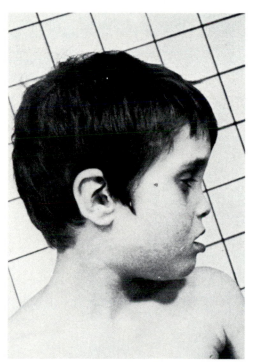

18.39

consideration. Several authors have noted the raucous and dissonant character of the voice, which has been described as a "husky voice."

Seizures have been reported.

Prognosis

In about 10 percent of the cases, the infant dies within the first few months of life. Otherwise, life expectancy seems normal. The first patient described died at the age of 14 years. Several patients have attained adult age.

CYTOGENETIC STUDY

In 80 percent of the cases, the deletion occurs *de novo.* The breakpoint is frequently located at 18q21.2.

In 10 percent of the cases, there is mosaicism and the patients may be less severely afflicted.

In 10 percent of the cases, the deletion results from a parental pericentric inversion or translocation.

A female carrier of 18q– has reportedly had six children, including a pair of twins, both with the deletion (Subrt and Pokorny, 1971).

DERMATOGLYPHICS

The only constant feature is an excess of whorls on the fingertips. Ten percent of the patients have 10 whorls, the excess occurring at the expense of arches. Of 40 patients examined, only one had an arch on a finger. This provides an excellent example of "type and countertype" when compared to trisomy 18.

The remainder of the palmar dermatoglyphics do not show any significant features.

LABORATORY FINDINGS

A deficiency in IgA has been observed in one-third of the cases.

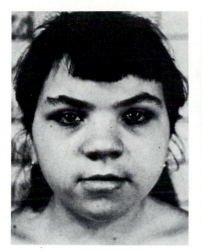
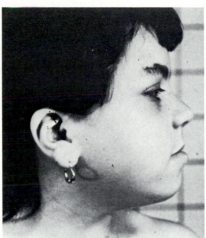
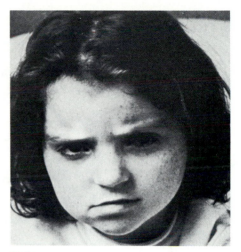

18.40   18.41

18.42

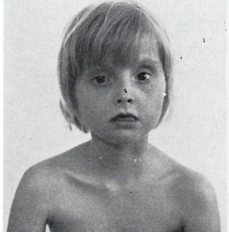
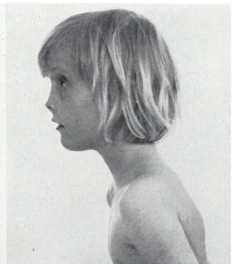

r(18)

Predominance of the features of 18q monosomy

REFERENCES

GROUCHY J. de: The 18p−, 18q− and 18r syndromes. *Birth Defects, Orig. Art. Ser., V-5*: 74−87, 1969.

RETHORE M.O.: Chromosome deletions and ring chromosome syndromes. *In* N.C. MYRIANTHOPOULOS, ed.: *Handbook of Clinical Neurology. Vol. 31, part II, Congenital malformations of the brain and skull.* Amsterdam, North Holland publ., pp. 549−620, 1977.

WANG H.C., MELNYCK J., McDONALD L.T., UCHIDA I.A., CARR D.H., GOLDBERG B.: Ring chromosomes in human beings. *Nature*, **195**:733−734, 1962.

The first observation of r(18) was reported in 1962 by Wang et al. The r(18) phenotype shares features of the 18p− syndrome and those of the 18q− syndrome. On the whole, features of the 18q− syndrome seem to predominate (Grouchy, 1969; Rethoré, 1977).

GENERALIZATIONS

- Sex ratio: 3F/2M
- Mean maternal age: 26.7 years
- Mean paternal age: 30.5 years
- Mean birth weight: 2,700 g

PHENOTYPE

Crown-heel length is usually below normal.

Craniofacial Dysmorphism

Microcephaly is noted in two-thirds of the cases.

Facial dysmorphism may be reminiscent of 18q monosomy because of a certain degree of depression of the midface, and especially the very characteristic carplike mouth.

Hypertelorism and bilateral epicanthus are present in half the cases. Ocular malformations are present in one-fourth to one-third of the cases, and include strabismus, ptosis, coloboma, anomalies of the fundi oculi, and nystagmus.

The ears are often strongly folded, as in 18q monosomy.

A high-arched or cleft palate is reported in nearly half the cases.

Neck, Thorax, and Abdomen

Features reminiscent of Turner's syndrome may be present: pterygium colli, shortness of the neck, and funnel-shaped thorax.

Limbs

Micromelia, clinodactyly of the fifth finger, and disorderly implantation of the toes may be observed.

Genitalia

The external genitalia are abnormal in one-fifth of the cases.

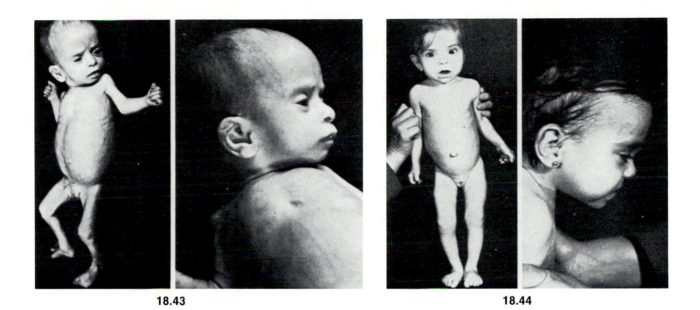

18.43

18.44

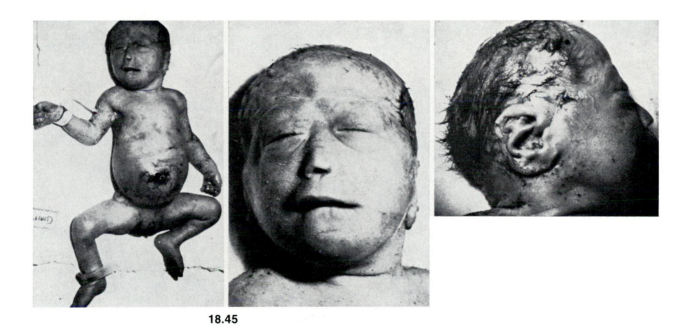

18.45

Malformations

Severe malformations have been reported, essentially including cebocephalia with arhinencephaly in 7 percent of the cases, and microphthalmia in 2 percent of the cases.

Mental Retardation

Mental retardation is constantly found. Its severity seems to lie between those observed in the 18q− and 18p− syndromes.

Prognosis

Life expectancy is usually normal.

CYTOGENETIC STUDY

In almost every case, r(18) occurs *de novo*. In one case, the anomaly was transmitted from mother to child. In one-tenth of the cases, there was mosaicism: 46,XX(Y)/ 46,XX(Y),r(18).

DERMATOGLYPHICS

The major feature is an excess of whorls: more than five were found in 40 percent of the cases.

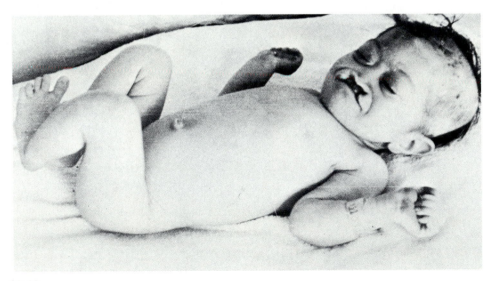

18.46

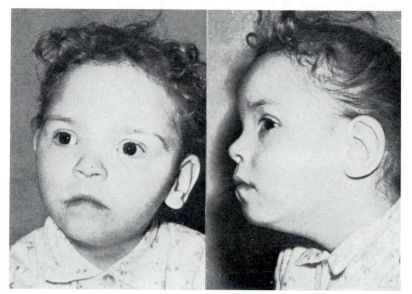

18.47

18.48

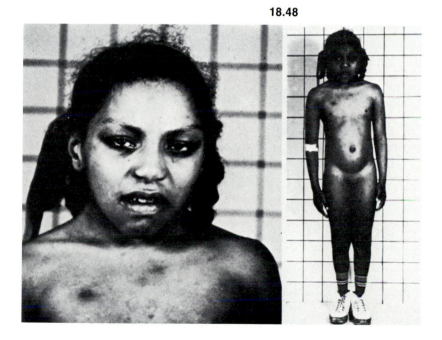

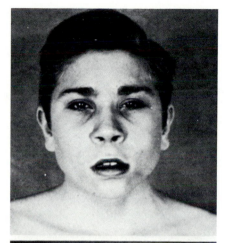

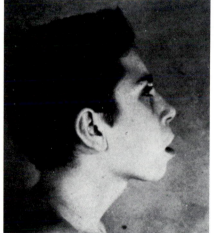

18.49

CHROMOSOME 19

Chromosome 19 is a small, but undoubtedly important chromosome, since very few cases of imbalance of this chromosome are known. Gene assignments include, among others, the gene coding for mannosidase whose mutation causes mannosidosis, and a site which is sensitive to the polio virus.

Chromosomal disorders are represented by 19q trisomy, of which only six observations are known.

A polymorphism of the centromere of 19 is visible in C-banding.

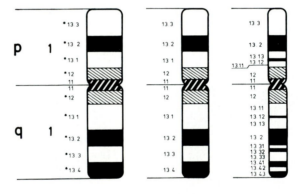

GENE MAP

19pter-19q13	MANB		Lysosomal alpha-D-mannosi-dase-B (24850) S
19pter-19q13	GPI		Glucose phosphate isomerase (17240) S, D
19q	PVS		Polio virus sensitivity (17385) S
19q	E11S	(L)	Echo 11 sensitivity (12915) S

19pter-19q13	PEPD		Peptidase D (17010) S
	M7VS1	(P)	Baboon M7 virus receptor (?same as virus RD114 receptor) (10918) S
	CB3S	(P)	Coxsackie B3 virus susceptibility (12005) S
	BCT2	(P)	Branched chain amino acid transaminase-2 (11353) S
19pter-19q13	DNL	(P)	Lysosomal DNAase (12635) S
	HC	(L)	Hypercholesterolemia (14389) F
	GUSM	(P)	Beta-glucuronidase, mouse, modifier of (23161) S
Probable order:	C3	(P)	Complement component-3, fibroblast (12070) S (?same as serum C3)
C3-Le-DM-Se-Lu	Le	(P)	Lewis blood group (11110) F (linked to serum C3)
	DM	(P)	Myotonic dystrophy (16090) F (in C3-Le-Se-Lu linkage group)
	Se	(P)	Secretor (18210) F
	Lu	(P)	Lutheran blood group (11120) ⎤
	Hh	(P)	Bombay phenotype (Hh) (21110) ⎦ (v,F)
	APOE	(P)	Apolipoprotein E (20776) F
	NF1	(P)	Neurofibromatosis (16220) F
	FEN	(P)	Ferritin (13479) S
	CGB	(P)	Chorionic gonadotropin, beta chain (11886) REa

MORPHOLOGY AND BANDING

QFQ　　GTG　　RHG　　RBA　　THA　　C

19q TRISOMY

REFERENCES

LANGE M., ALFI O.S.: Trisomy 19q. *Ann. Génét.,* **19**:17–21, 1976.

PANGALOS C., GHICA M., COUTURIER J.: Trisomy 19qter associated with partial trisomy 22q in two members of the same family. Analysis using the thymidine + RBA banding technique. *In: 6th International Congress of Human Genetics, Abstract, Jerusalem, September 13–18, 1981,* p. 171, 1981.

SCHMID W.: Trisomy for the distal third of the long arm of chromosome 19 in brother and sister. *Hum. Genet.,* **46**:263–270, 1979.

The description of the phenotype of 19q trisomy is based on the observations of two couples of sibs (Lange and Alfi, 1976; Schmid, 1979), and one couple that included an uncle and his niece (Pangalos et al., 1981).

GENERALIZATIONS

- Sex ratio: 3M/3F
- Mean birth weight: 2,240 g (4 children)
- Mean crown-heel length at birth: 41.25 cm (4 children)
- Mean head circumference at birth: 31 cm (4 children)

PHENOTYPE

Craniofacial Dysmorphism

These children are microcephalic and brachycephalic. The face is flat. Cranial sutures are widely gaping. Palpebral fissures slant downward and outward. Hypertelorism and bilateral palpebral ptosis are present, as well as protuberance of the glabella in two sibs.

The nose is small and snubbed.

The philtrum is short and prominent.

The mouth is fishlike; corners of the lips are down-turned. Cleft palate is found in three children.

The ears are low-set, in posterior rotation, with a sizable folding of the helix.

Neck, Thorax, and Abdomen

The neck is short with an excess of skin.

The thorax is barrel-shaped, with widely spaced nipples. A sacral dimple is noted in three children. Diastasis recti and kyphosis may also be noted.

Limbs

Hands and feet are small and pudgy. In two cases, clinodactyly of the Vth finger is present, and in one case, a unilateral bifid thumb. Two other infants have talipes valgus and laterally curved big toes.

Genitalia

In boys, cryptorchidism, hypospadias, and testicular hypoplasia are present.

Malformations

In the patients described by Lange and Alfi, the autopsy revealed severe heart disease, anomalies of the bronchial tree, hypoplasia of the gallbladder, bilateral diaphragmatic eventration, malrotation of the intestines (in one case), polycystic kidneys, and small gonads. The other patients are still alive; one of them has mesocardia

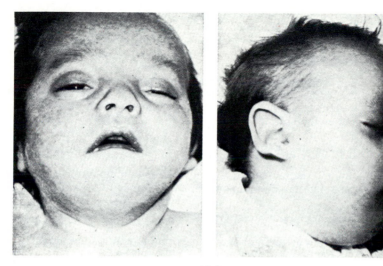

19.1

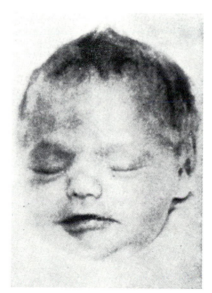

19.2

with a systolic murmur, renal difficulties, vertebral anomalies; another patient suffered from anal atresia.

Prognosis

The neonatal period is marked by feeding difficulties, seizures, stridor, and apnea. Growth is very retarded, falling well below the third percentile. Two sibs died, at 6 months and 2 years of age (Lange and Alfi, 1976). Mental retardation varies from very severe to slight.

CYTOGENETIC STUDY

Each case involved a familial translocation, with a 22 in 2 cases and a 20 in one case. The breakpoint is in q13.

DERMATOGLYPHICS

An excess of arches was reported in two of Schmid's patients.

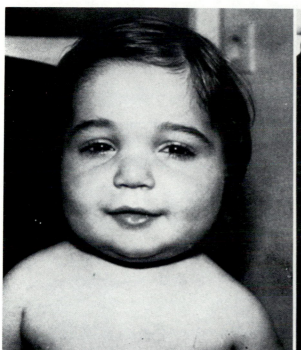
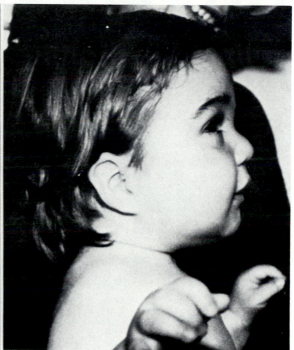

19.3

19.4

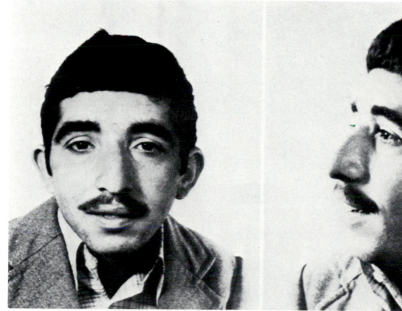

Gene assignments and chromosomal diseases are not very abundant. Two enzyme gene assignments are proven: inosine triphosphatase (ITPA) and adenosine deaminase (ADA).

The assignment for Sipple syndrome, or multiple neoplasia of the endocrine glands, is based on seven families in which carriers of a short intercalary deletion of 20p are found. Should this observation be confirmed, one will be confronted with a situation similar to that of retinoblastoma (chromosome 13) and of aniridia (chromosome 11).

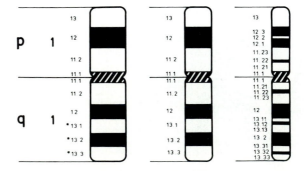

GENE MAP

20p	ITPA		Inosine triphosphatase (14752) S
20p11	FS		Fragile site (13658)
20p12.2	MEN2	(P)	Multiple endocrine neoplasia, type 2 (17140) Ch
20q13.2-qter	ADA		Adenosine deaminase (10270) S, D
	DCE	(P)	Desmosterol-to-cholesterol enzyme (12565) F
	SAHH	(P)	S-adenosylhomocysteine hydrolase (18089) S
	SRC		Protooncogene *src* (Rous sarcoma) (19009) REa

MORPHOLOGY AND BANDING

QFQ   GTG   RHG   RBA   THA   C

20p TRISOMY

Round face
High cheekbones
Elongated palpebral fissures
Normal growth

REFERENCES

ARCHIDIACONO N., TECILAZICH D., TONINI G., ROCCHI M., FILIPPI G.: Trisomy 20p from maternal t(3;20) translocation. *J. med. Genet.*, **16**:229–232, 1979.

CENTERWALL W., FRANCKE U.: Familial trisomy 20p. Five cases and two carriers in three generations. A review. *Ann. Génét.*, **20**:77–83, 1977.

DELICADO A., LOPEZ PAJARES I., VICENTE P., GRACIA R.: Partial trisomy 20. *Ann. Génét.*, **24**:54–56, 1981.

FRANCKE U.: Abnormalities of chromosomes 11 and 20. *In* J.J. YUNIS: *New Chromosomal Syndromes*, New York, Academic Press, pp. 245–271, 1977.

SCHINZEL A.: Trisomy 20pter→q11 in a malformed boy from a t(13;20) (p11;q11) translocation carrier mother. *Hum. Genet.*, **53**:169–172, 1980.

The first 13 observations of 20p trisomy were reviewed by Francke (1977) and Centerwall and Francke (1977). Four other cases have been published since then (Archidiacono et al., 1979; Schinzel, 1980; Delicado et al., 1981).

GENERALIZATIONS

- Sex ratio: 1M/1F
- Mean birth weight: 3,050 g

PHENOTYPE

Contrary to most carriers of a chromosomal anomaly, these patients have normal pre- and postnatal growth development.
Facial dysmorphism is relatively minor and variable.

Craniofacial Dysmorphism

The face is round with high cheekbones. Hair is thick.
Palpebral fissures slant upward and outward, with a fold in the upper eyelid and a slight enophthalmia. Hypertelorism is present, and strabismus is frequently encountered.
The nose is short and snubbed, with large nares.
The philtrum is somewhat short. Tooth decay and dental anomalies are frequent.
The chin is small.

Neck, Thorax, and Abdomen

Vertebral deformities may be present, as well as bone anomalies that show up on X-ray films.

Limbs

Abnormal implantation of the toes and clubfoot have, in rare cases, been reported.

Genitalia

These are generally normal, but hypogonadism and hypospadias may occasionally be seen.

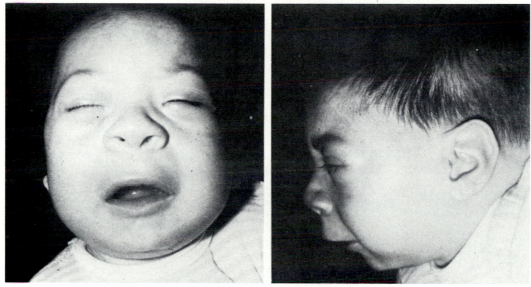

20.1

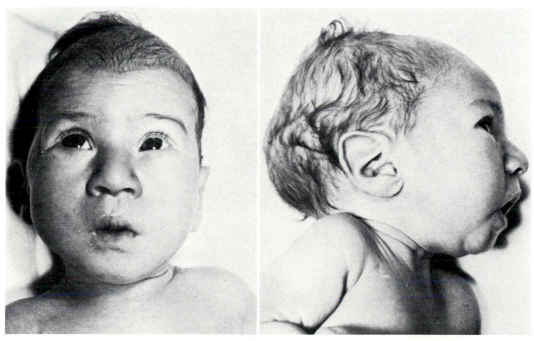

20.2

Malformations

Malformations are rare. Heart and kidney anomalies have been described. Radiological anomalies of the vertebrae were observed in most cases. Severe spondylar dysplasia was reported in one case.

Mental Retardation

Mental retardation is variable, with a mean IQ of 50. Language and coordination difficulties are usually present.

Prognosis

The vital prognosis is good. The oldest patient is 45. Early deaths are the exception.

CYTOGENETIC STUDY

Trisomy always results from a parental translocation. The breakpoint is variable, but is generally in p11.

DERMATOGLYPHICS

An excess of arches is noted.

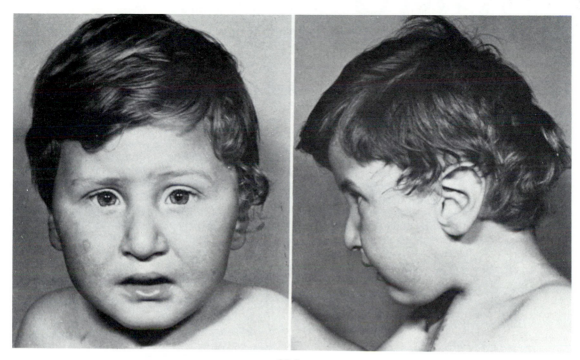

20.3
20.4

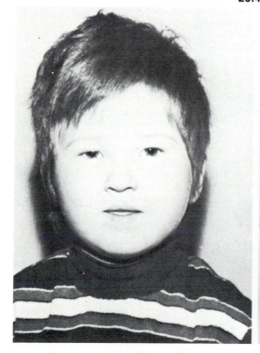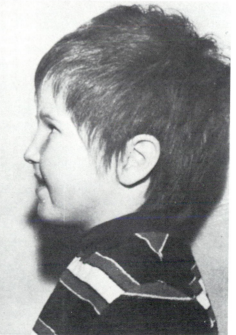

CHROMOSOME 21

REFERENCE

LEJEUNE J.: Investigations biochimiques et trisomie 21. *Ann. Génét.*, **22**:67–75, 1979.

Chromosome 21 has an important historical connotation, for it was the discovery of Lejeune et al. in 1959 of 21 trisomy which marked the beginning of human cytogenetics.

Known gene assignments include, among others, the gene for superoxide dismutase-1 (SOD1), the first enzyme for which gene dosage effect was demonstrated (Sinet et al., 1976), that of phosphoribosylglycinamide synthetase, and that of the hepatic (and erythrocytic) phosphofructokinase. These latter two enzymes also show a gene dosage effect. The gene for the interferon receptor, or antiviral protein, is also localized on 21, and also shows a gene dosage effect.

21 trisomy remains the most frequently encountered of the autosomal anomalies. It has served as a model for a number of etiological, epidemiological, and biochemical studies. It is possible that it will also be one of the first to respond to therapy based on a knowledge of biochemical imbalance (Lejeune, 1979).

Partial trisomies permitting the regional localization of both the "phenotype" for 21 trisomy and for marker genes are also known, as are monosomies.

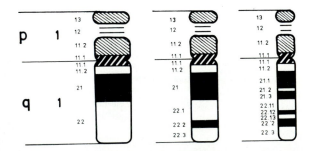

GENE MAP

21p12	RNR		Ribosomal RNA (18045) A
21q21-21qter	AVP		Antiviral protein (interferon receptor) (10745) S, D
21q21	SOD1		Superoxide dismutase-1 (soluble) (14745) S, D
	PRGS		Phosphoribosylglycinamide synthetase (13844) S, H
	PRAIS	(P)	Phosphoribosylaminoimidazole synthetase (17244) S
	S14	(P)	Surface antigen 14 (18559) S
	PFKL		Phosphofructokinase, liver type (17186) S, D
21q11-qter	THC	(P)	Primary thrombocytosis (18795) Ch
	AABT	(L)	Beta-amino acids, renal transport of (10966) Ch
	LDLR	(L)	Low density lipoprotein receptor (14389) S (on 5 or 21 or both)
	BAS	(L)	Beta adrenergic stimulation, response to (10967) D
	HTOR	(P)	5-hydroxytryptamine oxygenase regulator (14346) D

MORPHOLOGY AND BANDING

QFQ   GTG   RHG   RBA   THA   C

21 TRISOMY

Hypotonia
Round flat face
Palpebral fissures slanted upwards and outwards
Brushfield spots on the irides
Small ears
Flat nape

REFERENCES

BERGER R.: Trisomie 21. *Rev. Pédiat.,* **8**:287–301, 1972.

BOTT C.E., SEKHON G.S., LUBS H.A.: Unexpected high frequency of paternal origin of trisomy 21. *Proc. Amer. Soc. Hum. Genet. 27th Annual Meeting, Baltimore, Oct. 8–11, 1975,* p. 20A, 1975.

COX R.P.: Activity and regulation of alkaline phosphatase in cells from patients with abnormal chromosome complements. *Ann. N.Y. Acad. Sci.,* **166**:406–416, 1969.

DEKABAN A.: Twins probably monozygotic: one mongoloid with 48 chromosomes, the other normal. *Cytogenetics,* **4**:227–239, 1965.

DUTRILLAUX B.: Les aberrations chromosomiques transmissibles. In *Journées Pédiatriques Parisiennes,* pp. 13–21. Paris, Flammarion Médecine Sciences, 1972.

GERMAN J.: Mongolism, delayed fertilization and human sexual behaviour. *Nature,* **217**:516–518, 1968.

GIRAUD F., MATTEI J.F.: Aspects épidémiologiques de la trisomie 21. *J. Génét. hum.,* **23**:1–30, 1975.

GROUCHY J. de: Clinical cytogenetics. In *The Cell Nucleus, vol. 2.* New York, Academic Press, pp. 373–436, 1974.

HAMERTON J.L.: *Human Cytogenetics, Clinical Cytogenetics. Vol. II.* New York, Academic Press, 1971.

HAMERTON J.L., BRIGGS S.M., GIANELLI F., CARTER C.O.: Chromosome studies in detection of parents with high risk of second child with Down's syndrome. *Lancet,* **ii**:788, 1961.

HARA Y., SASAKI M.: A note on the origin of extra chromosomes in trisomies 13 and 21. *Proc. Japan Acad.,* **51**:295–299, 1975.

HOLT S.B.: Fingerprints in mongolism. *Ann. hum. Genet.,* **27**:279–282, 1964.

HSIA D.Y.Y., NADLER H.L., SHIH L.Y.: Biochemical changes in chromosomal abnormalities. *Ann. N.Y. Acad. Sci.,* **155**:716–736, 1968.

JEROME H., LEJEUNE J., TURPIN R.: Etude de l'excrétion urinaire de certains métabolites du tryptophane chez les enfants mongoliens. *C.R. Acad. Sci. (Paris),* **251**:474–476, 1960.

KAMOUN P., LAFOURCADE J., JEROME H.: Catabolism of serotonin in Down's syndrome (trisomy 21). *Biomedicine,* **21**:426–428, 1974.

Twenty-one trisomy was the first chromosomal aberration to be described in man (Lejeune et al., 1959). It is the most frequent autosomal anomaly.

The first clinical description of the disease appears to be that by Seguin who, in 1846, used the name "furfuraceous idiocy." In 1866, Down redescribed it as "mongolian idiocy."

The term "mongolism" has been justifiably criticized, particularly for its racial implications. Anglo-Saxon authors most frequently designate this disease by the term "Down's syndrome." Other authors prefer the more precise term, 21 trisomy.

GENERALIZATIONS

- Frequency: 1.45 per 1,000, or about 1 for every 700 births. The frequency seems constant, regardless of the ethnic or socioeconomic groups considered.
- Sex ratio: 3M/2F
- Mean maternal age: 34.4 years. The mean maternal age in the general population is 28.2 years. An elevated maternal age is thus clearly demonstrated.

The risk of a child being born with 21 trisomy increases exponentially with maternal age. Whereas it is on the order of 1 per 2,000 at 20 years, it increases slightly up to age 30; it is 1 per 300 at age 35, 1 per 100 between 40 and 45, and attains 1 per 50 after age 45 (Penrose, 1933; Lilienfield and Benesch, 1969; Mikkelsen, 1972).

The distribution curve of maternal ages is bimodal, with one peak near 28 years and the other near 36–37 years. The first corresponds to the maximum peak for births and includes the majority of sporadic or inherited translocations. The second peak seems strongly correlated with maternal age.

The reason for the maternal-age effect is not known. Several hypotheses have been proposed, in particular that of an "aging" of the ovum, due, for example, to the spacing of sexual relations in older couples (German, 1968), but, at present, no satisfactory hypothesis has been found.

- Paternal age: there is no increase in paternal age for a given maternal age.

Through the existence of chromosome 21 markers (21p− or 21p+), it is now possible to determine, in a limited number of cases, whether nondisjunction occurred in maternal or paternal gametogenesis. It appears that some 25 percent of nondisjunctions are of paternal origin (Bott, 1975; Hara and Sasaki, 1975; Mikkelsen, 1981). If one assumes that the rate of nondisjunction is similar in males and females, then elevated maternal age would be responsible for approximately 50 percent of the cases of nondisjunction.

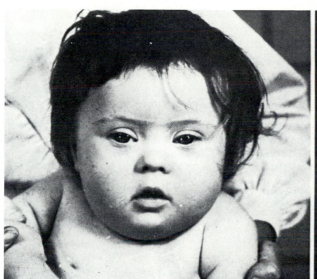
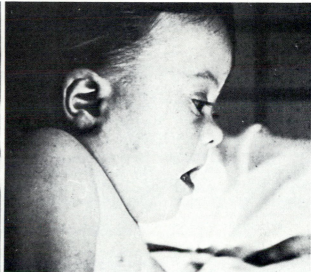
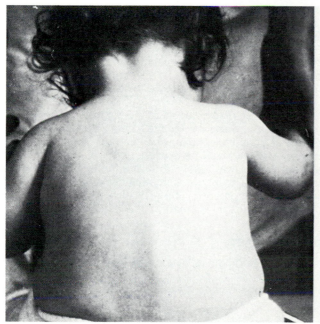
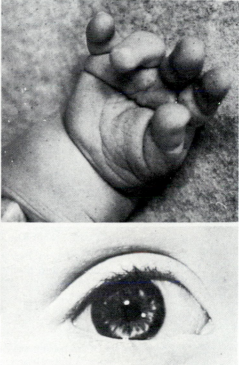

21.1

KJESSLER B., LA CHAPELLE A. de: Meiosis and spermatogenesis in two post-pubertal males, with Down's syndrome: 47,XY,G+. *Clin. Genet.*, **2**:50–57, 1971.

LAFOURCADE J., RETHORE M.O.: Les dermatoglyphes. In *Journées Parisiennes de Pédiatrie 1967*. Paris, Flammarion Médecine Sciences, 1967.

LEJEUNE J.: Réflexions sur la débilité de l'intelligence des enfants trisomiques 21. *C. R. Acad. Sci. Pontif.*, **3**:1–12, 1975.

LEJEUNE J., GAUTIER M., TURPIN R.: Les chromosomes humains en culture de tissus. *C. R. Acad. Sci. (Paris)*, **248**:602–603, 1959.

LEJEUNE J., LAFOURCADE J., BERGER R.: Les aberrations autosomiques. In *Maladies Humaines par Aberrations Chromosomiques. XXᵉ Congrès des Pédiatres de langue française, Nancy, Sept. 1965. Tome III*, Paris, Expansion Scientifique, 1965.

LEJEUNE J., LAFOURCADE J., SCHARER K., WOLFF E. de, SALMON C., HAINES M., TURPIN R.: Monozygotisme hétérocaryote, jumeau normal et jumeau trisomique 21. *C. R. Acad. Sci. (Paris)*, **254**:4404–4406, 1962.

LILIENFELD A.M., BENESCH C.H.: *Epidemiology of Mongolism*. Baltimore, The Johns Hopkins Press, 1969.

MIKKELSEN M.: Familial Down's syndrome. A cytogenetical and genealogical study of twenty-two families. *Ann. hum. Genet.*, **30**:125–146, 1966.

MIKKELSEN M.: The effect of maternal age on the incidence of Down's syndrome. *Humangenetik*, **16**:141–146, 1972.

MIKKELSEN M.: Epidemiology of trisomy 21: population, peri- and antenatal data. *In* G.R. BURGIO et al., eds: *Trisomy 21, An International Symposium*. Springer-Verlag, pp. 211–226, 1981.

NIEBUHR E.: Down syndrome: the possibility of a pathogenetic segment on chromosome n° 21. *Humangenetik*, **21**:99–101, 1974.

PENROSE L.S.: The relative effects of paternal and maternal age in mongolism. *J. Genet.*, **27**:219–224, 1933.

PENROSE L.S.: Fingerprints, palms and chromosomes. *Nature*, **197**:933–938, 1963.

PENROSE L.S., SMITH G.F.: *Down's anomaly*. Boston, Little, Brown, 1966.

PRIEUR M.: *Etude statistique du quotient intellectuel de 474 enfants atteints de trisomie 21*. Doctor's Thesis, Paris, 29 p., 1968.

PRIEST J.H.: Parental dermatoglyphics in age-independent mongolism. *J. med. Genet.*, **6**:304, 1969.

RAOUL O., DUTRILLAUX B., CARPENTIER S., MALLET R., LEJEUNE J.: Trisomies partielles du chromosome 21 par translocation maternelle t(15;21)(q26.2;21). *Ann. Génét.*, **19**:187–190, 1976.

RETHORE M.O., LAFOURCADE J., PRIEUR M., CRUVEILLER J., TANZY M., LEJEUNE J.: Mère et fille trisomiques 21 libres. *Ann. Génét.* **13**:42–45, 1970.

SCHARRER S., STENGEL-RUTKOWSKI S., RODEWALD-RUDESCU A., ERDLEN E., ZANG K.D.: Reproduction in a female patient with Down's syndrome. Case report of a 46,XY child showing slight phenotypical anomalies, born to a 47,XX,+21 mother. *Humangenetik*, **26**:207–214, 1975.

SINET P., COUTURIER J., DUTRILLAUX B., POISSONNIER M., RAOUL O., RETHORE M.O., ALLARD D., LEJEUNE J., JÉROME H.: Trisomie 21 et superoxyde dismutase-1 (IPO-A). Tentative de localisation sur la sous-bande 21q22.1. *Exp. Cell Res.*, **97**:47–55, 1976.

- Pregnancy: mean duration of 270 days (instead of 282 days)
- Mean birth weight: 2,900 g
- Mean crown-heel length at birth: slightly decreased
- Mean cranial circumference at birth: less than or equal to 32 cm in 40 percent of the cases

PHENOTYPE

The phenotype of 21 trisomy is now well known. Although clinical diagnosis in the child offers no difficulty, it can at times be uncertain in the newborn (Lejeune et al., 1965; Penrose and Smith, 1966; Berger, 1972; Grouchy, 1974).

The infants are always very hypotonic; this constant feature is often accompanied by ligamental hyperlaxity.

The skin is marbelized and rough. In cases of cardiac malformation, cyanosis of the lips and fingers can be observed.

Craniofacial Dysmorphism

The cranium is small and round; the occiput is flat. The nape of the neck is short, flat, and broad, with an excess of skin: this outstanding feature often enables diagnosis upon viewing the child from the back or in three-quarters profile.

The face is round and has a flat profile.

The forehead is normally convex.

Palpebral fissures are very clearly slanted upward and outward. Their inner angle is masked by epicanthal folds, which often give the false impression that hypertelorism is present. The eyelashes are sparse and short. Blepharitis is frequent. Strabismus is frequently present.

The irides, especially when they are blue, display an almost pathognomonic feature: the presence of Brushfield spots. These small, round, whitish spots, somewhat irregular, form a ring at the junction of the middle third and the external third of the iris.

The nasal bridge is flat because of hypoplasia of the subtending bones. The nose is short, and the nostrils are visible when the face is viewed from the front.

The mouth is small; the lips are thick and may be fissured. The tongue is often thick and protruding. Glossitis exfoliativa may be present.

The ears are small and round. The upper rim of the helix forms a horizontal fold. The height of the ear is reduced. The lobule is small and adherent. A very characteristic feature is the abnormal development of the root of helix, which extends entirely across the concha. The auditory canal is small.

Neck, Thorax, and Abdomen

Owing to hypotonia, the abdomen is distended, often with diastasis recti abdominis and umbilical hernia.

The pelvis is rather small, and iliac and acetabular angles are always decreased on X-ray films.

Limbs

The hands are broad and stocky. The fingers are short, especially the fifth digit and the thumb. Brachymesophalangia and clinodactyly of the fifth digit are very characteristic, and there is sometimes a single flexion crease.

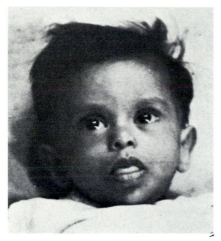

21.2

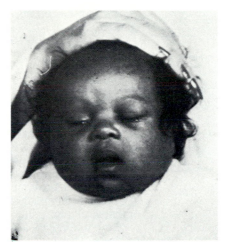

21.3

21.4

21.5

21.6

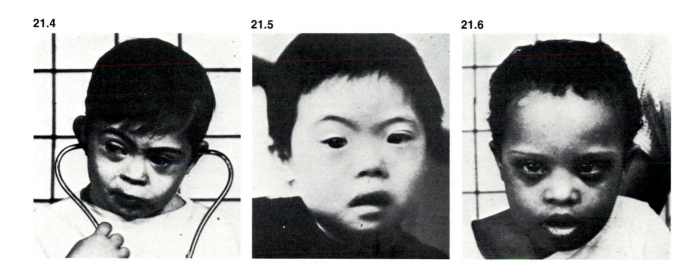

STENE T.: Detection of higher recurrence risk for age dependent chromosome abnormalities with an application to trisomy G (Down's syndrome). *Hum. Hered.*, **20**:112–122, 1970.

TURPIN R., BERNYER G.: De l'influence de l'hérédité sur la formule d'Arneth (cas particulier du mongolisme). *Rev. Hémat.*, **2**:189–206, 1947.

WILLIAMS J.D., SUMMITT R.L., MARTENS P.R., KIMBRELL R.A.: Familial Down syndrome due to t(10;21) translocation; evidence that the Down phenotype is related to trisomy of a specific segment of chromosome 21. *Amer. J. hum. Genet.*, **27**:478–485, 1975.

WOODFORD F.P., BEARN A.G.: A critical examination of some reported biochemical abnormalities in mongolism. *Ann. N.Y. Acad. Sci.*, **171**:551–558, 1970.

The feet are broad, small, and flat, with short toes. The first two toes are widely spaced.

Genitalia

The genitalia are normal.

Malformations

Visceral malformations are frequent; the most important are cardiac malformations, which afflict 40 percent of the patients. In order of decreasing frequency, these malformations are atrioventricular canal, ventricular septal defect, atrial septal defect, and patent ductus arteriosus.

Digestive malformations essentially include duodenal stenosis. Inversely, one-third of duodenal stenoses are seen in 21 trisomic patients. Annular pancreas, anal atresia, megacolon, and rectal prolapsus are also observed.

Bone anomalies are suggestive; those of the pelvis, already noted, are the most frequent. Also noteworthy is the shortness of the second phalange of the fifth digit, the presence of two ossification points of the sternal manubrium, the absence of the twelfth rib, inconstant microcephaly, absence of the frontal sinuses, and persistence of the metopic suture. Bone age is often slightly delayed.

No typical anomaly of the central nervous system has been demonstrated.

Mental Retardation

Mental deficiency in 21 trisomy varies widely and as a function of age. On the average, the IQ is 50 at the age of 5 years. It then degenerates progressively to a mean value of 38 at around the age of 15. Before the age of 5, the IQ is more difficult to determine; it degenerates to 50 between 2 and 3 years, increases to 58 between 3 and 4 years, and then degenerates again (Prieur, 1968).

Patients may be seen on either side of the mean: those with relatively high IQs, on the order of 70–80, and those condemned to a vegetative life.

Generally speaking, it is the abstract reasoning faculties that are most afflicted, whereas affectivity and social ability are fairly normal, at least in infants. In adults, the acquisition of a certain number of automatisms may mask the extent of mental retardation.

Prognosis

In weight and especially in stature the patients remain below normal. The mean height of the trisomic adult is 154 cm in the male, and 144 cm in the female.

Facial dysmorphism is modified with time. The "roundness" of the face is lessened, as is the epicanthus and flattening of the nasal bridge. A reddening of the cheeks appears, and the face takes on a wizened appearance. The voice is raucous.

The onset of puberty is normal in both sexes. Females are fertile, though their libido seems to be poor. Some 20 cases of maternity are known in 21 trisomic females. As predicted by genetic theory, an equal number of chromosome 21 disomic and trisomic infants are observed in their progeny (Rethoré et al., 1970; Scharrer et al., 1975).

No case of paternity has been proven. Testicular histology does not allow considering male patients as sterile. If decreased fertility exists, it is probably due to faulty postmeiotic maturation (Kjessler and La Chapelle, 1971).

21.7

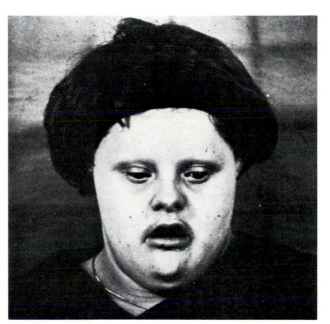

21.8

As the patient grows older, premature aging, frequent psychotic disorders, and senile cataracts appear.

Life expectancy depends on the presence of cardiopathies and digestive malformations, sensitivity to infections, and the increased risk of leukemia or cancers.

Twenty-five to 30 percent of the patients die within the first year, 50 percent before 5 years. In the other age groups, fatality remains five to six times higher than in the general population. Eight percent of patients survive beyond 40 years, and 2.6 percent beyond 50 years. Mean life expectancy at birth is 16.2 years. It increases to 22.7 years when the infant is 1 year old, and again increases to 26.7 years when the child is 5 to 9 years old.

CYTOGENETIC STUDY

Although cytogenetic examination usually confirms the clinical diagnosis, it is indispensable for genetic counseling.

The majority of 21 trisomies are free trisomies, which can be considered as isolated accidents. A minority of trisomies are due to translocations, which may be familial.

Free 21 Trisomy

Free 21 trisomy occurs in 92.5 percent of the cases. It results from nondisjunction, occurring at the time of the first or second meiotic division, whether maternal or paternal (see above).

Translocations

The frequencies indicated are those given by Giraud and Mattei (1975) based on 4,760 cases of 21 trisomy.

Translocations account for 4.8 percent of 21 trisomies.

The most frequent are t(Dq21q), which represent 54.2 percent of all translocations. The next most frequent are t(21qGq), with a frequency of 40.9 percent. The other translocations account for the remaining 4.9 percent.

In order of frequency, t(Dq21q) comprise t(14q21q): 58.5 percent; t(13q21q): 22 percent; t(15q21): 19.5 percent.

In 55 percent of the cases t(Dq21q) arise *de novo*.

In 45 percent a balanced translocation is present in one of the parents.

The t(21qGq) comprise t(21q21q): 83.3 percent and t(21q22q): 16.6 percent.

The t(21qGq) occur *de novo* in 96 percent of the cases. In only 4 percent are they inherited.

Other translocations (reciprocal translocations, tandem translocations, etc.) are *de novo* in 22 percent of the cases and transmitted in 78 percent of the cases.

Mosaicisms

These represent 2.7 percent of all cases of 21 trisomy. The only cases considered here are those detected because of a trisomic phenotype. It is known that at the extreme of the distribution there exist subjects with a normal phenotype and a very reduced trisomic cellular population. These subjects may eventually be discovered among the parents of 21 trisomy patients (see below).

In mosaic individuals a great variability in symptomatic expression can be seen, both in morphology and psychology. Generally, a mosaic 47,XY(X),+21/46,XY(X) is involved. Cases of more complex mosaicism are extremely rare.

Partial 21 Trisomies

Niebuhr (1974) was the first to support the hypothesis that the phenotype of 21 trisomy results from trisomy for the distal segment of chromosome 21.

Several observations of partial trisomies are now known, either for the proximal segment q1q21 or for the distal segment q22. These observations confirm that only the latter are concomitant with the "classical" 21 trisomy phenotype, the former having a practically normal phenotype, with moderate mental retardation (Williams et al., 1975; Raoul et al., 1976; Wahrman et al., 1976).

The same observations confirm the localization of the locus for SOD-1 on the distal segment of chromosome 21 (Sinet et al., 1976).

Associations

The association of 21 trisomy and aneuploidy of the sexual chromosomes (monosomy X, XXY, XYY) occurs more frequently than can be accounted for by chance. There is no clear explanation for this increased frequency of double aneuploidies.

The association of a different autosomal trisomy is also relatively frequent (18 trisomy in particular).

A *de novo* translocation t(DqDq) may also be associated with 21 trisomy.

21 Trisomies and Twinning

Dizygotic twins are most often discordant for 21 trisomy, whereas monozygotic twins are concordant. Moreover, there seems to be a significant lack of monozygotic twins among 21 trisomy patients.

Exceptional observations of monozygotic twins have been described under the name of heterokaryotic monozygotism. These observations concern pairs of monozygotic twins differing by a single pair of chromosomes. They result from the concomitant production of twinning and mosaicism. Such pairs of twins, one of which is normal, the other trisomic, are known for chromosome 21 (Lejeune et al., 1962; Dekaban, 1965).

RISK OF RECURRENCE AND GENETIC COUNSELING

In close to 5 percent of the cases, 21 trisomy is correlated with a translocation. In slightly more than half of the cases, it occurs *de novo* and may be considered as an accident that will probably not recur. In other cases, a balanced translocation has been demonstrated in one of the parents, and the risk of recurrence can be estimated. For t(Dq21q) and t(21q22q) the risk varies, depending on whether the translocation is transmitted by the mother or the father. If it is the mother, the risk is 16 percent, and if the father, it is lower than 5 percent. These figures differ from the theoretically expected frequency (33 percent). The reasons for this difference and for the difference in risk due to the sex of the carrier parent of the translocation are unknown.

For t(21q21q) the chances of either a trisomic child or a spontaneous abortion are 100 percent (Hamerton, 1970; Dutrillaux, 1972).

Taking all translocations into account, they are transmitted by the mother in 95 percent of the cases, and by the father in 5 percent of the cases.

In theory, the risk of recurrence of free 21 trisomy should not increase for a given maternal age. In practice, however, an excess of families with more than one free 21 trisomic subject is observed. In several cases, a very weak parental mosaic can be demonstrated, most often maternal. In other cases, a translocation not involving chromosome 21 may be discovered in one of the parents. The most notable is t(DqDq), in which case the risk of birth of a 21 trisomic child seems as great as the risk of birth of a 13 trisomic child.

Nonetheless, the cause for recurrence of free 21 trisomy has not been discovered in one-half of the cases. Several hypotheses have been proposed: germinal mosaicism, family tendencies toward nondisjunction, possibly of genetic origin (diverse aneuploidies can be observed in certain families), and the possible role of minor rearrangements (marker chromosomes).

Some authors assume that for women having had their first trisomic child before the age of 30, the risk of recurrence would be 1 to 2 percent, and for women having had their first trisomic child after 30, the risk of recurrence would not be increased (translocations and mosaics excluded) (Hamerton, 1961; Mikkelsen, 1966; Stene, 1970). Fortunately, those families with more than one child with free 21 trisomy are exceptional, and it is still difficult to accept the reality of these estimates.

DERMATOGLYPHICS

On the whole, the hand of a 21 trisomic patient has stereotyped dermatoglyphic features of great diagnostic value (Penrose, 1963; Holt, 1964; Priest, 1969; Lafourcade and Rethoré, 1967).

In the newborn, the ridges are usually immature and difficult to analyze, especially in the hypothenar region.

The principal dermatoglyphic features are:

- Single transverse palmar crease, bilateral in 26.3 percent of the cases, and unilateral in 15 percent of the cases.
- Axial triradius in t″ in more than 70 percent of the cases. Its occurrence in t′ is rare.
- Hypothenar patterns more frequent than normal.
- Increased index of transversality: above 31 in 75 percent of the cases.
- Increased frequency of interdigital patterns.
- Excess of ulnar loops, at the expense of other patterns on the fingertips.
- Clinodactyly and a single flexion crease on the fifth digit.

LABORATORY FINDINGS

Twenty-one trisomy has been the object of a great number of biological studies, primarily with two goals: (1) to reveal a defined biochemical imbalance, possibly leading to some form of therapy (Lejeune, 1975); (2) to try to establish the gene map of chromosome 21.

Metabolic Disorders

Tryptophan metabolism: the rate of urinary excretion of xanthurenic acid is significantly decreased, which suggests a disorder in tryptophan metabolism by acceleration of the cynurenin pathway (Jérôme et al., 1960).

The rate of captation of blood platelets for serotonin is decreased, although they catabolize this amine normally. The activity of platelet monoamine oxidase is decreased (Kamoun et al., 1974).

Hyperuricemia is one of the most constant metabolic disorders (Woodford and Bearn, 1970).

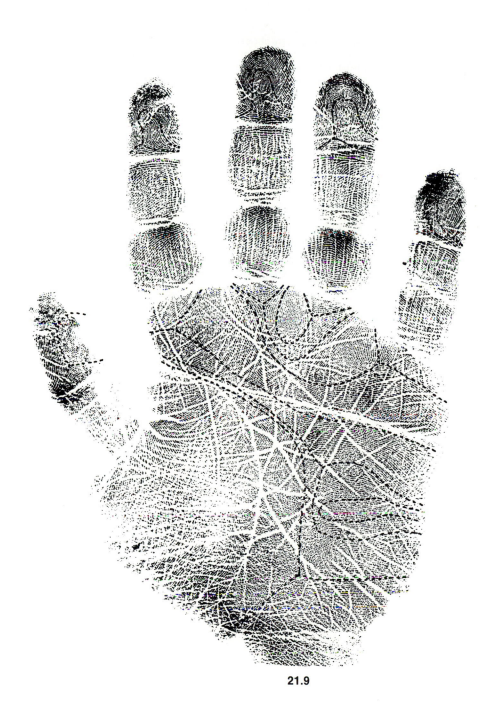

21.9

Enzymatic Disorders

The activity of leukocyte alcaline phosphatase is increased, although no quantitative gene dosage effect can be established (Cox, 1969). Other enzymatic activities are also increased, such as that of G6PD (for which it is known with certainty that the structural gene is localized on the X chromosome); those of leukocyte galactose-1-phosphate uridyl transferase; erythrocyte galactokinase; erythrocyte phosphofructokinase and phosphohexokinase; erythrocyte lactate dehydrogenase; leukocyte acid phosphatase; leukocyte 5-nucleotidase; leukocyte fructose aldolase; and glutamo-oxalo-acetic transaminase (Hsia et al., 1968).

In fact, the only enzymes for which we are certain of the gene localization on 21 are superoxide dismutase 1 (SOD1), the phosphorybosilglycinamide synthetase (PRGS), and hepatic phosphofructokinase (PFKL). A gene dosage effect exists for all three of these enzymes.

Particular Pharmacological Sensitivity

Twenty-one trisomy patients have an exaggerated pupillary sensitivity to atropine, pilocarpine, and ephedrine, which seems to correspond to a cholinergic deficiency.

Interferon Receptor

The interferon receptor (antiviral protein) is localized on 21 and demonstrates a gene dosage effect.

Immunological Disorders

Anomalies of the gammaglobulins (decreased and abnormal electrophoretic migration of IgG) and an excessively high frequency of Australia antigen have been reported, but their significance remains uncertain.

Hematological Disorders

Nuclear lobulation of polynuclear neutrophils is lowered, resulting in a decrease in the index of segmentation, hence a leftward deviation in Arneth's formula (Turpin and Bernyer, 1947).

According to some authors, the mean age of red cells is decreased with an increase in the rate of replacement. A turnover of blood components different from that of normal subjects could partially explain the deviations from the norm observed in enzymatic activities.

21.10

r(21)

Hypertonia
Protuberant nasal bridge
Palpebral fissures slanted downward and outward
Protuberant occiput
Large ears

REFERENCES

ARMENDARES S., BUENTELLO L., CANTU-GARZA J.M.: Partial monosomy of a G group chromosome (45,XY,G−/46,XY,Gr). Report of a new case. *Ann. Génét.*, **14**:7–12, 1971.

CRANDALL B.F., WEBER F., MULLER H.M., BURWELL J.K.: Identification of 21r and 22r chromosomes by quinacrine fluorescence. *Clin. Genet.*, **3**:264–270, 1972.

LEJEUNE J., BERGER R., RETHORE M.O., ARCHAMBAULT L., JEROME H., THIEFFRY S., AICARDI J., BROYER M., LAFOURCADE J., CRUVEILLER J., TURPIN R.: Monosomie partielle pour un petit acrocentrique. *C.R. Acad. Sci. (Paris)* **259**:4187–4190, 1964.

MAGENIS R.E., ARMENDARES S., HECHT F., WELEBER R.G., OVERTON K.: Identification by fluorescence of two G rings: (46, XY,21r) G deletion syndrome I and (46,XX, 22r) G deletion syndrome II. *Ann. Génét.*, **15**:265–266, 1972.

RETHORE M.O.: Chromosome deletions and ring chromosome syndromes. *In* N.C. MYRIANTHOPOULOS, ed.: *Handbook of Clinical Neurology. Vol. 31, part II, Congenital malformations of the brain and skull.* Amsterdam, North Holland publ., pp. 549–620, 1977.

RICHMOND H.G., MAC ARTHUR P., HUNTER D.: A «G» deletion syndrome antimongolism. *Acta paediat. Scand.*, **62**:216–220, 1973.

SHIBATA K., WALDENMAIER C., HIRSCH W.: A child with a 21 ring chromosome, 45,XX,−21/46,XX,21r, investigated with the banding technique. *Humangenetik*, **18**:315–319, 1973.

WARREN R.J., RIMOIN D.L.: The G deletion syndromes. *J. Pediat.*, **77**:658–663, 1970.

WARREN R.J., RIMOIN D.L., SUMMITT R.L.: Identification by fluorescent microscopy of the abnormal chromosomes associated with G-deletion syndromes. *Amer. J. hum. Genet.*, **25**:77–81, 1973.

The first observation of partial monosomy of a G chromosome involved a ring r(G). It was reported in 1964 by Lejeune et al. These authors considered that the patient's phenotype was in countertype to that of 21 trisomy; thus, they interpreted the rearrangement as an r(21).

Other observations of r(G) were subsequently published, but with no possibility of identifying the rearranged chromosome other than by clinical features. Thus, a distinction was made between syndrome I, corresponding to the countertype of 21 trisomy, and syndrome II, corresponding, as was believed, to an r(22).

Only the development of banding techniques permitted the rearranged chromosome to be identified with certainty.

The r(21) syndrome will be described based on five observations subjected to banding (Warren and Rimoin, 1970 and Warren et al., 1973; Armendares et al., 1971, and Magenis et al., 1972; Crandall et al., 1972; Shibata et al., 1973; Richmond et al., 1973). Six previous observations that can be considered as observations of r(21) because of the clinical features will be briefly referred to (Rethoré, 1977).

GENERALIZATIONS

Eleven cases are considered here:

- Sex ratio: 1M/1F
- Maternal age: 26.8 years
- Paternal age: 32.4 years
- Mean birth weight: 2,400 g

PHENOTYPE

Several features of the phenotype are in countertype to 21 trisomy. The infants are hypertonic and have significant growth retardation.

Craniofacial Dysmorphism

Microcephaly is the most constant feature. The occiput is protuberant, the forehead high and bossing with a high-set hairline.

The nasal bridge is broad and most often protuberant. Palpebral fissures are slanted downward and outward, or horizontal. A cutaneous fold of the lower eyelid can redirect the lashes toward the eyeball (epiblepharon). The lashes are long.

The tip of the nose is wide, with a medial ridge. The nostrils are broad and horizontal.

The upper lip is long, with a narrow philtrum and a vermilion visible only in the medial portion. The mouth is small.

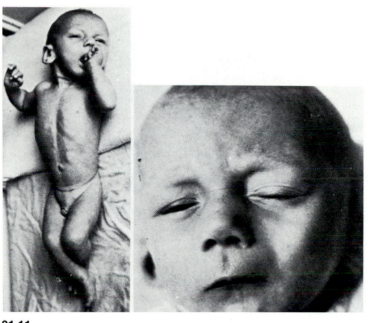

21.11

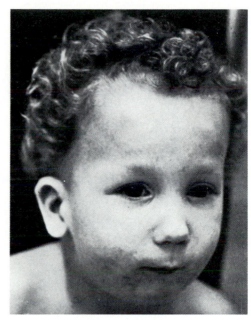

21.12

21.13

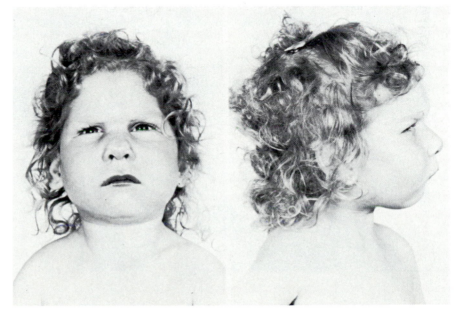

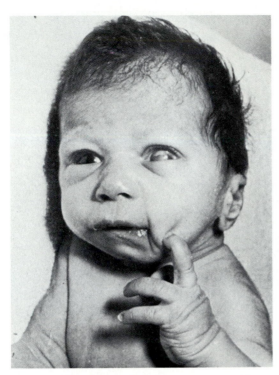

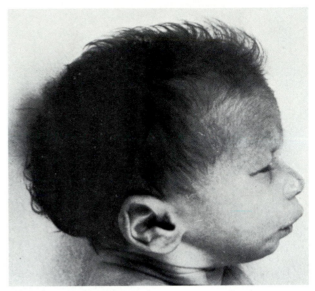

21.14

21.15

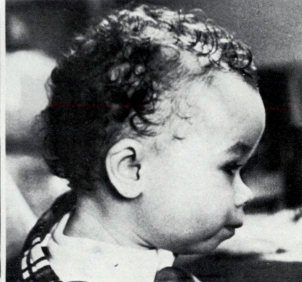

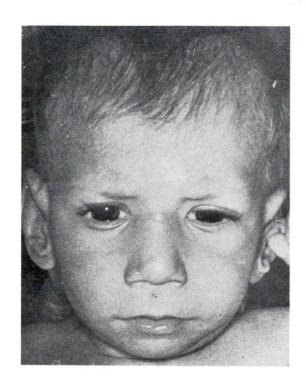

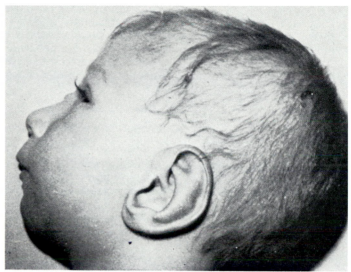

21.16

21.17

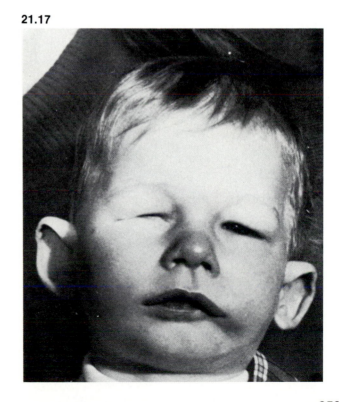

Microretrognathia is constant.

The ears are most often set normally, rarely low-set. They are large with a protuberant anthelix, broad concha, and a broad external auditory canal. The lobe, usually well developed, may at times be small and adherent.

Other features have been observed in isolated cases: choanal stenosis, bifid uvula, cleft palate, and pretragial nodule.

Malformations

Malformations are considerable. They include:

- Ocular malformations: microphthalmia, keratitis, blepharochalasis, persistence of the pupillary membrane, cataract, corneal opacities; and nystagmus
- Digestive malformations: pyloric stenosis, inguinal hernia, hiatal hernia, Meckel's diverticulum, and a common mesentery
- Cardiac malformations: ventricular septal defect, coarctation of the aorta, malpositioning of the large blood vessels, and systolic murmur
- Renal malformations: renal agenesia, and ureteral anomaly
- Genital malformations: almost constant hypospadias, cryptorchidism
- Skeletal malformations: a thirteenth rib, hemivertebrae, supernumerary vertebrae, narrow iliac wings, and hip dysplasia

Blood anomalies are also noted: thrombocytopenia, eosinophilia, hypogammaglobulinemia, normal or lowered alkaline phosphatase activity.

The 5-hydroxyindolacetic/cynurenine ratio is elevated in one case.

Prognosis

Mental retardation is severe and fatality is high.

DERMATOGLYPHICS

- Normal flexion creases
- Axial triradius in t or t′

Type and Countertype

	21 Trisomy	r(21)
Tonus	Hypotonia	Hypertonia
Cranium	Flat occiput	Protuberant occiput
Nasal bridge	Aplastic	Protuberant
Orientation of the palpebral fissures	Upward and outward	Downward and outward
Nostrils	Small and oriented forward	Broad and horizontal
Ears	Small with small concha, small auditory canal	Large with large concha, large auditory canal
Dermatoglyphics	Immature ridges	Hypermature ridges
Palmar creases	Single transverse palmar crease	Normal

21 MONOSOMIES

21q1 Monosomy

Discrete dysmorphism

REFERENCES

DUTRILLAUX B., JONASSON J., LAUREN K., LEJEUNE J., LINDSTEN J., PETERSEN G.B., SALDANA-GARCIA P.: An unbalanced 4q/21q translocation identified by the R but not by the G and Q chromosome banding techniques. *Ann. Génét.*, **16**:11–16, 1973.

HOLBEK S., FRIEDRICH V., BROSTROM K., PETERSEN G.B.: Monosomy for the centromeric and juxtacentromeric region of chromosome 21. *Humangenetik*, **24**:191–195, 1974.

LAURENT C., DUTRILLAUX B., BIEMONT C.L., GENOUD J., BETHENOD M.: Translocation t(14q–;21q+) chez le père. Trisomie 14 et monosomie 21 partielles chez la fille. *Ann. Génét.*, **16**:281–284, 1973.

RETHORE M.O., DUTRILLAUX B. BAHEUX G., GERBEAUX J., LEJEUNE J.: Monosomie pour les régions juxta-centromériques d'un chromosome 21. *Exptl. Cell Res.*, **70**:455–456, 1972.

RETHORE M.O., DUTRILLAUX B., LEJEUNE J.: Translocation 46,XX,t(15;21) (q13;q221) chez la mère de deux enfants atteints de trisomie 15 et de monosomie 21 partielles. *Ann. Génét.*, **16**:271–275, 1973.

RETHORE M.O.: Chromosome deletions and ring chromosome syndromes. *In* N.C. MYRIANTHOPOULOS, ed.: *Handbook of Clinical Neurology. Vol. 31, part II, Congenital malformations of the brain and skull.* Amsterdam, North Holland publ., pp 549–620, 1977.

SCHMIDT R., MUNDEL G., ROSENBLATT M., KATZNELSON M.B.: Apparent G monosomy, G deletion, and incomplete Down's syndrome in a single family. *J. med. Genet.*, **9**:457–461, 1972.

WAHRMAN J., GOITEIN R., RICHLER C., GOLDMAN B., AKSTEIN E., CHAKI R.: The mongoloid phenotype is due to triplication of the distal pale G band of chromosome 21. *In* P.L. PEARSON, K.R. LEWIS: *Chromosomes Today, vol. 5* (Proc. Leiden Chromosome Conference, July 15–17, 1974). New York, John Wiley and Sons and Jerusalem, Israel Universities Press, 1974, pp. 241–248, 1974.

The first observation of proximal 21 monosomy was reported in 1972 by Rethoré et al. Eight cases of monosomy are now known, all involving the same segment (Schmidt et al., 1972; Rethoré et al., 1973; Dutrillaux et al., 1973; Laurent et al., 1973; Holbek et al., 1974; Wahrman et al., 1974). They constitute a pathological entity delineated by Rethoré (1977).

Other varieties of 21 monosomy are also known. They represent isolated cases, and the exact limits of the monosomy are usually not known. They will not be considered here.

GENERALIZATIONS

In the eight cases that will be referred to, monosomy results from a translocation, parental or *de novo,* of band q22 onto another chromosome with loss of the remaining portion of chromosome 21. For this reason it is often accompanied by partial monosomy or trisomy of another chromosome, usually of little phenotypic significance.

- Sex ratio: 5F/3M
- Mean birth weight: 2,300 g

PHENOTYPE

Craniofacial Dysmorphism

The forehead is high.
Palpebral fissures are horizontal, the eyeballs small and sometimes deeply set.
The nasal bridge is aplastic, and in infants the nose is flat.
The nasolabial sulcus is very deep; the maxillae may be slightly protuberant; the philtrum is broad with blunt borders.
The vermilion of the upper lip is visible only in its medial portion.
The mandible is wide.
The ears are low-set, with large pinnae.

Neck, Thorax, and Abdomen

The neck is short.

Limbs

Joint stiffness is noted, and in the infant an abnormal positioning of the fingers which is similar to that of chromosome 18 trisomy.

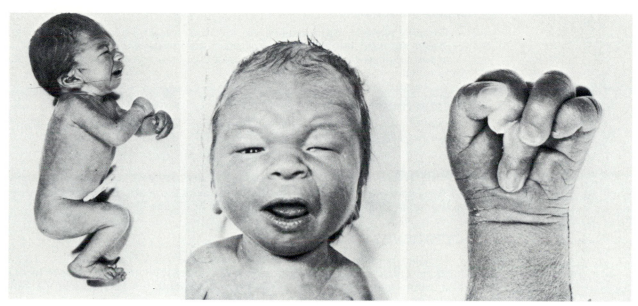

21.18

21.19

21.20

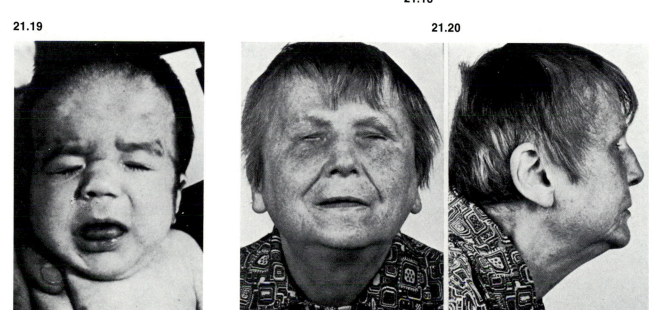

Malformations

Primarily skeletal malformations are involved: supernumerary vertebrae, dorsal cyphosis, hip dislocation, clubfoot, aplasia of the radius, and bone fragility.

In two autopsied patients, cerebral, cardiac, pulmonary, and renal malformations were noted.

Mental Retardation

Mental retardation is considerable. The IQ is always below 50. Tonus disorders have been reported in several cases.

Prognosis

Two infants died, one at 4½ months, the other at 2½ years. The other patients were examined in childhood. Only one was seen at adult age (54 years).

CYTOGENETIC STUDY

Proximal monosomy 21pter→q21 is in every case due to malsegregation of a translocation, which is usually parental, but may in rare instances occur *de novo* (with loss of a centromere). In one case (the 54-year-old patient), the karyotype of the parents could not be studied.

Dermatoglyphics

The dermatoglyphics do not have characteristic features. Interdigital patterns (pelotes) are noted in 7 and 9, as well as hypothenar patterns and an excess of whorls.

Other 21 Monosomies

REFERENCES

GRIPENBERG U., ELFVING J., GRIPENBERG L.: Case report. A 45,XX,21 − child: attempt at a cytological and clinical interpretation of the karyotype. *J. med. Genet.*, **9**:110–115, 1972.

HALLORAN K.H., BREG W.R., MAHONEY M.J.: 21 monosomy in a retarded female infant. *J. med. Genet.*, **11**:386–389, 1974.

KANEKO Y., IKEUCHI T., SASAKI M., SATAKE Y., KUWAJIMA S.: A male infant with monosomy 21. *Humangenetik*, **29**:1–7, 1975.

MIKKELSEN M., VESTERMARK S.: Karyotype 45,XX,−21/46,XX,21q− in an infant with symptoms of G-deletion syndrome I. *J. med. Genet.*, **11**:389–392, 1974.

SUMMITT R.L., MARTENS P.R., WILROY R.S.: X-autosome translocation in normal mother and effectively 21 monosomy daughter. *J. Pediat.*, **84**:539–546, 1974.

WEBER F.M., SPARKES R.S., MULLER H.: Double monosomy mosaicism (45,X/45,XX,−21) in a retarded child with multiple congenital malformations. *Cytogenetics*, **10**:404–412, 1971.

Some cases of complete 21 monosomy, homogeneous or in mosaic, have been reported (Weber et al., 1971; Gripenberg et al., 1972; Halloran et al., 1974; Mikkelsen and Vestermark, 1974; Summitt et al., 1974; Kaneko et al., 1975). They are of little value because of their small number and the uncertainty of the cytogenetic diagnoses and especially of the associated chromosomal rearrangements.

In the only case of total monosomy in mosaic (with no other anomaly) the phenotype is very close to that of the r(21) syndrome.

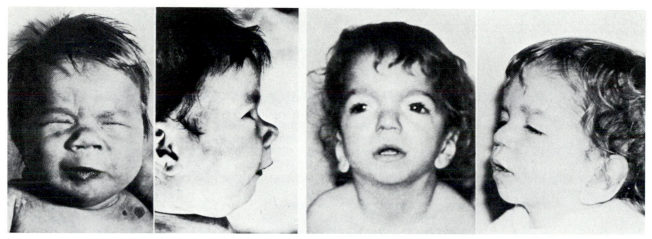

21.21   21.22

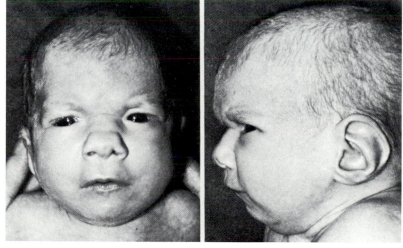

21.23

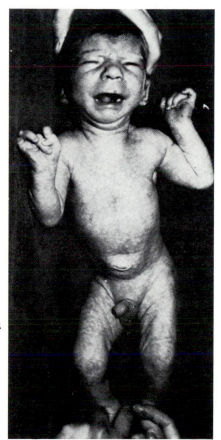

21.24

CHROMOSOME 22

REFERENCE

LA CHAPELLE A. de, HERVA R., KOIVISTO M., AULA P.: A deletion in chromosome 22 can cause Di George syndrome. *Hum. Genet.*, **57**:253–256, 1981.

Chromosome 22 includes the gene family responsible for synthesis of the lambda light chain of immunoglobulins. In addition, it carries the genes for β-galactosidase 2; arylsulfatase A, whose mutation causes delayed infantile metachromatic leucodystrophy; mitochondrial aconitase; and α-galactosidase B. These enzymes merit our attention in that their dosage could be envisaged in suspected cases of 22 trisomy. Recently, the DiGeorge syndrome was, at least in certain cases, associated with an intercalary deletion of 22q11 (La Chapelle et al., 1981).

Chromosomal disorders, however, are relatively scarce. We have considered here the rings and the "cat-eye" syndrome, though the latter has not been definitively proven to result from a partial trisomy 22.

Complete trisomy mosaicism has not been considered, nor have partial distal trisomies and partial deletions, because of their extreme rareness.

A polymorphism, due to the variation of the short arm length and of the satellite size, is present. Chromosomes with satellites in duplicate or triplicate may also be observed.

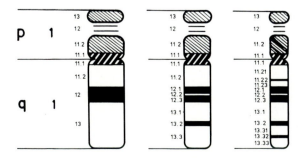

GENE MAP

22p12	RNR		Ribosomal RNA (18045) A
22pter-22q11	CES	(L)	Cat eye syndrome (11547) Ch
22q11	DGS	(P)	DiGeorge syndrome (18840) Ch
22q13-22qter	GALB		Beta-galactosidase-2 (10968) S
22q11	CML		Chronic myeloid leukemia (15141) Ch
22q13-22qter	ARSA		Arylsulfatase A (25010) S
22q13-22qter	DIA1		NADH-diaphorase-1 (25080) S
22q11-22q13	ACO2		Aconitase, mitochondrial (10085) S
22q13	GLB2		N-acetyl-alpha-D-galactosamini-dase (alpha-galactosidase B) (10417) S
	S22	(P)	Surface antigen 22 (18558) S
	IGL		IMMUNOGLOBULIN LAMBDA LIGHT CHAIN GENE FAMILY REa,
	IGKV		Variable region of lambda light chains (many genes) (14724)
	IGKJ		J region of lambda light chains (several genes) (14723)
	IGLC		Gene (multiple) for constant region of lambda light chains (14722)
	IDA	(P)	Alpha-L-iduronidase (25280) S
	SIS		Onc gene: simian sarcoma virus (19004) S

MORPHOLOGY AND BANDING

QFQ   GTG   RHG   RBA   THA   C

22 TRISOMY

REFERENCES

SCHINZEL A.: Incomplete trisomy 22. II. Familial trisomy for the distal segment of chromosome 22q in two brothers from a mother with a translocation, t(6;22) (q27;q13). *Hum. Genet.*, **56**:263–268, 1981.

SCHINZEL A.: Incomplete trisomy 22. II. Mosaic trisomy 22 and the problem of full trisomy 22. *Hum. Genet.*, **56**:269–273, 1981.

G trisomies, or "non-21 trisomies" for a long time formed a heterogenous group some of which were considered as 22 trisomies. Extensive use of banding techniques have led to the opinion that so-called "22 trisomy" is, in fact, an 11qter trisomy resulting from 3:1 segregation of the reciprocal translocation, t(11;22) (see Chapter on chromosome 11).

It is presently held that complete, homogeneous 22 trisomy is incompatible with survival. In this regard, it should be noted that there exists no 22 trisomic child born to a parent carrier of a Robertsonian translocation involving a 22. In the special case of the very rare t(22q;22q), all pregnancies end in early, spontaneous abortions.

Several rare observations report the birth of children with 22 trisomy mosaicism. All of these infants died soon after birth (Schinzel, 1981).

On the other hand, 22 trisomy is frequently observed in spontaneous abortions (Boué et al., 1976).

Insofar as partial 22 trisomies are concerned, these are rather exceptional, apart from trisomies resulting from a t(11;22), and possibly the "cat-eye" syndrome (Schinzel, 1981). A special chapter is devoted to this latter syndrome.

THE CAT-EYE SYNDROME

Ocular coloboma
Anal atresia
Preauricular tags

REFERENCES

BUHLER E.M., MEHES K., MULLER H., STALDER G.R.: Cat-eye syndrome. A partial trisomy 22. *Humangenetik,* **15**:150–162, 1972.

GUANTI G.: The etiology of the cat eye syndrome reconsidered. *J. med. Genet.,* **18**:108–118, 1981.

SCHACHENMAN G., SCHMID W., FRACCARO M., MANNINI A., TIEPOLO L., PERONA G.P., SARTORI E.: Chromosomes in coloboma and anal atresia. *Lancet,* **2**:290, 1965.

SCHINZEL A., SCHMID W., FRACCARO M., TIEPOLO L., ZUFFARDI O., OPITZ J.M., LINDSTEN J., ZETTERQVIST P., ENELL H., BACCHICHETTI C., TENCONI R., PAGON R.A.: The «cat eye syndrome». Dicentric small marker chromosome probably derived from a n° 22 (tetrasomy 22qter-q11) associated with a characteristic phenotype. *Hum. Genet.,* **57**:148–158, 1981.

In 1965, Schachenmann et al. reported the observation of four patients having a tiny metacentric supernumerary chromosome, and a comparable phenotype, which is now known as the cat-eye syndrome.

The first familial observation of the syndrome due to a familial 22pter→q11 trisomy (Buhler et al., 1972) led a number of authors to believe that the supernumerary chromosome was a 22q−. This origin, however, remains open to debate, as we will see later. For the sake of convenience, we have chosen to describe the cat-eye syndrome in this chapter devoted to chromosome 22.

Around 60 cases have been described in the literature. A chromosome aberration anomaly was observed in around five out of six cases (Guanti, 1981; Schinzel et al., 1981).

PHENOTYPE

The complete syndrome includes the following symptoms:

- Moderate mental retardation
- A set of congenital malformations, including:
    Anal atresia
    Ocular coloboma
    Palpebral fissures slanting downward and outward
    Microphthalmia
    Pretragian tags and/or pits
    Congenital heart disease especially atrial and ventricular septal defect
    Anomalies of the urinary tract
    Skeletal anomalies

In fact, however, the complete syndrome is observed in little more than a third of the cases. At least one of the major features, either coloboma or anal atresia, is necessary in order to make the diagnosis.

CYTOGENETIC STUDY

The characteristic chromosomal aberration in this syndrome consists of a small supernumerary metacentric chromosome whose size and morphology are variable. It is generally shorter than, or the same size as, chromosome 22, and is present in mosaic.

Several hypotheses have been advanced to determine the origin of this supernumerary chromosome.

According to Schinzel (1981), the chromosome is derived from duplication of the centromeric portion of a 22: 22pter→q11. Patients would therefore be tetrasomic for a short segment of chromosome 22.

Another hypothesis was formulated by Guanti (1981), in which the supernumerary chromosome is a "compact" 13 resulting from an intercalary deletion involving the central region of 13q.

363

r(22)

Doe's eyes
Low-set eyebrows

REFERENCES

ALLER V., ABRISQUETA J.A., DE TORRES M.L., MARTIN-LUCAS M.A., PEREZ-CASTILLO A., DEL MAZO J.: An r(22) (p11 q13) in a moderately mentally retarded girl. *Hum. Genet.*, **51**:157–162, 1979.

FOWLER G., KAISER-McCAW B., HECHT F.: The use of sequential silver and quinacrine staining to determine the parental origin and breakpoints of a ring-22 human chromosome. *Clin. Genet.*, **18**:274–279, 1980.

FRYNS J.P., VAN DEN BERGHE H.: Ring chromosome 22 in a mentally retarded child and mosaic 45,XX,−15, −22, +t(15;22)(p11;q11)/46, XX,r(22)/46,XX karyotype in the mother. *Hum. Genet.*, **47**:213–216, 1979.

HUNTER A.G.W., RAY M., WANG H.S., THOMPSON D.R.: Phenotypic correlations in patients with ring chromosome 22. *Clin. Genet.*, **12**:239–249, 1977.

MAGENIS R.E., ARMENDARES S., HECHT F., WELEBER R.G., OVERTON K.: Identification by fluorescence of two G rings: (46,XY,21r) G deletion syndrome I and (46,XX,22r) G deletion syndrome II. *Ann. Génét.*, **15**:265–266, 1972.

RETHORE M.O., NOEL B., COUTURIER J., PRIEUR M., LAFOURCADE J., LEJEUNE J.: Le syndrome r(22). A propos de quatre nouvelles observations. *Ann. Génét.*, **19**:111–117, 1976.

WELEBER R.G., HECHT F., GIBLETT E.R.: Ring-G chromosome, a new G deletion syndrome? *Amer. J. Dis. Child.*, **115**:489–494, 1968.

The first observation of an r(22) is that of Weleber et al. in 1968, verified by Q-banding by Magenis et al. in 1972. Rethoré et al. (1976) considered 14 observations to describe the r(22) syndrome. Approximately 20 observations are currently known (Hunter et al., 1977; Aller et al., 1979; Fryns and Van den Berghe, 1979; Fowler et al., 1980).

GENERALIZATIONS

- Sex ratio: 4F/3M
- Mean maternal age: 27.6 years
- Mean paternal age: 30.7 years
- Mean birth weight: 3,100 g

PHENOTYPE

Craniofacial Dysmorphism

The head circumference is below normal in two-thirds of the cases. The face is round in young children, oval in older patients.

The palpebral fissures are horizontal and broadly almond-shaped, thus evoking "doe's eyes." Epicanthus is constant at birth. It disappears with age. Ptosis of the eyelids is noted in one-third of the cases. The eyebrows are low-set and sometimes converging. Hypertelorism is present.

The nose has a bulbous tip in infants, but subsequently becomes normal in appearance.

The upper lip is normal, sometimes a bit long. The palate is high-arched.

The ears are large and usually normally set. A small adherent lobule and a poorly folded helix are often observed.

Neck, Thorax, and Abdomen

No anomaly in particular is noted.

Limbs

Syndactylies are frequent on either the hands or feet. Clubfeet have been reported.

Genitalia

Genital organs are normal.

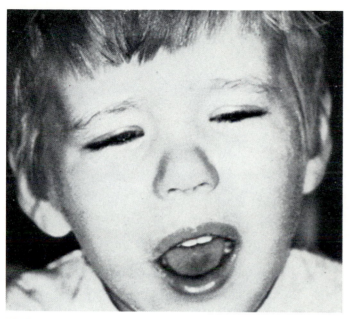

22.1

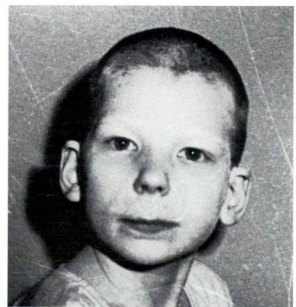

22.2

22.3

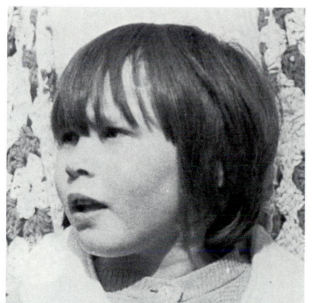

22.4

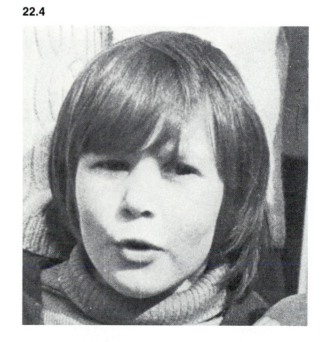

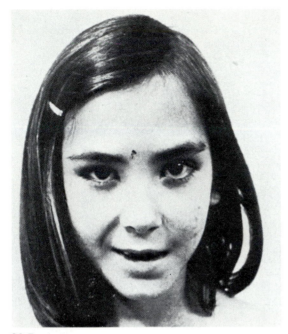

22.5

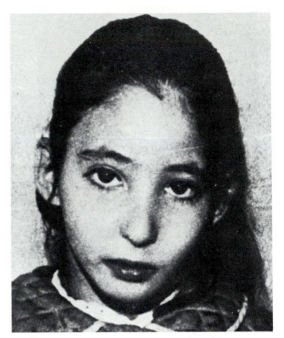

22.6

22.7

22.8

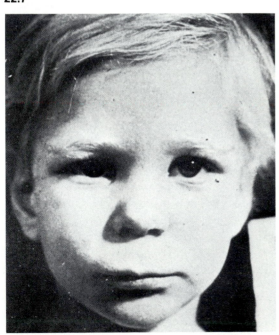

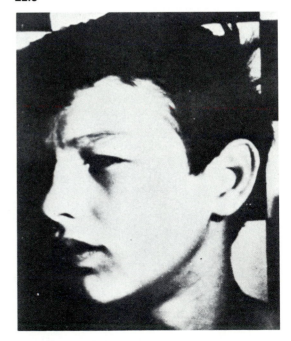

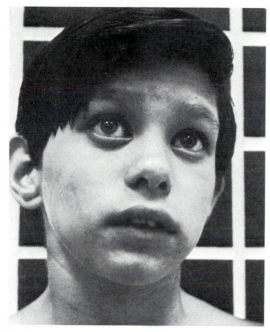

22.9

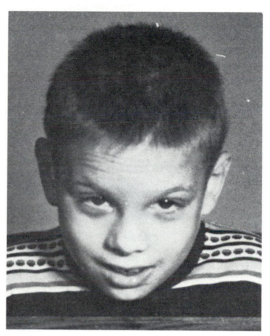

22.10

22.11

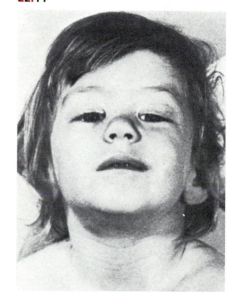

Malformations

Malformations seem to be exceptional, as none have been mentioned in the cases reported.

In one case, enteropathy with an immunological deficiency was mentioned.

Mental Retardation

Mental retardation is always severe and is aggravated with age. The mean IQ degenerates from 50 before two years of age to 30 after four years of age.

Hypotonia is constant, except in one case with hypertonia. Motor incoordination is present, especially in walking, as well as excitability and instability.

Prognosis

Growth is usually normal, but may in rare cases be delayed. Bone age is normal. Diagnosis is usually made late, between one and fourteen years. No child is known to have died.

CYTOGENETIC STUDY

The karyotype of the parents is normal in every case. In one observation the sister of the patient is a carrier of the translocation t(15q22q).

Overall, ring 22 seems stable, since in more than half of the cases it is present in all observed cells, without mosaicism.

DERMATOGLYPHICS

Palmar creases are usually normal.

In several cases, two axial triradii are reported. Also noted are hypothenar patterns and an excess of whorls.

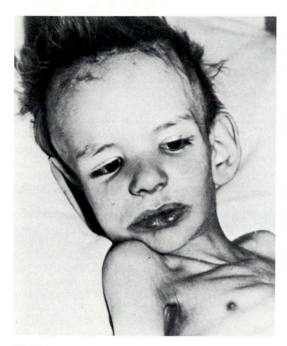

22.12

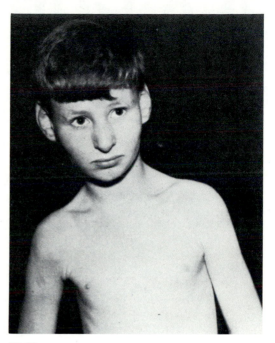

22.13

22.14

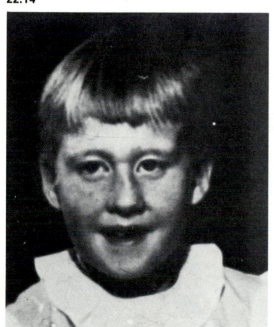

22.15

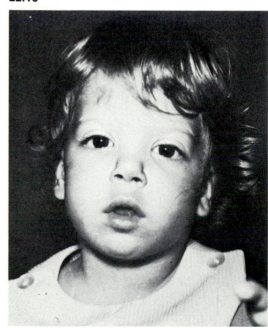

SEX CHROMOSOMES

X CHROMOSOME

From the characteristics of X-linked inheritance and, in two instances, from studies of somatic cell hybrids, over 115 genetic loci have been assigned to the X chromosome. Three assigned by somatic cell genetics are tyrosine aminotransferase regulator (TAT1, 31435), surface antigen, X-linked (SAX, 31345), and temperature sensitive complement (31365). The others were assigned by pedigree pattern (in the first instance, at least); confirmation by other methods, such as study of cell hybrids, and mapping to the X chromosome in other mammals (Ohno's law), is available for some. In other instances (Lesch-Nyhan syndrome, testicular feminization, ornithine transcarbamylase deficiency, PRPP synthetase, etc.) confirmation was provided by demonstration of lyonization in cell cultures or other material from heterozygotes. (From Victor A. McKusick, *The Human Gene Map Newsletter*, March 1, 1983.)

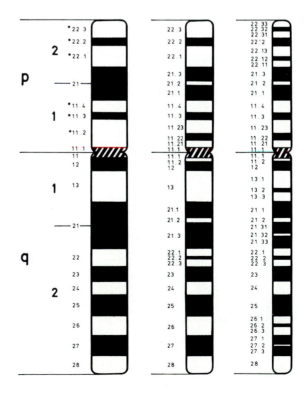

GENE MAP OF THE X CHROMOSOME

A. *Regional assignments on the X chromosome by study of cell hybrids*

Xp22.3	Xg	Xg blood group (31470)]
Xp22.3	STS	Steroid sulfatase (ichthyosis) (30810)
(L) Xp22		Aicardi syndrome (30405) Ch
(L) Xp22.32	CDP	Chondrodysplasia punctata (30295)
(L) Xp21-Xp22	HYB	Gonadal dysgenesis, XY female type (?deficiency of gonad-specific H-Y receptor) (30610)
(P) Xp22.3	HYC	H-Y regulator (or repressor) (30697) (may be the same)
Xp21-Xp22	DMD	Duchenne muscular dystrophy (31020)
(P) Near centromere (p11-q13) (proximal Xq, prob. q11)	TFM	Testicular feminization (31370)
(L) Xq11.2-Xq21.1		X chromosome controlling element (31467)
(L) Xq12		Aarskog-Scott syndrome (30540)
(L) Xq12		Anhidrotic ectodermal dysplasia (30510)
Xq13	PGK	Phosphoglycerate kinase (31180)
Near PGK		Monoamine oxidase A (30985)
Xq22-Xq24	GLA	Alpha-galactosidase A (Fabry disease) (30150)
Xq22-Xq26	PRPS	Phosphoribosylpyrophosphate synthetase (31185)
Xq26-Xq27	HPRT	Hypoxanthine-guanine phosphoribosyltransferase (30800)
Near HPRT		Temperature-sensitive mutation, mouse and hamster, complementation of (31365)
Xq (between HPRT and G6PD)	SAX	Surface (or species) antigen, X linked (31345)
Xq27-28	FS	Fragile site (at interface between Xq27 and Xq28) (30955)
Xq28	G6PD	Glucose-6-phosphate dehydrogenase (30590)

Radiation-induced segregation (Goss-Harris) supports order:

PGK—GALA—PRPS—HPRT—SAX—G6PD

B. *Regional mapping by family linkage studies*
    1. The Xg cluster

	Xg		Xg (31470)**
Xp22	STS		Ichthyosis (steroid sulfatase deficiency) (30810)
	Xk,CGD		Xk (and chronic granulomatous disease) (31485, 30640)
	OA1		Ocular albinism, Nettleship-Falls type (30050)
	OA2	(P)	Ocular albinism, Forsius-Eriksson type (30060)
	RS		Retinoschisis (31270)
		(L)	Mental retardation, ? type (?30950)

**Ropers et al. (HGM6) suggested order: Xg—H-Y repressor—
  H-Y receptor—STS—Xk
Ferguson-Smith (HGM6) suggested order: STS—11cM—Xg—?2cM—Xk—OA

    2. The G6PD cluster (linkages to G6PD unless otherwise indicated)

Xq28	G6PD		Glucose-6-phosphate dehydrogenase (30590)
	CBD		Deutan color blindness (30380)
	CBP		Protan color blindness (30390)
	HEMA		Hemophilia A (30670)
	ALD		Adrenoleukodystrophy (30010)
		(L)	Blue-mono-cone-monochromatic color blindness (30370)
		(P)	Manic-depressive psychosis (30920)
			Xm (31490) (linked to color blindness)
		(P)	Becker muscular dystrophy (31010) (? linked to color blindness)
		(P)	Emery muscular dystrophy (31030) (? linked to color blindness)
		(P)	Hunter syndrome (30990) (? linked to Xm)
Xq27-28			Mental retardation with large testes and fragile site (30955) (linkage to G6PD or color blindness has been demonstrated)

C. *X-specific DNA fragments with regional assignment*
(A considerable and increasing number of cloned X-specific DNA fragments
  have been localized to specific regions.)

MORPHOLOGY AND BANDING OF THE X CHROMOSOME

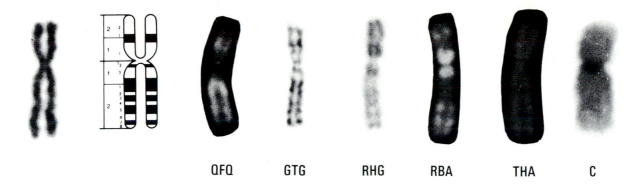

| QFQ | GTG | RHG | RBA | THA | C |

Y CHROMOSOME

REFERENCES

BISHOP A, BLANK CE, HUNTER H: Heritable variation in the length of the Y chromosome. *Lancet* **2**:18–20, 1962.

COHEN MM, SHAW MW, McLUER JW: Racial differences in the length of the human Y chromosome. *Cytogenetics* **5**:34–52, 1966.

GENEST P, LABERGE C, POTY J, GAGNÉ R, BOUCHARD M: Transmission d'un petit "Y" durant onze générations dans une lignée familiale. *Ann. Génét.* **13**:233–238, 1970.

GROUCHY J de: Annulation de reconnaissance de paternité fondée sur la non identité des chromosomes Y. Un jugement du Tribunal de Paris. *Ann. Génét.* **20**:133–135, 1977.

LIN CC, GEDEON MM, GRIFFITH P, SMINK WK, NEWTON DR, WILKIE L, SEWELL LM: Chromosome analysis on 930 consecutive newborn children using quinacrine fluorescent banding technique. *Hum Genet* **31**:315–328, 1976.

NUZZO F, CAVIEZEL F, CARLI L de: Y chromosome and exclusion of paternity. *Lancet* **2**:260–262, 1966.

SOUDEK D, LANGMUIR V, STEWART DJ: Variation in the nonfluorescent segment of long Y chromosome. *Humangenetik* **18**:285–290, 1973.

UNNERUS V, FELLMAN J, LA CHAPELLE A de: The length of the human Y chromosome. *Cytogenetics* **6**:213–227, 1967.

The distal segment of Yq, brightly fluorescent after quinacrine staining, is remarkable for its polymorphism.

This segment is generally considered inactive with regard to sexual differentiation. Its function remains unknown.

Its size varies considerably from one race to another. It is large in Semites and Japanese (Bishop et al., 1962; Cohen et al., 1966; Soudek et al., 1973).

It also varies from individual to individual.

It is stable from one generation to another and constitutes an excellent familial marker transmitted from father to son (Genest et al., 1970).

Its value as a proof of paternity is not yet recognized by jurisprudence but might well be in the future (Nuzzo et al., 1966; Unnerus et al., 1967; Grouchy, 1977).

The figure on p. 374 shows this polymorphism of the Y chromosome (Lin et al., 1976).

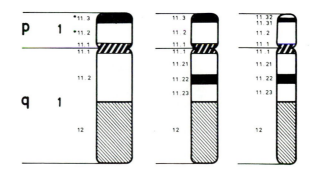

GENE MAP OF THE Y CHROMOSOME

Yp	HYA	Y-histocompatibility antigen. Locus A (TDF: testis determining factor probably the same as HYA).
	STA	Stature
Yq11(?)	SP3	Spermatogenesis controlling factor-3 (azoospermia 3rd factor).
Yq11(?)	TS	Tooth size

MORPHOLOGY AND BANDING OF THE Y CHROMOSOME

QFQ   GTG   RHG   RBA   THA   C

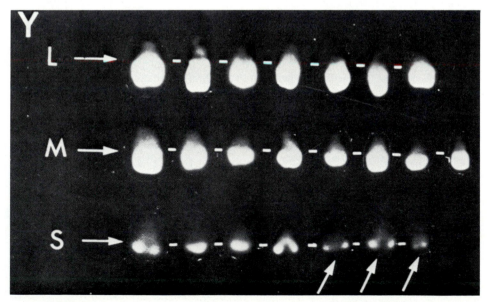

23.1

TURNER'S SYNDROME

Short stature
Lack of pubertal development
Ovarian agenesis
Characteristic somatic malformations

REFERENCES

FORD C.E., JONES K.W., POLANI P.E., AL-MEIDA J.C. de, BRIGGS J.H.: A sex chromosomal anomaly in a case of gonadal dysgenesis (Turner's syndrome). *Lancet*, I:711–713, 1959.

GROUCHY J. de: Clinical Cytogenetics. In *The Cell Nucleus. Vol. 2*, New York, Academic Press, 373–436, 1974.

HAMERTON J.L.: *Human Cytogenetics. Clinical Cytogenetics, Vol. II*, New York, Academic Press, pp. 65–112, 1971.

LAMY M., JOSSO N., GROUCHY J. de, BITAN A.: Anomalies des gonosomes. *XXe Congrès de l'Association des Pédiatres de langue française, Nancy, 1965, Tome 3*. Paris, Expansion scientifique, 1965.

POLANI P.E.: Turner's syndrome and allied conditions. *Brit. med. Bull.*, **17**:200–205, 1961.

TURNER H.H.: A syndrome of infantilism, congenital webbed neck, and cubitus valgus. *Endocrinology*, **23**:566–574, 1938.

In 1938 Turner reported seven observations of a new syndrome characterized by a female phenotype, small stature, impuberism with primary amenorrhea and sterility, and somatic malformations, primarily including bilateral pterygium colli and cubitus valgus. Ullrich had already described these somatic anomalies in 1930, and compared his observations with those of Bonnevie on the *my* mouse, although he did not relate them to the absence of puberty.

In 1954, Polani et al. observed the absence of the Barr body in certain patients afflicted with ovarian agenesis (Polani, 1961).

In 1959, Ford et al. discovered the 45,X karyotype of these patients, thus describing the first anomaly of the sexual chromosomes.

The clinical syndrome correlated with the 45,X karyotype was soon well defined. Consequently, a great variety of anomalies in the number and structure of the X chromosome and even of the Y chromosome were identified in patients with a more or less complete Turnerian phenotype (Lamy et al., 1965; Hamerton, 1971; Grouchy, 1974).

Terminology of Turner's syndrome admittedly remains poorly defined; for example, the designation "Bonnevie-Ullrich syndrome" was used for a long time. It is now mainly reserved for the syndrome observed in the infant and associated with lymphedema of the hands and feet and pterygium colli. More ambiguous is the use of the terms ovarian (or gonadal) agenesis (or dysgenesis). Although basically correct, these terms are generally used to designate certain specific anomalies of sexual development: true gonadal dysgenesis and mixed gonadal dysgenesis (see further below).

Alternatively, the somatic features of Turner's syndrome can be seen piecemeal in the female and also in the male (the "male" Turner's syndrome). This syndrome is now generally designated by the name Noonan's syndrome (see further below).

GENERALIZATIONS

- Frequency: at birth the frequency is 0.4 per 1,000 females. In fact, it is quite likely that the frequency of the 45,X karyotype at conception is much higher: 39 of 40 zygotes with a 45,X karyotype seem to abort early.
- Mean parental ages: normal.
- Pregnancy: normal duration.
- Mean birth weight: 2,500 to 2,900 g, depending on the authors.
- Mean crown-heel length at birth: 45 to 47.5 cm, depending on the authors. Small size at birth is often evident.

PHENOTYPE

The phenotype described here is comprehensive; in reality, it is rare that all features are present in one patient. There is also great phenotypic variability as a function of karyotypic variants.

The disease is generally diagnosed either at birth, because of suggestive congenital malformations, or in young girls, based on delayed growth, or in adolescents because puberty fails to occur.

In the Infant

Turner's syndrome can be diagnosed at birth when the following features occur together: small size, lymphedema of the hands and feet, and an "excess of skin" on the nape.

Lymphedema is very characteristic. It is restricted to the backs of the hands and feet and does not extend to the distal portions of the limbs. It is hard and noninflammatory. Though it disappears during the second year, it sometimes leaves a puffiness on the backs of the fingers.

The excess of skin on the nape may be very impressive in some cases. It is often more subdued, evidenced by simple laxity of the skin (cutis laxa). It eventually becomes transformed into pterygium colli.

These combined features constitute the Bonnevie-Ullrich syndrome, which corresponds to Turner's syndrome in about 30 percent of the cases. Inversely, 70 to 80 percent of patients with Turner's syndrome have lymphedema at birth.

In the Young Girl and the Adolescent

Small stature and impuberism are the basic features of Turner's syndrome. Small stature is a prerequisite condition of the syndrome, which cannot be diagnosed in a patient with a height greater than 150–153 cm. Subnormal size is often apparent at birth, and the child's growth is slow: height remains at a level 3 to 4 SD below the mean for a given age.

Craniofacial Dysmorphism

Craniofacial dysmorphism may be very suggestive. The face is triangular. Palpebral fissures are slanted downward and outward, possibly with uni- or bilateral epicanthus, and ptosis.

The mouth may be downturned. The teeth are poorly inserted and the palate very high-arched.

Hypoplasia of the mandible and retrognathia are the most characteristic features of facial dysmorphism.

The ears are often low-set.

Neck, Thorax, and Abdomen

The hairline is very low on the nape and may extend to the upper region of the shoulders.

The neck is short and thickened by pterygium colli—that is, a webbing extending from the mastoid to the acromial area.

The thorax is deformed in a very characteristic way: the biacromial diameter is excessively large in relation to that of the pelvis. The thorax is broad and shield-shaped, with wide-spaced nipples. The latter are often hypoplastic.

Limbs

Cubitus valgus is a classic feature, as is the very characteristic shortening of the fourth and fifth metacarpals (brachymetacarpia).

The lower limbs are deformed by flattening of the inner tibial plate (Kosowicz' sign).

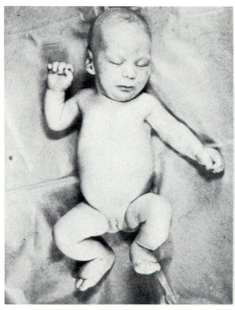

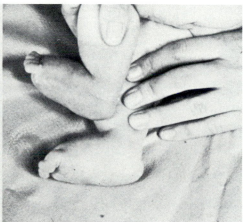

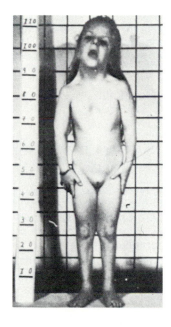
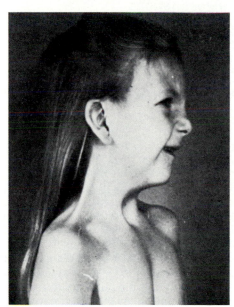
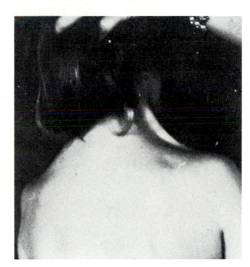

23.3

Skin and Nails

The presence of nevi is an important diagnostic feature; they may be seen on the face, the forearms, the thorax, and on any other part of the body.

The nails are often hypoplastic and exaggeratedly convex.

The marked tendency for keloid scars raises the question of whether to perform esthetic surgery.

Genitalia

External genitalia remain infantile.

Pubic hair is very scant or even nonexistent. Axillary hair is absent. A certain degree of virilization may occasionally be observed, manifested primarily by clitoral hypertrophy, and secondarily by posterior labial fusion.

Celioscopy and laparotomy reveal atrophy of the gonads, which are reduced to "streak gonads." The uterus is hypoplastic; it may be bifid or reduced in its cervical portion. The fallopian tubes are normal or slender.

In typical cases, histology of the gonads reveals neither germinal cells nor follicular formation, but a fibrous tissue resembling the cortical ovarian stroma.

Secondary sexual characteristics do not appear. The nipples remain infantile. Occasionally, a beginning of mammary development may be noted, although it is possible to be misled by fatty accumulation, especially in plump girls.

Hormonal Investigations

Urinary hormonal elimination is compatible with impuberism of gonadal origin. It is characterized by the absence of estrogens and of pregnandiol as well as by an increased level of FSH above 40 MU. In some cases, these appear in the urine long after the usual age of puberty. Inversely, the level of FSH may be normal in adult patients having no sign of estrogenic impregnation.

The 17-ketosteroid level is often low. Thyroid function is normal.

Malformations

- *Heart and blood vessels:* cardiovascular malformations are frequent, especially in the classic form 45,X (at least 20 percent of the cases). Coarctation of the aorta exists in more than half the cases.
- *Kidneys:* the high frequency of renal malformations (40 to 60 percent) and the usual absence of clinical manifestation justify systematic intravenous pyelography. Most frequent are horseshoe kidney, anomalies of rotation, hypoplasia or agenesia, hydronephrosis, and reno-ureteral bifidity. Hypertension, when present, does not seem to be correlated with renal malformations.
- *Skeleton:* X-ray films confirm the anomalies of the knees, wrists, and hands.

  The knee: The inner tibial plate is low, slightly slanted downward and inward, and projects beyond the subtending metaphysis (Kosowicz' sign). It can be observed especially after the child reaches age 7.

  The wrist: A raised semilunar bone causes an ovalization of the carp. The lower extremity of the radius is often somewhat excessively slanted downward and outward, and, at its worst produces Madelung's deformity.

  The hand: Shortness of the fourth and occasionally of the fifth metacarpal is very characteristic.

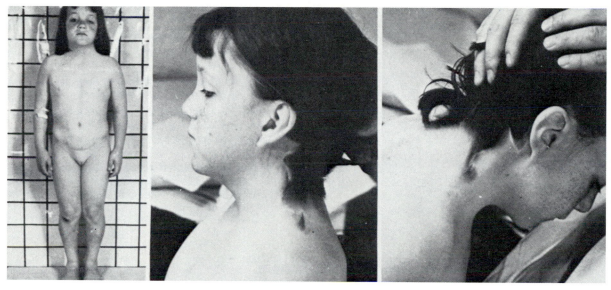

23.4

23.5

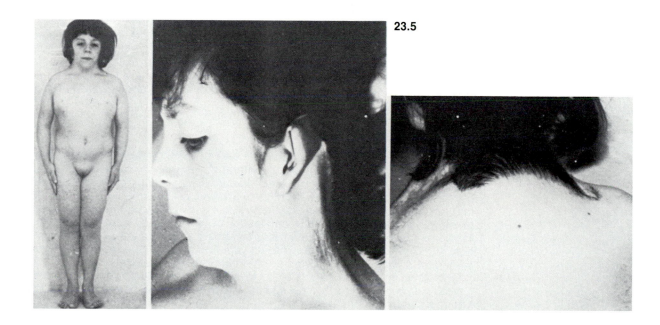

Apart from the malformations described above, X-ray examination may reveal:

Delayed bone age, especially evident after the age of puberty.

Excessive transparency of the skeleton, but with neither severe osteoporosis nor fractures.

A sella turcica small in size, hypoplasia of the first cervical vertebra, and unequal development of the dorso-lumbar vertebral plates.

- *Sensory organs:* severe myopia, congenital cataracts, congenital deafness, or hearing difficulties are frequent complications of Turner's syndrome.
- *X-linked characteristics and diseases:* it is particularly interesting to note that the same frequency of sex-linked characteristics and anomalies are found in 45,X patients and normal 46,XY males, as predicted by genetic theory. The most frequent anomaly is daltonism. Patients with hemophilia and muscular dystrophy are also known. In a certain number of cases, these anomalies allow the determination of the paternal or maternal origin of the X chromosome present in the patients, in much the same way that normal markers such as the Xg and Xm blood groups are used.

Psychomotor Development

Psychomotor development is extremely variable from one patient to another. Slight retardation is not uncommon, but the IQ may also attain high levels.

The frequency of 45,X subjects seems no higher in specialized institutions than in the general population.

Prognosis

Somatic malformations may be severe enough to cause death in the neonatal period, but such cases are rare. Overall, malformations are rather subdued and the survival rate is normal, with the disease usually remaining undiagnosed until the age of puberty.

Treatment

Until recently, treatment was not undertaken before the age of 16, with the intention of avoiding "blocking" possible natural development. Currently, it is held that treatment may be undertaken at age 14 years, and possibly earlier if warranted by psychological considerations. It primarily involves estrogen treatment, which usually induces satisfactory breast development, pubic and axillary hair growth, development of internal and external genitalia, and finally menstruation. This treatment has considerable psychological value in allowing mammary development as well as menstruation. Nonetheless, there exists a certain variation in individual response to the treatment, and it is advisable to avoid undue optimism at the outset.

The usual therapy does not seem to affect growth. This is also true of human growth hormone or anabolizing agents, which have a temporary effect but do not seem to modify the final height.

Another aspect of treatment consists of prevention of the gonadal tumors (gonadoblastoma, dysgerminoma). These are exceptional in the course of "classic" Turner's syndrome and do not justify a gonadectomy. Conversely, when the patient is a carrier of a karyotypic variant with a Y chromosome, gonadectomy is expressly indicated.

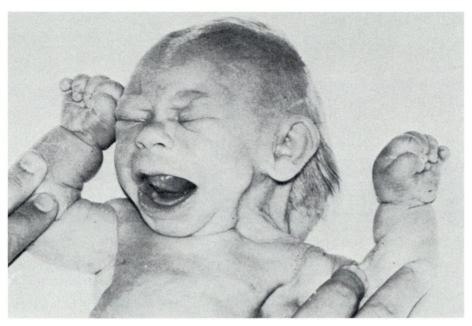

23.6

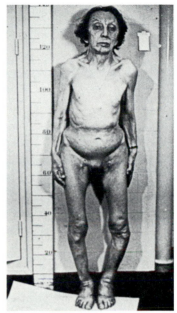

23.7

23.8

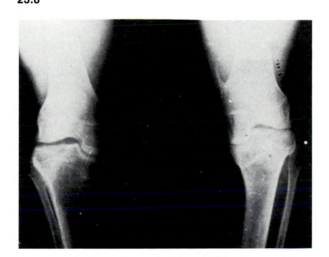

23.9

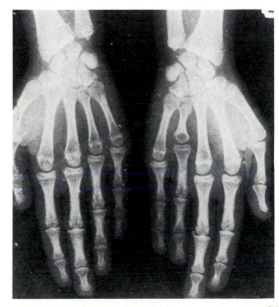

DERMATOGLYPHICS

Palmar and digital dermatoglyphics do not allow diagnosis of Turner's syndrome, but nonetheless present some characteristic features:

- The axial triradius is most often in the normal t position, more rarely in t' or t".
- The frequency of complex hypothenar patterns is increased whereas the frequency of thenar patterns is decreased.
- Subdigital triradius b is shifted toward the ulnar margin. Line A terminates in zone 1 rather than in zone 3, and line T in zone 11 rather than in zone 13.
- The index of transversality is lowered.
- The total ridge count is increased.

CYTOGENETIC STUDY

It is difficult to determine the actual frequency of the 45,X karyotype with respect to the karyotypic variants observed in Turner's syndrome, and especially mosaicism. Much depends on the quality of the cytogenetic studies and on the number of different cells and tissues studied in a given patient.

In this respect, the frequencies given by various authors vary somewhat. It can be estimated that the homogenous 45,X karyotype is observed in about 55 percent of the patients, where study of the buccal smear reveals the absence of Barr bodies. The study of genetic markers, in particular of blood group Xg, reveals that in 76 percent of the cases the X chromosome present is of maternal origin.

In 45 percent of the cases there is a karyotypic variation in number or structure, which may be accompanied by the presence of one or more Barr bodies. These variants appear in Table 1. It can be noted that the most frequent mosaicism without structural anomaly is the 46,XX/45,X mosaicism. The most frequent structural anomaly is an isochromosome of the long arm of X: i(Xq). Banding techniques show that certain anomalies of the X chromosomes (maybe on the order of half) previously considered to be i(Xq) are actually translocations between two X chromosomes. Likewise, it is now becoming possible to distinguish the rare cases of i(Xp) from those of Xq−.

Techniques that allow study of the replication of the X chromosome (autoradiography or, better, BrdU incorporation) show that in the great majority of cases it is the rearranged X that is "inactive" or late replicating (see technical appendix).

CLINICAL FORMS

The phenotype associated with the 45,X karyotype is relatively variable and especially in the intensity of the dysmorphic syndrome.

This variability is more pronounced in cases of karyotypic variants. Small stature must be considered as a constant and essential feature. However, relatively eumorphic development can be observed, as well as a beginning of puberty with primary mammary development and both axillary and pubic pilosity, or more rarely, spontaneous menstruations (the reality of which is not always certain). Cases of pregnancy have been reported, but the reality of a homogeneous 45,X karyotype remains uncertain.

In other patients, and often in the presence of a Y chromosome, a certain degree of virilization can be seen, in particular, hypertrophy of the clitoris.

Both the macroscopic and microscopic aspects of the gonads are also variable. Although the classic presence of a streak gonad is most frequent, either total aplasia (proven by serial sectioning of the parametrium) or rudimentary development, some-

TABLE 1   **Karyotypes in Turner's Syndrome**

Karyotypes	Barr Bodies	Frequency (%)
Monosomy X		
45,X	0	55
Mosaicism		
46,XX/45,X	1 ⎫	
47,XXX/45,X	2 ⎬	10
47,XXX/46,XX/45,X	2 ⎭	
Isochromosome X		
46,X,i(Xq)	1 ⎫	
46,X,i(Xq)/45,X	1 ⎪	
47,X,i(Xq),i(Xq)/46,X,i(Xq)/45,X	2 ⎬	20
46,X,i(Xp)	1 ⎪	
46,X,i(Xp)/45,X	1 ⎭	
Deletion of X		
46,X,del(Xp)	1 ⎫	
46,X,del(Xq)	1 ⎬	5
46,X,del(Xp)/45,X	1 ⎪	
46,X,del(Xq)/45,X	1 ⎭	
X ring		
46,X,r(X)/45,X	1	5
Y chromosome		
46,XY	0 ⎫	
46,XY/45,X	0 ⎪	
46,XYY/45,X	0 ⎬	5
46,X,del(Yq)/45,X	0 ⎪	
46,X,i(Yq)	0 ⎪	
46,X,dic(Y)	0 ⎭	
Normal karyotype		
46,XX	1	except.

times reduced to the presence of several primordial follicles, sometimes with hypoplastic but recognizable ovaries, can be observed. Finally, "large, hard" ovaries as in Stein-Leventhal's syndrome, may be present.

TURNER'S SYNDROME AND PREGNANCY

Since the initial observation of fertility in a patient with Turner's syndrome (Bahner et al., 1960), more than 80 cases of beginning pregnancies have been found in women with a 45,X homogeneous or numerical mosaicism, or who have a structural anomaly of the X (see Grouchy et al., 1981). The genetic risk in such cases is then high (on the order of 80%), and includes spontaneous abortions, stillbirths, the birth of children with abnormal karyotypes (21 trisomy, aberrations in the Turner series), severe malformations (congenital heart disease, spina bifida, omphalocele, and intestinal malformations). Only slightly more than one-third of all children born alive can be considered normal (Mavel et al., 1980).

REFERENCES

BAHNER F., SCHWARZ G., HARNDEN D.G., JACOBS P.A., HIENZ H.A., WALTER K.: A fertile female with XO sex chromosome constitution. Lancet II:100, 1960.

GROUCHY J. de, THIBAUD E., TURLEAU C., ROUBIN M., ROSE F., CACHIN O., RAPPAPORT R.: Femme 45,X/46,X,del (X)(q25) accouchée d'une fille normale 46,XX. Ann. Génét., 24:229–230, 1981.

MAVEL A., TURC C., FELDMAN J.P., MICHIELS Y., NIVELON-CHEVALIER A., KOSLOWSKI J.P., GERVAIS G.: La fonction gonadique des femmes à caryotypes XO homogène ou en mosaïque numérique. J. Gyn. Obstet. Biol. Reprod., 9:875–886, 1980.

NOONAN'S SYNDROME

REFERENCES

NOONAN J.A., EHMKE D.A.: Associated noncardiac malformations in children with congenital heart disease. *J. Pediat.* **63**:468–470, 1963.

The term "Noonan's syndrome" (Noonan and Ehmke, 1963) now tends to be used instead of the older terms "male Turner's syndrome," "Turnerian phenotype with normal karyotype," "Ullrich's syndrome," and "pseudo-Turner female syndrome."

Defined by a "Turnerian" phenotype and a normal karyotype, it may be observed in the male as well as in the female. It is transmitted as an autosomal dominant character. Gonadal function is variable, ranging from gonadal agenesia to normal gonadal development and fertility.

THE 47,XXX KARYOTYPE

REFERENCES

JACOBS PA, BAIKIE AG, COURT-BROWN WN, MacGREGOR TN, MacLEAN N, HARNDEN DG: Evidence for the existence of the human superfemale. *Lancet* **2**:423–425, 1959.

LAMY M, JOSSO N, GROUCHY J de, BITAN A: Anomalies des gonosomes. *Rapport au XXè Congrès de Pédiatrie de Langue Française, Nancy,* 1965.

SINGER J, SACHDEVA S, SMITH GF, HSIA DYY: Triple X female and a Down's syndrome offspring. *J. Med. Genet.* **9**:238–239, 1972.

TENNES K, PUCK M, BRYANT K, FRANKENBURG W, ROBINSON A: A developmental study of girls with trisomy X. *Am. J. Hum. Genet.* **27**:71–80, 1975.

ZIZKA K, BALICEK P, NIELSEN J: XXYY son of a triple-X mother. *Humangenetik* **26**:159–160, 1975.

The 47,XXX karyotype was described for the first time in 1959 by Jacobs et al. under the name "superfemale" by analogy with observations made in Drosophila, but the term was quickly abandoned. It is now known that the anomaly is frequent, but not correlated with a true clinical syndrome (Lamy et al., 1965; Tennes et al., 1975).

GENERALIZATIONS

- Frequency: 0.8 per 1,000 females
- Mean maternal age: 32.8 years
- Mean paternal age: 36.5 years
- Mean birth weight: 2,350 g

PHENOTYPE

In most patients, phenotype, puberty, and fertility are normal. In certain patients, menstrual disorders can be seen: secondary amenorrhea, spaniomenorrhea, and often early menopause.

Intellectual Development

In one-third of the cases intellectual development is considered to be normal, and in two-thirds of the cases slightly below normal, with schizophrenic tendencies. The frequency of 47,XXX patients is in fact elevated in psychiatric hospitals.

CYTOGENETIC STUDY

The karyotype is either homogeneous 47,XXX or a mosaic 47,XXX/46,XX.

The progeny of 47,XXX females should, in theory, be one-half normal and one-half 47,XXX or 47,XXY. In fact, the progeny of such women are usually normal. There have nonetheless been reports of several cases of 47,XXY or mosaicism (47,XXY/46,XY and 47,XXX/46,XX), as well as children who are carriers of diverse anomalies (21 trisomy, 48,XXYY) (Singer et al., 1972; Zizka et al., 1975). It is possible that the genetic risk involved is greater than was thought following the first observations of births in women trisomic for chromosome X.

DERMATOGLYPHICS

There is an excess of radial loops and arches. The total ridge count is reduced.

48,XXXX SYNDROME

Oval face
Vague resemblance to 21 trisomy

REFERENCES

CARR D.H., BARR M.L., PLUNKETT E.R.: An XXXX sex chromosome complex in two mentally defective females. *Canad. Med. Ass. J.,* **84**:131–137, 1961.

GARDNER R.J.H., VEALE A.M.O., SANDS V.E.: XXXX syndrome: case report and a note on genetic counselling and fertility. *Humangenetik,* **17**:323–330, 1973.

GROUCHY J. de, BRISSAUD H.E., RICHARDET J.M., REPESSE G., SANGER R., RACE R.R., SALMON C., SALMON D.: Syndrome 48,XXXX chez une enfant de six ans. *Ann. Génét.,* **11**:120–124, 1968.

GROUCHY J. de, VIALATTE J., CHAVIN-COLIN F., ROUBIN M., TURLEAU C.: Le syndrome 48,XXXX. *Arch. Fr. Pédiat.* (supplt), **36**:XLI–XLVII, 1979.

HAMERTON J.L. *Human Cytogenetics. Clinical Cytogenetics, Vol. II,* New York, Academic Press, pp. 99–112, 1971.

PENA S.D.J., RAY M., DOUGLAS G., LOADMAN E., HAMERTON J.L.: A 48,XXXX female. *J. med. Genet.,* **11**:211–215, 1974.

The first description of a 48,XXXX karyotype was reported in 1961 by Carr et al. and concerned two patients with normal female phenotype and normal menstruation, but who were mentally retarded.

Although not all 48,XXXX females have necessarily been detected due to the fact that some have normal morphology and intelligence, this constitution can be considered rare. Thirty cases have been reported in the literature (Gardner et al., 1973; Pena et al., 1974; Grouchy et al., 1979).

GENERALIZATIONS

- Frequency: poorly known, but undoubtedly quite low
- Parental ages: maternal: 26.5 years
  paternal: 31.3 years

PHENOTYPE

Although a characteristic phenotype concomitant with the 48,XXXX constitution cannot be described, dysmorphisms are nonetheless more pronounced than in cases of 47,XXX females.

Craniofacial Dysmorphism

The dysmorphic features most frequently observed are heavy oval face, slight hypertelorism, bilateral epicanthal folds, palpebral fissures slightly slanted upward and outward and flattening of the bridge of the nose. These combined features can be vaguely reminiscent of 21 trisomy.

Other features are sometimes encountered: strabismus, coloboma iridis, nystagmus, myopia, cataracts, anomalies of the ears, and microcephaly.

Neck, Thorax, and Abdomen

The neck is short, and pterygium colli is occasionally present. There may be osteoarticular anomalies, dorsal kyphoses, and hip dislocations.

Stature is variable. The mean height of adult patients is 170 cm for a weight of 70 kg.

Limbs

Radioulnar synostosis, shortness of the fingers (involving the last phalange), and clinodactyly of the fifth digit have been reported in several cases.

Genitalia

External genitalia are usually normal.
Menstrual disorders are very frequent.
Mammary development is usually normal.
Only one 48,XXXX woman is known to have had two children, and one of these was 21 trisomic.

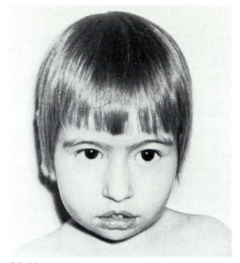

23.10

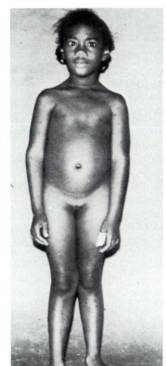

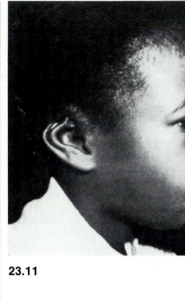

23.11

23.12

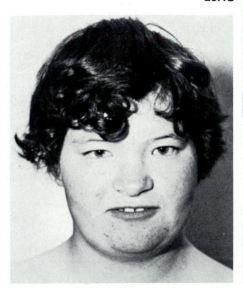

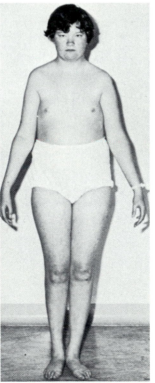

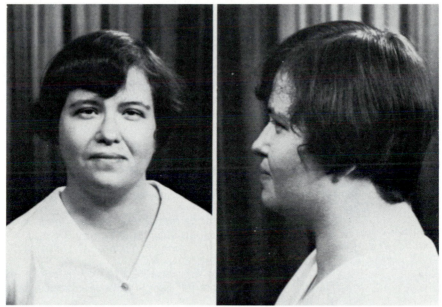

23.14

23.13

23.15

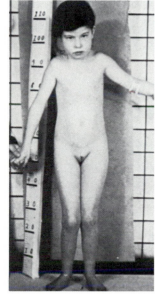

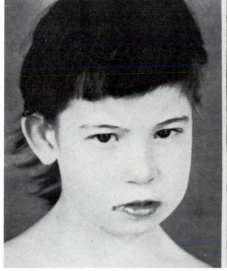

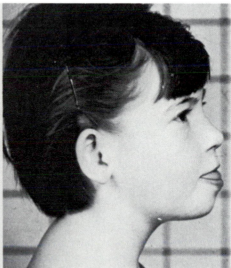

Mental Retardation

IQ varies between 50 and 100. Normal or subnormal intelligence is nonetheless rare (one-fifth of the cases).

Language difficulties, including difficulties in articulation rather than in expression, or even logorrhea, would seem to be fairly characteristic of this syndrome. A case of aggressive tendencies has been reported, but does not appear to be typical. No schizophrenic or psychotic tendencies have been observed as in 47,XXX syndrome.

Prognosis

Life expectancy is not affected. The eldest patient is 66.

CYTOGENETIC STUDY

Apart from homogeneous forms, several cases of mosaicism can be observed. In one case, study of Xg blood groups demonstrated the maternal origin of the four X chromosomes, the father not having supplied any sexual chromosome (Grouchy et al., 1968). Similar observations in 49,XXXXY and 49,XXXXX patients suggest that 48,XXXX individuals are in fact false 49,XXXXX (or less often 49,XXXXY) subjects, owing to the loss of the paternal sex chromosome at fertilization of a 26,XXXX ovule (Hamerton, 1971).

DERMATOGLYPHICS

The total ridge count is decreased, as compared with both normal and 47,XXX women.

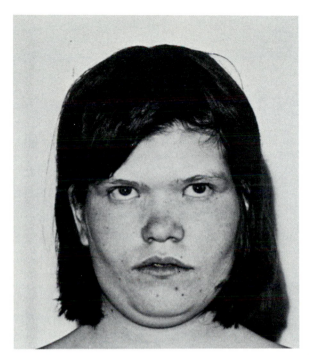
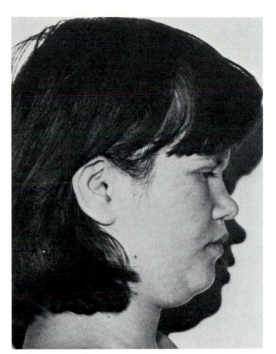

23.16

23.17

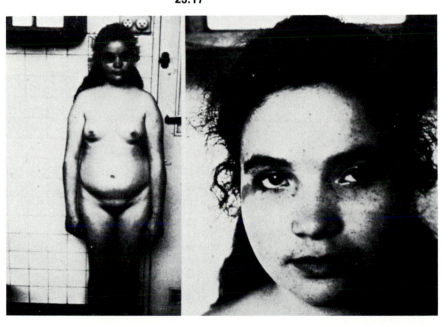

49,XXXXX SYNDROME

REFERENCES

FUNDERBURK S.J., VALENTE M., KLISAK I.: Pentasomy X: report of patient and studies of X-inactivation. *Am. J. med. Genet.*, **8**:27–33, 1981.

KESAREE N., WOOLLEY P.V.: A phenotypic female with 49 chromosomes, presumably XXXXX. *J. Pediat.*, **63**:1099–1103, 1963.

MONHEIT A., FRANCKE U., SAUNDERS B., JONES K.L.: The penta-X syndrome. *J. med. Genet.*, **17**:392–395, 1980.

Heavy set face
Hypertelorism
Vague resemblance to 21 trisomy

In 1963, Kesaree and Wooley observed a 49,XXXXX karyotype in a 2-year-old female with minor malformations and delayed psychomotor development.

Such a genetic constitution is rarely observed. Mosaicisms included, around 20 cases are known (Monheit et al., 1980; Funderburk et al., 1981).

GENERALIZATION

- Mean parental ages:
  maternal: 28.4 years
  paternal: 28.9 years
- Mean birth weight: 2,030 g

PHENOTYPE

Diverse dysmorphic features have been described. However, the corresponding syndrome remains rather poorly defined.

Craniofacial Dysmorphism

The head circumference is small in half of the cases. The face is heavy set and elongated, giving the impression of flabbiness. The most frequent features are hypertelorism, palpebral fissures slanted upward and outward, and a flattened nasal bridge. These features produce a vague resemblance to 21 trisomy.

Dental malformations and low-set ears are frequently encountered.

Neck, Thorax, and Abdomen

The neck can be short, with a low hairline.

The appearance of the thorax and the narrowness of the shoulders can be evocative.

Scoliosis, uni- or bilateral hip dislocation, as well as minor and multiple anomalies of the bones and articulations can be observed.

Limbs

The most suggestive malformations are radio-cubital synostosis and campto-clinodactyly of the Vth finger. Club foot, micromelia, genu valgum, and a retardation in bone age may also be observed.

Genitalia

The external genitalia are usually normal. A majority of the observations involve children. Nevertheless, ovarian dysgenesis, absence of an ovary, and atrophic uterus have all been reported.

Malformations

Cardiac malformations have been reported in half the patients, especially patent ductus arteriosus.

Incoordination of ocular movements is noted in some cases.

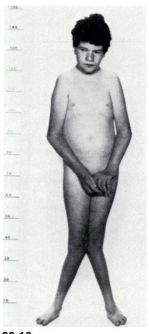

23.18

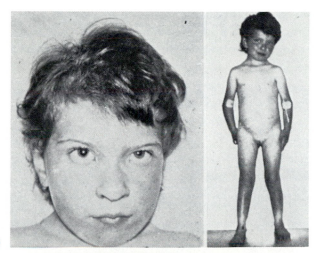

23.19

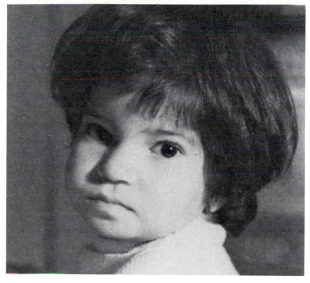

23.20

Mental Retardation

The degree of mental retardation is variable, but appears to be comparable with 48,XXXX patients.

CYTOGENETIC STUDY

In more than two-thirds of the cases the karyotype seems homogeneous. Somewhat complex mosaics are observed in the other cases.

DERMATOGLYPHICS

A decrease in the total ridge count is reencountered here. The ridges may be immature.

KLINEFELTER'S SYNDROME

Gynecomastia
Testicular atrophy
Sterility

REFERENCES

BRADBURY J.T., BUNGE R.G., BOCCA-BELLA R.A.: Chromatin in Klinefelter's syndrome. *J. Clin. Endocr.*, **16**:689, 1956.

JACOBS P.A., STRONG J.A.: A case of human intersexuality having a possible XXY sex-determining mechanism. *Nature*, **183**:302–303, 1959.

KLINEFELTER H.F., REIFENSTEIN E.C., ALBRIGHT F.: Syndrome characterized by gynecomastia, aspermatogenesis, without aleydigism and increased excretion of follicle stimulating hormone. *J. Clin. Endocr.*, **2**:615–627, 1942.

POLANI P.E., BISHOP P.M.F., LENNOX B., FERGUSSON-SMITH M., STEWART J.S.S., PRADER A.: Colour vision study and the X chromosome constitution of patients with Klinefelter's syndrome. *Nature*, **182**:1092–1093, 1958.

In 1942, Klinefelter et al. described a syndrome involving gynecomastia, testicular atrophy with azoospermia but without atrophy of Leydig cells, and increased excretion of FSH.

In 1956, Bradbury et al. as well as Plunkett and Barr demonstrated the presence of Barr bodies in the cell nuclei of patients.

In 1958, Polani et al. investigated the frequency of daltonism in these patients and suggested the presence of two X chromosomes. The following year, Jacobs and Strong demonstrated the karyotype to be 47,XXY.

GENERALIZATIONS

- Frequency: 1.18 per 1,000 male births (mosaics included)
- Mean maternal age: 31.3 years
- Mean paternal age: 35.5 years

PHENOTYPE

Klinefelter's syndrome entails no significant dysmorphism apart from the symptoms described above.

Only in exceptional cases of systematic newborn population studies can the disease be diagnosed at birth.

Equally rare is the diagnosis of the syndrome in young children because of the observation of nonspecific associated malformations of the genitalia, such as testicular ectopy, hypospadias, and hypoplasia of the penis and scrotum.

The disease is usually diagnosed at puberty. Gynecomastia may appear at around the age of 12–13 years. Involved is true gynecomastia of moderate volume, often asymmetric at the outset, but symmetric in the adult, and without hyperpigmentation. It is far from being a constant feature and is present in only one-fourth to one-third of the patients.

Testicular atrophy contrasts with the normal development of the penis and an otherwise normal puberty. It is constant. The testicles are small, soft, and often insensitive to pressure. The scrotum is normal in size and pigmentation.

Morphology of the subjects is variable and not a necessary diagnostic criteria. Whereas some subjects are dolichomorphic with long limbs and a gynecoid appearance, others have a normal masculine morphology and may even be small in size. Diagnosis can, in fact, only be made because of sterility.

Other features are equally inconstant: poor musculature, delicate skin, and scarceness of hair.

Histology

In the child, testicular histology is almost normal.

At puberty, the seminiferous tubules are arranged in a very irregular manner, alone

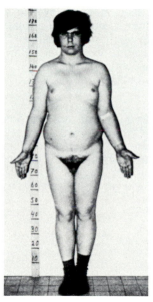

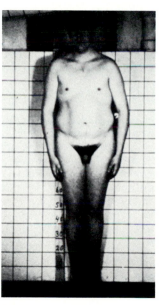

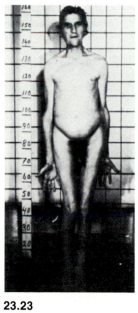

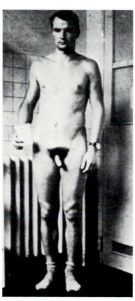

23.21

23.22

23.23

23.24

or in bundles. Atrophic and hyalinosed, they contain almost nothing but Sertoli cells. Several elements of germinal development can be observed, but their maturation rarely goes beyond the spermatocyte stage. Azoospermia is a practically constant feature.

Leydig cells are present as large clusters between the atrophied tubes. Tubulo-interstitial sclerohyalinosis increases after puberty.

Endocrinology

Before puberty, hormonal excretions are unmodified. An increase in the rate of excretion of FSH can be observed at puberty. The excretion of ketosteroids is variable. It may be normal or lowered in half the cases.

Intellectual Development

A majority of 47,XXY subjects have normal intellectual development. However, it is known that a certain number have more or less severe mental debility or psychotic disorders. They may experience difficulties in social adaptation and sometimes find themselves condemned by the courts for minor offenses or abnormal sexual behavior.

On the whole, libido and sexual activity of 47,XXY subjects is reduced.

CYTOGENETIC STUDY

In 80 percent of the cases, the 47,XXY karyotype is homogeneous. It corresponds to the "classic" Klinefelter's syndrome.

In 20 percent of the cases, a chromosomal variant is observed, such as mosaicism; the most frequent are indicated in Table 2. In the 47,XXY/46,XY mosaic fairly complete spermatogenesis can occasionally be observed.

The 48,XXXY karyotype and corresponding mosaics are accompanied by more severe hypogonadism and mental retardation.

TABLE 2 **Karyotypes in Klinefelter's Syndrome**

47,XXY	80%
48,XXXY	
47,XXY/46,XY	
47,XXY/46,XX	20%
47,XXY/46,XY/45,X	
47,XXY/46,XY/46,XX	

DERMATOGLYPHICS

Palmar and digital dermatoglyphics are not very characteristic. The following tendencies may be noted:

- Decreased total ridge count with an increased frequency of arches
- Increased frequency of the single transverse palmar crease
- Axial triradius displaced toward the ulnar margin of the hand

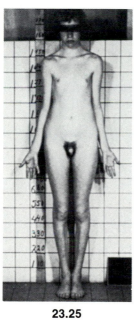 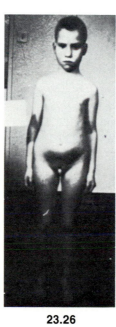 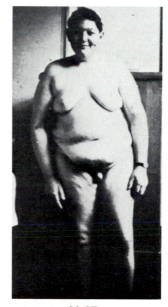

23.25   **23.26**   **23.27**

46,XX MALES

Gynecomastia
Testicular atrophy
Sterility

REFERENCES

EVANS H.J., BUCKTON K.E., SPOWART G., CAROTHERS A.D.: Heteromorphic X chromosomes in 46,XX males: evidence for the involvement of X-Y interchange. *Hum. Genet.,* **49**:11–31, 1979.

FERGUSON-SMITH M.A.: X-Y chromosomal interchange in the aetiology of true hermaphrodism and of XX Klinefelter syndrome. *Lancet,* **ii**:475–476, 1976.

GROUCHY J. de, JOSSO N., LAMY M., FREZAL J., NEZELOF C., FEINTUCH G.: Syndrome de Klinefelter chez un nourrisson hypospade. Caryotype à 46 chromosomes. *Ann. Pédiat.,* **39**:173–177, 1963.

LA CHAPELLE A. de, HORTLING H., NIEMI M., WENNSTROM J.: XX sex chromosomes in a human male. First case. *Acta med. Scand., Suppl.* **412**:25–38, 1964.

LA CHAPELLE A. de: Nature and origin of males with XX sex chromosomes. *Amer. J. hum. Genet.,* **24**:71–105, 1972.

LA CHAPELLE A. de: The aetiology of human XX males. *Vth International Congress of Human Genetics, Mexico.* Amsterdam, Excerpta Medica. International Congress Series, no 397, 1976.

LA CHAPELLE A. de, KOO G.C., WACHTEL S.S.: Recessive sex determining genes in human XX male syndrome. *Cell,* **15**:837–842, 1978.

LA CHAPELLE A. de, SIMOLA K., SIMOLA P., KNUUTILA S., GAHMBERG N., PAJUNEN L., LUNDQVIST C., SARNA S., MURROS J.: Heteromorphic X chromosomes in 46,XX males? *Hum. Genet.,* **52**:157–167, 1979.

The first observation of an apparently normal 46,XX female karyotype in a subject with a male phenotype was reported by La Chapelle et al. in 1964.

However, in 1963, Grouchy et al., in a boy afflicted with hypogonadism, described a supposed Y/autosome translocation in the presence of two normal X chromosomes. A new examination after banding showed that what was involved was, in fact, a 46,XX male with a Gp+ and a Dp+.

46,XX males constitute a clinical entity very closely related to the 46,XXY Klinefelter's syndrome (La Chapelle, 1972).

GENERALIZATIONS

- Frequency: still not entirely known; it seems to be ten to fifteen times less frequent than 47,XXY—i.e., about 1/10,000 male births
- Mean maternal age: 27.8 years
- Mean paternal age: 30.9 years

PHENOTYPE

The phenotype is very similar to that of Klinefelter's syndrome. However, the following points can be noted:

- Mean stature (168.2 cm) is below that of 47,XXY subjects (177.4 cm). It is intermediate between that of normal 46,XY males and normal 46,XX females.
- Contrary to certain 47,XXY subjects, a disproportion between the trunk and limbs is usually not observed.
- Gynecomastia is slightly less frequent than in Klinefelter's syndrome.

Genitalia

The external genitalia are morphologically identical to those of 47,XXY subjects.

Histology

The appearance of the gonads is intermediate between that observed in Klinefelter's syndrome and in Del Castillo's syndrome. Seminiferous tubules are normal or slightly small in size and contain only Sertoli cells, with no hyperplasia of Leydig cells, and with subdued sclerohyalinosis.

Hormonal Investigations

The modifications of hormonal secretions are similar to those observed in Klinefelter's syndrome.

Psychological Development

As in Klinefelter's syndrome, different psychological disorders are described, psychasthenia in particular, although neither aggressivity nor psychotic tendency are encountered.

On the whole, intelligence is normal. There is no concentration of 46,XX subjects in mentally retarded populations.

CYTOGENETIC STUDY

Two cytogenetic theories have been proposed to explain the development of male gonadal tissue in the absence of a Y chromosome:

1. A reciprocal translocation between homologous portions of the X and Y chromosomes (Ferguson-Smith, 1966; La Chapelle et al., 1979; Evans et al., 1979).

2. An undetected mosaicism or an elimination of the Y chromosome (La Chapelle et al., 1964).

The development of banding techniques has not yet allowed discrimination of the correct hypothesis.

A third theory, that of a gene mutation, did not seem compatible with the absence of consanguinity in the parents, nor with the absence of familial cases. Recent observations are however compatible with this hypothesis (La Chapelle, 1976).

DERMATOGLYPHICS

These do not show anything in particular. The characteristics observed in Klinefelter's syndrome are not reencountered here.

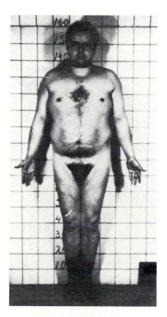

23.28

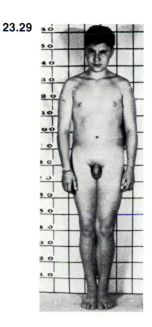

23.29

47,XYY MALES

REFERENCES

BENEZECH M., BOURGEOIS M., NOEL B.: Les hommes XYY. Aspects psychiatriques et criminologiques. *Ann. médico-psychol.* 2:365–394, 1974.

HAUSCHKA T.S., HASSON J.E., GOLDENSTEIN M.N., KOEPS G.F., SANDBERG A.A.: One XYY man with progeny indicating familial tendency to nondisjunction. *Amer. J. hum. Genet.*, **14**:22–30, 1962.

HOOK E.B.: XYY genotype. *Science,* **189**:1044–1045, 1975.

JACOBS P.A.: XYY genotype. *Science,* **189**:1044–1045, 1975.

JACOBS P.A., BRUNTON M., MELVILLE M.M., BRITTAIN R.P., McCLEMONT W.F.: Aggressive behaviour, mental subnormality and the XYY male. *Nature,* **208**:1351–1352, 1965.

MELNYCK J., THOMPSON H., RUCCI A.J., VANASEK F., HAYES S.: Failure of transmission of the extra chromosome in subjects with 47,XYY karyotype. *Lancet,* ii:797–798, 1969.

SANDBERG A.A., KOEPF G.F., ISHIHARA T., HAUSCHKA T.S.: An XYY human male. *Lancet,* ii:488–489, 1961.

SUNDEQUIST U., HELLSTROM E.: Transmission of 47,XYY karyotype. *Lancet,* ii:1367, 1969.

The 47,XYY constitution was first observed in 1961 by Sandberg et al. in a man of normal intelligence who had had numerous progeny by two women, including an amenorrheic daughter, a pair of twins, one a "blue baby" and the other with 21 trisomy, and two spontaneous abortions (Hauschka et al. 1962).

As early as 1963, investigations conducted in centers for violent criminals and subjects of subnormal intelligence demonstrated a concentration of male subjects with Barr bodies; one-third of them were 48,XXYY.

Following these observations, Jacobs et al. (1965) suggested that the presence of two Y chromosomes could produce a predisposition toward abnormally aggressive behavior. Examination of 197 incarcerated subjects, mentally retarded and with violent or criminal tendencies, revealed seven 47,XYY males, one 48,XXYY, and one 47,XYY/46,XY mosaic, as well as three carriers of autosomal structural anomalies. These authors concluded that there could be a relationship between the presence of a supernumerary Y chromosome and a tendency toward aggressivity. They also noted that the subjects involved were unusually tall.

This publication immediately caused a sensation, and with the involvement of the news media, the additional Y chromosome became the "chromosome of crime."

With the passage of time and the accumulation of data, a proper perspective has been gained. It is now known that the 47,XYY anomaly is one of the most frequent and the majority of 47,XYY subjects do not stand out from the general population.

GENERALIZATIONS

- Frequency: 1.1 per 1,000 male births.
- Parental ages: these do not seem modified. There is, however, no satisfactory investigation of this point.

PHENOTYPE

The phenotype of 47,XYY males is normal. However, the subjects involved are often tall. Mean height varies from 180 to 186 cm, depending on the reports. The frequency of 47,XYY subjects in a given population increases rapidly as a function of height. It attains the 10 percent level for subjects taller than 200 cm.

EEG modifications have been reported, but seem to be minor and without significance.

There is no significant modification in the level of hormonal secretions.

The majority of 47,XYY males are usually fertile. Histological lesions of the gonads have nonetheless been described, the most extreme being more or less complete arrest of spermatogenesis. Genital hypoplasia, testicular ectopy, and hypospadias can be seen, but are not the rule.

The investigation of meiosis shows the formation of a normal sexual bivalent containing a single Y chromosome. The supernumerary Y chromosome is apparently eliminated by selective nondisjunction before spermatogenesis. Rare cases (probably less than five) of subjects having procreated a 47,XYY son are known (Melnyck et al., 1969; Sundequist and Hellstrom, 1969).

SOCIAL RISK

There is a definite correlation between the 47,XYY karyotype and the chance of internment in an institution for mentally retarded delinquents. The exact nature and significance of this association remain to be determined.

The frequency of 47,XYY males is in the vicinity of 0.11 percent in the general population. It attains 2 percent in institutions for dangerous mentally ill subjects—an increase on the order of eighteen times. Moreover, this concentration is not restricted to 47,XYY males but also involves 47,XXY subjects and especially 48,XXYY subjects, for whom the increase may be as much as a hundredfold. Whereas the frequency of males with an abnormal chromosomal constitution is significantly higher in centers for retarded delinquents, this frequency is also higher, though less so, in either exclusively mentally retarded, or exclusively delinquent populations.

Psychiatric examination of "deviant" 47,XYY males has primarily revealed the following: the precocity of character disorders and of first offenses; adequate stability in a protected environment contrasted with poor social integration; emotional instability; low tolerance to frustration; impulsiveness with poor self-control; and aggressivity (Bénézech and Noël, 1975).

The nature of the offenses committed is usually limited to objects (thefts, swindling, and arson). Only very rarely is bloodshed involved.

The proportion of "deviant" subjects among 47,XYY males is difficult to determine, the majority of them remaining undetected. The discovery of a 47,XYY karyotype in a child or young adult is often fortuitous, the examination perhaps being indicated by the presence of diverse somatic or genital anomalies, or by anomalies of progeny. Such a discovery does not allow prediction of subsequent social adaptation of the subject. Moreover, such situations raise questions that are difficult to answer. In particular, should the parents of such a child be informed, or should the individual himself be informed later? Or again, should the condition be tracked down systematically in a given human population? At present, doctors, experts, and psychologists are widely separated on these issues (Hook, 1975; Jacobs, 1975).

23.30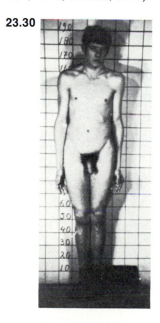

48,XXYY MALES

Hypogonadism
Severe mental retardation
Aggressivity

REFERENCE

MULDAL S., OCKEY C.H: The «double male»: a new chromosome constitution in Klinefelter's syndrome. *Lancet*, II:492, 1960.

The 48,XXYY karyotype was reported for the first time by Muldal and Ockey (1960) under the name of "double male."

Because of the association of the phenotypic characteristics of Klinefelter's syndrome with those of 47,XYY individuals, the 48,XXYY males constitute a very special group.

GENERALIZATIONS

• Frequency: at birth, it is on the order of 0.04 per 1,000 male births. It is 50 to 100 times greater in mental/security institutions.

PHENOTYPE

The phenotype resembles that of Klinefelter's syndrome sometimes in the gynecoid appearance, hypogonadism with its histological modifications and hormonal characteristics, and gynecomastia.

However, it resembles the 47,XYY phenotype in above-normal stature and aggressivity.

Mental retardation is more pronounced and more frequent than for either of these two constitutions.

The association of mental debility and aggressivity explains the significant concentration (50 to 100 times) of 48,XXYY subjects in mental/security institutions. This significant increase, as compared to that of either 47,XXY or 47,XYY subjects, suggests a "multiplication" of the phenotypic effects correlated with these karyotypes.

DERMATOGLYPHICS

These show more simple digital patterns with an excess of arches, and a lower total ridge count than in Klinefelter's syndrome.

49,XXXXY SYNDROME

Oval face
Hypertelorism
Palpebral fissures slanted upward and outward
Aplastic nasal bridge
Radio-ulnar synostosis

REFERENCES

FRACCARO M., LINDSTEN J.: A child with 49 chromosomes. *Lancet*, ii:1303, 1960.

KUSHNICK T., COLONDRILLO M.: 49XXXXY patient with hemifacial microsomia. *Clin. Genet.*, **7**:442–448, 1975.

SHAPIRO L.R., HSU L.Y.F., CALVIN M.E., HIRSHHORN K.: XXXXY boy. A 15 month old child with normal intellectual development. *Amer. J. Dis. Child.*, **119**:79–81, 1970.

Described in 1960 by Fraccaro and Lindsten, the 49, XXXXY karyotype corresponds to a relatively well-defined phenotype. More than 60 cases had been reported in 1970 (Shapiro et al.). Since that time, isolated cases are only rarely subject to publication (Kushnick and Colondrillo, 1975).

GENERALIZATIONS

- Frequency: poorly known, but seems at least ten times greater than that of 49,XXXXX women
- Mean maternal age: 28.5 years
- Mean paternal age: 30 years
- Mean birth weight: 2,600 g

PHENOTYPE

The phenotype of 49,XXXXY children is undoubtedly more distinctive than that of 49,XXXXX females. Some features reminiscent of 21 trisomy are reencountered.

Craniofacial Dysmorphism

The shape and size of the cranium offer no special characteristic. The face is oval.
Most striking are the significant hypertelorism and the palpebral fissures slightly slanted upward and outward. The latter are larger, often edged with epicanthal folds. The orbits seem wide. Strabismus is frequent.
The nasal bridge is aplastic; the nose is broad and flat, often upturned.
The mouth is large, and well defined.
Prognathism may be present.
The ears are large and poorly folded.

Neck, Thorax, and Abdomen

The neck is short, sometimes broad with pterygium colli.
Minor malformations of the vertebral column are often observed.

Limbs

Anomalies of the upper limbs are very characteristic: bending movements of the elbow and supination are restricted by a radio-ulnar synostosis or, more rarely, by a dislocation without synostosis.
Clinodactyly of the fifth digit is nearly constant.
The most frequent anomaly of the lower limbs is coxa valga.

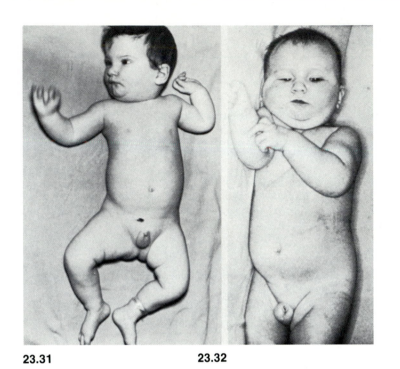

23.31   23.32

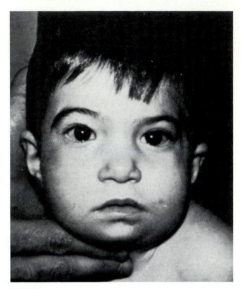

23.33

23.34

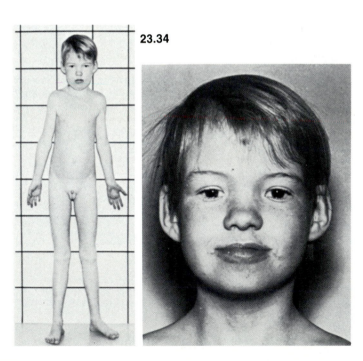

23.35

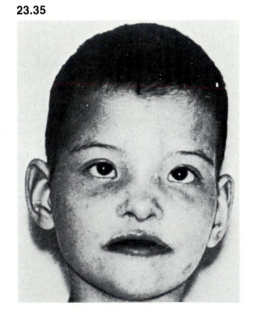

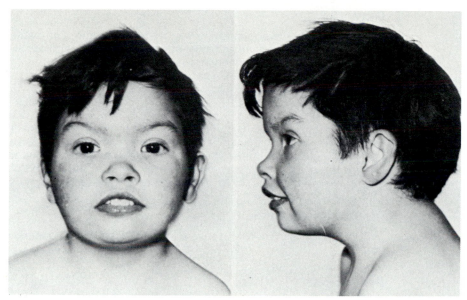

23.36

23.37

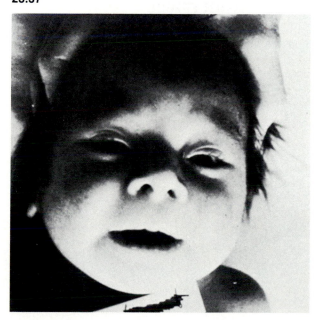

Genitalia

The degree of hypogonadism is extremely variable. The penis and scrotum are small. The testes are small and most often ectopic. In several cases a testicular biopsy revealed atrophy of the seminiferous tubules, more precocious than in Klinefelter's syndrome. Leydig cells are absent.

Radiology

X-ray examination shows that bone age is often very much delayed, and confirms the different malformations already described. Other anomalies can be revealed: an enlarged ulnar head, an elongation of the distal extremity of the ulna, pseudoepiphyses of the carpus and metacarpus, a shortening of the second phalange of the big toe, sclerosed cranial sutures, and vertebral anomalies.

Mental Retardation

Retardation is present and severe in every case. The IQ generally falls between 20 and 50.

Prognosis

This is primarily a function of the severity of mental retardation.

CYTOGENETIC STUDY

The karyotype is usually homogeneous 49,XXXXY. Mosaicism has been reported in about 15 percent of the cases.

DERMATOGLYPHICS

The total ridge count is decreased, as in the other cases of karyotypes involving more than two X chromosomes.

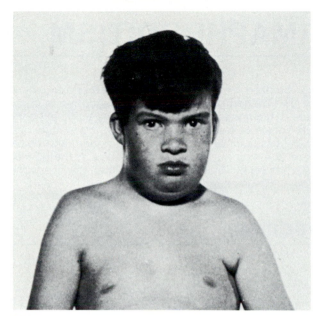

23.38

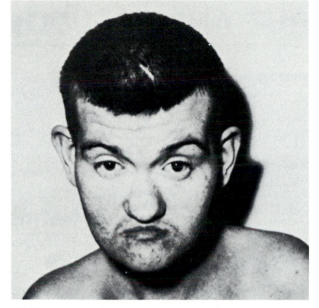

23.39

23.40

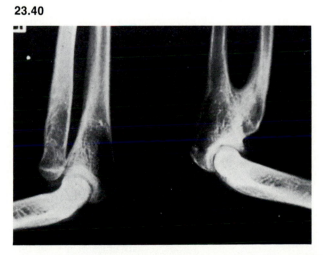

TRUE AND PSEUDOHERMAPHRODITISM

REFERENCES

GARTLER S.M., WAXMAN S.H., GIBLETT E.: An XX/XY human hermaphrodite resulting from double fertilization. *Proc. Nat. Acad. Sci.,* **48**:332–335, 1962.

GROUCHY J. de, MOULLEC J., SALMON C., JOSSO N., FREZAL J., LAMY M.: Hermaphrodisme avec caryotype XX/XY. Etude génétique d'un cas. *Ann. Génét.,* **7**:25–30, 1964.

HAMERTON J.L.: *Human Cytogenetics. Vol. I: General Cytogenetics.* New York, Academic Press, 1971.

JOSSO N.: The Intersex Child. *Pediatric and Adolescent Endocrinology,* vol. 8, Basel, S. Karger, p. 273, 1981.

JOSSO N., FREZAL J., LAMY M.: Aspects génétiques du pseudo-hermaphrodisme masculin. *Ann. Génét.,* **9**:176–180, 1966.

LA CHAPELLE A. de, SCHRÖDER J., RANTANEN P., THOMASSON B., NIEMI M., TIILIKAINEN A., SANGER R., ROBSON E.R.: Early fusion of two human embryos. *Ann. hum. Genet.,* **38**:63–75, 1974.

LAFOURCADE J.: La différenciation sexuelle dans l'espèce humaine. Etude des principales anomalies. In: *Biologie Génétique,* p. 73 (Monographie Annuelle de la Société Française de Biologie Clinique), Paris, Expansion scientifique, 1972.

The intersexual states are characterized by ambiguity of the internal and external genitalia. The distinction between true hermaphroditism and pseudohermaphroditism is based solely on the nature of the gonads. One necessary and sufficient condition is a prerequisite to the diagnosis of true hermaphroditism: the simultaneous presence of both kinds of gonadal tissue, male and female. Pseudohermaphroditism is characterized by the presence of a single type of gonadal tissue. If the gonad is male, the pseudohermaphroditism is said to be male; if female, then the pseudohermaphroditism is said to be female. (See for review Josso, 1981.)

True Hermaphroditism

PHENOTYPE

Genitalia

The ambiguity of the genitalia can assume all degrees between male and female. In two-thirds of the cases the genitalia are more masculine in nature. The newborn is most often considered to be of the male sex, with cryptorchidism and/or hypospadias.

At puberty, breasts develop eight times out of ten and menses become established five times out of ten. Yet consultations can arise for a young girl because of amenorrhea and hypertrophy of the clitoris, while a boy may consult because of gynecomastia or periodic hematuria.

Numerous classifications of true hermaphroditism have been proposed to describe the degree of sexual ambiguity. Each has its merits.

Examination of the internal genitalia most often shows the persistence of the Müllerian as well as Wolffian structures. Fallopian tubes, a normal or most often hypoplastic uterus, epididymis and seminal vesicles can coexist in multiple combinations.

Gonads

The gonads must contain male *and* female structures. The presence of "undifferentiated tissue" or a "stroma of the ovarian kind" is insufficient to diagnose the condition. Seminiferous tubules as well as follicles, must be clearly identifiable.

The following three forms are most often described: lateral or alternate hermaphroditism, with an ovary on one side and a testicle on the other; bilateral hermaphroditism, with an ovotestis on both sides; unilateral hermaphroditism, with an ovary or testicle on one side and an ovotestis on the other. These three forms are encountered with the respective frequencies of 40, 20, and 40 percent.

CYTOGENETIC STUDY

Different karyotypes can be observed in true hermaphroditism. These are indicated in Table 3.

The karyotype 46,XX is the most frequent, corresponding to half of the cases. As for 46,XX males, it is not known how these patients produce testicular tissue in the absence of a Y chromosome.

The 46,XY/46,XX mosaic is probably the result of double fertilization of a female gamete with two nuclei by two spermatozoa (Gartler et al., 1962; Grouchy et al., 1964). About twenty observations are known. The possibility of precocious fusion between two zygotes is not excluded in certain cases (La Chapelle et al., 1974).

Male Pseudohermaphroditism

Male pseudohermaphroditism is defined by an ambiguity of the external and internal genitalia associated with the exclusive presence of male gonadal tissue (for review see Josso et al., 1966; Hamerton, 1971; Lafourcade, 1972).

MALE PSEUDOHERMAPHRODITISM DUE TO CHROMOSOMAL ABERRATION

Male pseudohermaphroditisms are, in rare cases, due to chromosomal aberration. The most frequent of such cases is the 46,XY/45,X mosaicism. Others can, occasionally, be seen and are indicated in Table 4.

From a clinical point of view, the ambiguity of the external genitalia is distinct at birth. The infant is most often declared to be of the male sex. At puberty the appearance of pubic and axillary hair and a deepening of the voice are noted. Mammary development of the female type can be seen.

At adult age, morphology often develops as in males with muscular masses and fatty distribution of the masculine sort. More rarely seen is a female morphology with delicate skin and abundant scalp hair.

The genitalia can be more or less masculinized with a penile organ of variable size. A more or less differentiated urogenital sinus is always present.

Clinical and anatomical diagnosis as well as the choice of the sex in which the child is to be raised must be determined early. These are always of basic importance. Surgery can greatly facilitate the decision.

The danger of malignant degeneration of the gonads imposes preventive gonadectomy in numerous cases.

MALE PSEUDOHERMAPHRODITISM OF GENIC ORIGIN

There are several forms of familial male pseudohermaphroditism due to gene mutations and transmitted as autosomal recessive or sex-linked recessive conditions.

THE SO-CALLED SYNDROME OF TESTICULAR FEMINIZATION

Involved here is a genic mutation, probably X-linked, and responsible for the lack of synthesis of the tissular testicular hormone proteinic receptor. Patients have a female phenotype, usually normal. The motive for consultation is usually primary amenorrhea, or more rarely an inguinal hernia in a young girl.

The karyotype is always that of a normal 46,XY male.

TRUE GONADAL DYSGENESIS

True gonadal dysgenesis represents a particular form of gonadal dysgenesis, at the limit of male pseudohermaphroditism. The patients are of female phenotype, sometimes eunuchoid, amenorrheic, often with only slightly developed secondary sexual characteristics and rudimentary streak gonads.

The karyotype is most often 46,XY, but there can be a 46,XY/45,X mosaicism, or more rarely another kind of mosaicism (Table 5).

MIXED GONADAL DYSGENESIS

Mixed gonadal dysgenesis represents an extreme form of male pseudohermaphroditism. It is characterized by a female phenotype of the Turnerian sort, or more or less android, an ambiguity of the external genitalia with hypertrophy of the clitoris, and signs of virilization at puberty. The gonads are intraabdominal and asymmetric: on one side there is a streak gonad, on the other a more or less dysgenesic testicle. The intraabdominal structures are also asymmetrical, of Müllerian or Wolffian type depending on the gonad present on each side. The possibility of malignant transformation of the gonads imposes their preventive ablation.

The karyotype is often 46,XY, but here again a 46,XY/45,X mosaic or other mosaics can be seen (Table 6).

Female Pseudohermaphroditism

In the very great majority of cases, female pseudohermaphroditism is due to secondary virilization of a fetus of the female sex. Virilization is usually due to congenital hyperplasia of the adrenal glands, more rarely to a virilizing hormonal therapy during pregnancy, or to a masculinizing ovarian tumor in the mother. The karyotype is always normal: 46,XX.

Essential female pseudohermaphroditism is the exception, and anomalies of the karyotype are extremely rare.

TABLE 3   **Karyotypes in True Hermaphroditism**

Karyotype	Frequency
46,XX	50
46,XY	20
46,XY/45,X 46,XX/46,XY	20
47,XXY/46,XX 47,XXY/46,XY/46,XX 48,XXYY/46,XX 49,XXYYY/47,XXY/46,XX 47,XX,i(Yq)/46,XX/45,X 47,XXX/46,XX	10

TABLE 4   **Karyotypes in Male Pseudohermaphroditism Due to Chromosomal Rearrangement**

In most cases:
   46,XY/45,X
Exceptionally:
   46,X,del(Yq)/45,X
   46,XX/46,XY/45,X
   47,XY,i(Yp)/46,X,i(Yp)/45,X
   47,XXY/46,XX/46,XY/45,X
   47,XXY
   47,XXY/47,XYY/46,XY
   46,XX/45,X
   46,XX
   45,X

TABLE 5   **Karyotypes in Pure Gonadal Dysgenesis**

46,XY
46,XY/45,X
47,XYY/46,XY/45,X
46,XX/46,XY
46,XX
46,XX/45,X
47,XXX/46,XX/45,X
46,X,del(Xq)

TABLE 6   **Karyotypes in Mixed Gonadal Dysgenesis**

46,XY/45,X
45,X
46,XY
46,XX/46,XY/45,X
48,XXXY/45,X
47,XXY/46,XX/46,XY/45,X
47,XY,i(Yp)/46,X,i(Yp)/45,X
46,X,del(Yq)/45,X
48,XXX,del(Yq)/46,X,del(Yq)/45,X

MENTAL RETARDATION AND fra(X)(q28)

REFERENCES

CANTU J.M., SCAGLIA H.E., MEDINA M., GONZALEZ-DIDDI M., MORATO T., MORENO M.E., PEREZ-PALACIOUS G.: Inherited congenital normofunctional testicular hyperplasia and mental deficiency. *Hum. Genet.*, **33**:23–33, 1976.

FOX P., FOX D., GERRARD J.W.: X-linked mental retardation: Renpenning revisited. *Am. J. med. Genet.*, **7**:491–496, 1980.

GIRAUD F., AYME S., MATTEI J.F., MATTEI M.G.: Constitutional chromosomal breakage. *Hum. Genet.*, **34**:125–136, 1976.

GIRAUD F., MATTEI J.F.: Retard mental et fragilité du chromosome X: fra(X)(q28). *Arch. fr. Pédiatr.*, **38**:387–388, 1981.

GLOVER T.W.: FUdR induction of the X chromosome fragile site: evidence for the mechanism of folic acid and thymidine inhibition. *Am. J. hum. Genet.*, **33**:234–242, 1981.

HARVEY J., JUDGE C., WIENER S.: Familial X-linked mental retardation with an X chromosome abnormality. *J. med. Genet.*, **14**:46–50, 1977.

HERBST D.S.: Nonspecific X-linked mental retardation I: a review with information from 24 new families. *Am. J. med. Genet.*, **7**:443–460, 1980.

HERBST D.S., MILLER J.R.: Nonspecific X-linked mental retardation II: the frequency in British Columbia. *Am. J. med. Genet.*, **7**:461–470, 1980.

HOWARD-PEEBLES P.N.: Fragile sites in human chromosomes II: demonstration of the fragile site Xq27 in carriers of X-linked mental retardation. *Am. J. med. Genet.*, **7**:497–502, 1980.

JACKY P.B., DILL F.J.: Expression in fibroblast culture of the satellited-X chromosome associated with familial sex-linked mental retardation. *Hum. Genet.*, **53**:267–269, 1980.

JACOBS P.A., MAYER M., RUDAK E., GERRARD J., IVES E.J., SHOKEIR M.H.K., HALL J., JENNINGS M., HOEHN H.: More on marker X chromosomes, mental retardation and macro-orchidism. *N. Engl. J. Med.* **300**:737–738, 1979.

LEJEUNE J., MAUNOURY C., RETHORE M.O., PRIEUR M., RAOUL O.: Site fragile Xq27 et métabolisme des monocarbonés. Diminution significative de la fréquence de la lacune chromosomique par traitement in vitro et in vivo. *C.R. Acad. Sci. (Paris)*, **292** (série III):491–493, 1981.

LUBS H.A.: A marker X chromosome. *Am. J. hum. Genet.*, **21**:231–244, 1968.

PENROSE L.S.: A clinical and genetic study of 1280 cases of mental retardation. *Special Rep., med. Res. Council (Lond.)*, Ser. No 229, 1938.

RENPENNING H., GERRARD J.W., ZALESKI W.A., TABATA T.: Familial sex-linked mental retardation. *Can. med. Assoc. J.*, **87**:954–956, 1962.

SUTHERLAND G.R.: Fragile sites on human chromosomes: demonstration of their dependence of the type of tissue culture medium. *Science*, **197**:265–266, 1977.

Penrose (1938) had noted that there were about 25% more males than females in a population of mentally retarded individuals. Other surveys carried out between 1943 and 1963 confirmed the existence of mental deficiency due to X-linked genes (see Giraud and Mattei, 1981). Renpenning et al. (1962) described a Mennonite family of Dutch origin living in Canada, in which mental retardation was sex-linked and recessively transmitted.

Lubs (1968) described the association between X-linked mental retardation and a fragile site in the distal extremity of the long arm of an X, giving the impression of a satellite similar to that of acrocentric chromosomes. This observation was confirmed by Giraud et al. (1976), Harvey et al. (1977), and Sutherland (1977).

Turner et al. (1975) and Cantu et al. (1976) demonstrated the association between mental retardation and macroorchidism.

Turner et al. (1978), Sutherland and Ashford (1979), and Jacobs et al. (1979) gathered these disparate elements and attempted to characterize the association between sex-linked mental retardation, the fragile X, and macroorchidism. Fox et al. (1980) carried out a study on the family that Renpenning et al. (1962) had previously described, and demonstrated that the mentally retarded males in this family had neither the fragile X nor macroorchidism. It is possible, therefore, that at least two forms of mental retardation exist which are X-linked, depending on whether or not they are accompanied by macroorchidism and a fragile X. But it must be noted that these two signs are not always recognized. Paradoxically, macroorchidism may go unnoticed, and the fragile X requires special culture conditions in order to be demonstrated.

GENERALIZATIONS

The frequency of this illness is difficult to evaluate due to the absence of studies concerning normal populations. Research carried out on mentally retarded populations, or among siblings of patients, led Turner et al. (1975, 1978) to estimate the frequency of female carriers at 0.74 per 1,000, and that of the patients at 0.53 per 1,000. Herbst (1980) and Herbst and Miller (1980), on the basis of a wider investigation, estimated the frequency of heterozygous females at 2.4 per 1,000 and that of hemizygous males at 1.8 per 1,000. X-linked mental retardation is thus found more frequently than 21 trisomy in males, and would account for 25% of mentally retarded males.

It is difficult to evaluate the epidemiology of the syndrome and especially its ethnic distribution.

PHENOTYPE

We shall describe the complete syndrome; as we have stated, however, this may be dissociated.

Facial Dysmorphism

Facial dysmorphism is considered to be characteristic. The forehead is high, and the face long, with hypoplasia of the mid-face, a wide bridge of the nose, prognathism, a large mouth with thick lips and high-arched palate, and, in particular, detached ears which are large and poorly folded. This facial dysmorphism has been compared to that of 8 trisomy (Turleau et al., 1979).

SUTHERLAND G.R., ASHFORTH P.L.C.: X-linked mental retardation with macro-orchidism and the fragile site at Xq27 or 28. *Hum. Genet.*, **48**:117–120, 1979.

TURLEAU C., CZERNICHOW P., GORIN R., ROYER P., GROUCHY J. de: Débilité mentale liée au sexe, visage particulier, macro-orchidie et zone de fragilité de l'X. *Ann. Génét.*, **22**:205–209, 1979.

TURNER G., EASTMAN C., CASEY J., McLEAY A., PROCOPIS P., TURNER B.: X-linked mental retardation associated with macro-orchidism. *J. med. Genet.*, **12**:367–371, 1975.

TURNER G., TILL R., DANIEL A.: Marker X chromosomes, mental retardation and macro-orchidism. *N. Engl. J. Med.*, **296**:1472, 1978.

Genitalia

The question remains as to whether or not macrogenitalism is already present before puberty or even at birth.

In pubescent boys, the penis may be large in volume. Macroorchidism is the principle feature, and is considered to be present if testicular volume exceeds 25 ml. Normal testicular volume at age 19–20 is 18.6 ± 4.0 ml. The formula for calculating testicular volume is: $\pi/6 \times$ width$^2 \times$ length. In certain patients, this macroorchidism may attain 120 ml or more.

Macroorchidism may vary in size in patients from the same family.

Testicular histology shows no particular anomaly. Spermatogenesis may be normal or decreased. Cases of fertility have been found.

Mental Retardation

In typical patients, mental retardation is moderate, with an IQ of around 50, thus permitting a certain amount of autonomy and, occasionally, the possibility of working at odd jobs. The patients' behavior is fairly stereotyped; they have been described as "jocular" in nature. Their speech is irregular and halting, with poor articulation.

Epilepsy is found frequently. Certain patients are autistic.

CYTOGENETIC STUDY

The chromosomal aberration is characterized by a fragile site of the X chromosome at the interface of bands q27 and q28. This is observed in the lymphocytes, with extreme variations in percentages ranging from 5 to 60%. In the same patient, this percentage may vary from one examination to another. The role of the exogenous supplement of folic acid has been suggested in order to explain such variations (Lejeune et al., 1981).

The role played by the culture medium is, in effect, essential: it must be poor in folate, or must contain a folate antagonist such as methotrexate (Sutherland et al., 1977) or 5-fluorodeoxyuridine (FUdR), an inhibitor of thymidylate synthetase (Glover, 1981).

Detection of the fra(X)(q28) in fibroblasts or amniotic cells remains a highly uncertain process (Jacky and Dill, 1980).

HETEROZYGOTIC FEMALES

The chromosome aberration is observed only in a small percentage of cells (2–3%) and is sometimes not observed at all, especially in older women (Howard-Peebles, 1980).

Female carriers often have a normal phenotype. However, around 20% of carriers have more or less severe mental retardation (Herbst, 1980).

GENETIC COUNSELING

The anomaly is transmitted in the recessive sex-linked mode, and the question of genetic counseling for sisters of patients often arises. Due to the difficulty in detection of heterozygotes, only positive results can be considered as valid. The nondetection of a fra(X)(q28) does not necessarily mean that there is no risk.

The possibility of prenatal diagnosis will depend upon an improvement in research techniques on amniotic cells.

POLYPLOIDY

Polyploidy includes triploidy (3N = 69 chromosomes) and tetraploidy (4N = 92 chromosomes). It is without doubt the most frequently occurring accident of fecundation in the human species. It is estimated that 2–3% of fertilized eggs are polyploids. The majority of them are aborted, triploidy representing 20% of all chromosomal anomalies observed in the products of spontaneous abortions, and tetraploidy representing 6% (Boué et al., 1975).

Several polyploid pregnancies do, however, result in live births, but these infants die very early, during the first few hours or days.

TRIPLOIDY

REFERENCES

AL SAADI A., JULIAR J.F., HARM J., BROUGH A.J., PERRIN E.V., CHEN H.: Triploidy syndrome. A report on two live-born (69,XXY) and one still-born (69,XXX) infants. *Clin. Genet.*, **9**:43–50, 1976.

BÖÖK J.A., SANTESSON B.: Malformation syndrome in man associated with triploidy (69 chromosomes). *Lancet*, 1:858, 1960.

BOUE J., BOUE A., LAZAR P.: The epidemiology of human spontaneous abortions with chromosomal anomalies. In: *Aging Gametes, International Symposium, Seattle, 1973*, Basel, Karger, pp. 330–348, 1975.

CASSIDY S.B., WHITWORTH T., SANDERS D., LORBER C.A., ENGEL E.: Five month extra-uterine survival in a female triploid (69,XXX) child. *Ann. Génét.*, **20**:277–279, 1977.

COUILLIN P., HORS J., BOUE J., BOUE A.: Identification of the origin of triploidy by HLA markers. *Hum. Genet.*, **41**:35–44, 1978.

FRYNS J.P., Van de KERCKHOVE A., GODDEERIS P., Van den BERGHE H.: Unusually long survival in a case of full triploidy of maternal origin. *Hum. Genet.*, **38**:147–155, 1977.

FRYNS J.P., VINKEN L., GEUTJENS J., MARIEN J., DEROOVER J., Van den BERGHE H.: Triploid-diploid mosaicism in a deeply mentally retarded adult. *Ann. Génét.*, **23**:232–234, 1980.

GROUCHY J. de, ROUBIN M., RISSE J.C., SARRUT S.: Enfant triploïde (69,XXX) ayant vécu neuf jours. *Ann. Génét.*, **17**:283–286, 1974.

GROUCHY J. de: *Jumeaux, Mosaiques, Chimères et autres Aléas de la Fécondation Humaine*. Paris, Medsi, p. 206, 1980.

JACOBS P.A., ANGELL R.R., BUCHANAN I.M., HASSOLD T.J., MATSUYAMA A.M., MANUEL B.: The origin of human triploids. *Ann. hum. Genet.*, **42**:49–57, 1978.

The first observation of triploidy was that of Böök and Santesson (1960). About 40 cases are now known (see Wertelecki et al., 1976; Al Saadi et al., 1976; and Grouchy, 1980).

GENERALIZATIONS

- Pregnancy is often marked by hemorrhaging, polyhydramnios, pre-eclampsia, or post-partum accidents.
- Mean birth weight: 1,900 g in males and 1,500 g in females
- Mean crown-heel length at birth: 43.1 cm (males), 38 cm (females)

The fetus/placenta ponderal ratio is modified in favor of a voluminous placenta. Hydaditiform degenerescence is often noted.

PHENOTYPE

Craniofacial Dysmorphism

The skull is deformed, with hypoplasia of the occiput and parietals; the fontanel is widely gaping.

Ocular anomalies are found frequently, including anophthalmia, microphthalmia, coloboma, and cataracts.

Harelip, cleft palate, and macroglossia are often present.

Severe micrognathia also exists.

The ears are low-set and poorly folded.

Neck, Thorax, and Abdomen

The abdomen is deformed by diastasis recti, hepatomegalia, and occasionally by omphalocele.

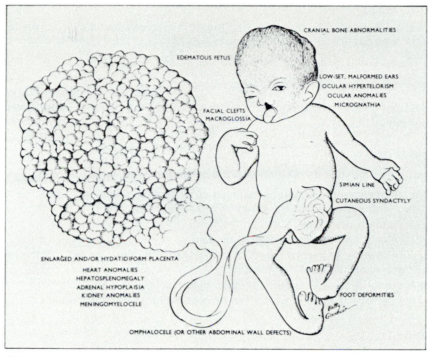

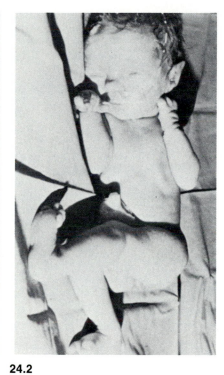

CRANIAL BONE ABNORMALITIES

EDEMATOUS FETUS

LOW-SET, MALFORMED EARS
OCULAR HYPERTELORISM
OCULAR ANOMALIES
MICROGNATHIA

FACIAL CLEFTS
MACROGLOSSIA

SIMIAN LINE

CUTANEOUS SYNDACTYLY

ENLARGED AND/OR HYDATIDIFORM PLACENTA
HEART ANOMALIES
HEPATOSPLENOMEGALY
ADRENAL HYPOPLAISIA
KIDNEY ANOMALIES
MENINGOMYELOCELE

FOOT DEFORMITIES

OMPHALOCELE (OR OTHER ABDOMINAL WALL DEFECTS)

24.1

24.2

24.3

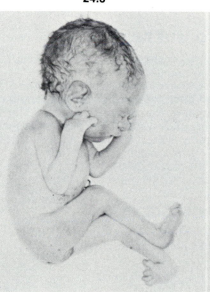

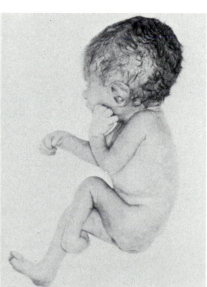

KAJII T., NIKAWA N.: Origin of triploidy and tetraploidy in man: 11 cases with chromosome markers. *Cytogenet. Cell Genet.*, **18**:109–125, 1977.

WERTELECKI W., GRAHAM J.M., SERGOVICH F.R.: The clinical syndrome of triploidy. *Obstet. Gynecol.*, **47**:69–76, 1976.

Limbs

Limbs are deformed, with thenarian and thumb hypoplasia, syndactyly of the toes and fingers, long spatular fingers, varus equin, hypoplasia of the big toe, and poor toe implantation.

Genitalia

The external genitals are abnormal in boys, and include micropenis, hypospadias, and cryptorchidism.

Malformations

Internal malformations are severe, and are of the following types:

- Cerebral: absence of the corpus callosum, holoproencephaly, aplasia of the falx cerebri, ventricular anomalies, encephalo- and meningocele
- Ocular: already mentioned
- Cardiac: atrial and ventricular septal defects
- Digestive: tracheobronchial fistula, atrophy of the bladder, agenesis of the pancreas, and malrotation of the intestine
- Kidney: renal dysplasia, hydronephrosis, hypoplasia of the adrenals
- Internal genitals: testicular or ovarian aplasia, pseudohermaphroditism

Prognosis

The vital prognosis is very poor; the infants usually die at birth or within a few hours or days. In rare cases, infants survived several weeks (Grouchy et al., 1974; Fryns et al., 1977; Cassidy et al., 1977).

CYTOGENETIC STUDY

Three karyotypes can be observed as a function of the sex chromosomes: 69,XXX, 69,XXY, and 69,XYY.

A number of mechanisms may result in formation of a triploid zygote, according to the extent of paternal or maternal contribution, and the moment at which the accident occurred in the course of meiosis. The study of chromosomal and hematological markers (mainly the HLA system) show that the most frequently occurring mechanism is diandry (80% of cases), probably through dispermia (Kaji and Nikawa, 1977; Jacobs et al., 1978; Couillin et al., 1978).

In the case of 2N/3N mosaicism, the clinical picture is less severe, and survival may be prolonged (Fryns et al., 1980).

DERMATOGLYPHICS

An increase in the number of whorls is the main distinctive feature as well as a single palmar flexion crease.

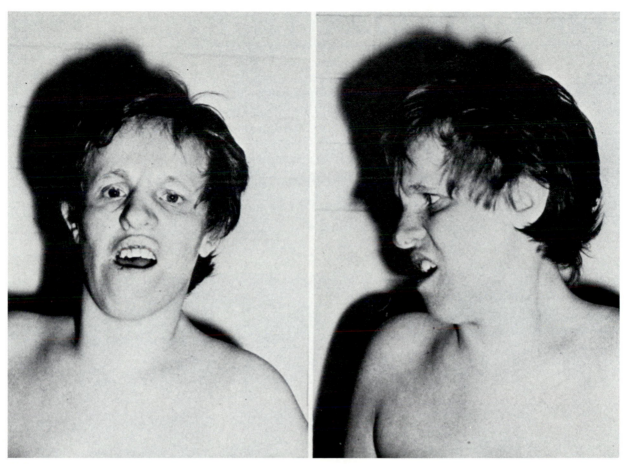

24.4

TETRAPLOIDY: 92,XXYY

REFERENCES

GOLBUS M.S., BACHMAN R., WILTSE S., HALL B.D.: Tetraploidy in a liveborn infant. *J. med. Genet.*, **13**:329–332, 1976.

PITT D., LEVERSHA M., SINFIELD C., CAMPBELL P., ANDERSON R., BRYAN D., ROGERS J.: Tetraploidy in a liveborn infant with spina bifida and other anomalies. *J. med. Genet.*, **18**:309–311, 1981.

Only two observations of tetraploidy in infants born alive are known (Golbus et al., 1976; Pitt et al., 1981). The principal features in common are:

- Growth retardation, with microcephaly, and a narrow forehead
- Facial dysmorphism, with a birdlike nose and short philtrum, and ear anomalies with an absence of cartilage
- Malpositioning of the limbs, hands, and feet
- At autopsy: renal urinary tract and brain anomalies
- A rapidly fatal evolution

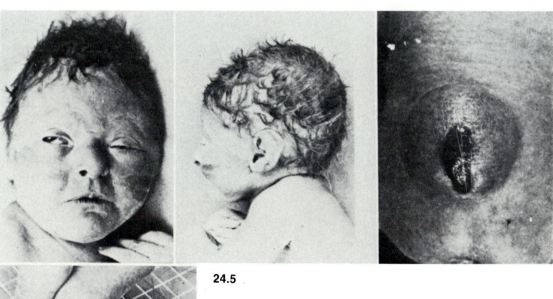

24.5

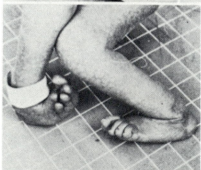

APPENDIXES

APPENDIX I
TECHNIQUES

Cell Cultures and Cytogenetic Preparations

Karyotypic investigations are generally initiated by the culture of blood lymphocytes stimulated by phytohemagglutinin (PHA). Mitoses at the prophase-metaphase stage are accumulated by exposure to colchicine or another mitotic poison. Swelling of the nuclei is obtained by use of hypotonic solutions in order to disperse the chromosomes. The cells are fixed with an acid-alcohol solution and spread on glass slides. These preparations can be stained immediately or else be subjected to various treatments resulting in banding of the chromosomes.

Most of the steps involved are subject to considerable variation. One reason is that each laboratory follows its own technique, and there are no "standard" recipes. Another is that the choice of a given technique, that of the hypotonic solution in particular, depends on the desired type of banding.

The cells used are usually blood lymphocytes because of their easy availability and culture. Other cells may also be used, such as fibroblasts originating from cutaneous and connective tissues; gonadal biopsies; fetal cells; cancerous cells; and bone-marrow cells in cases of malignant or nonmalignant hemopathies. Techniques must be adapted to these different tissues.

The principal techniques proposed here are the ones currently employed in our laboratory. They are proven and reliable. Variations will also be described, but these have not necessarily been tested in our laboratory.

BLOOD LYMPHOCYTE CULTURES

References

Dutrillaux and Couturier, 1972; Hamerton, 1971.

Blood Sampling

The blood sample should be taken from a vein, or from a heel or finger prick in young babies. It must be done aseptically.

Culture

In order to avoid unnecessary transfers, collect the blood directly in centrifuge tubes with a conical tip. Use a hermetic and sterilizable stopper (Teflon disk).

For a 10 ml tube, inoculate 5–10 drops of whole blood into 6.5 ml of culture medium:

TC 199 medium	5	ml
Human AB serum	1.5	ml
PHAC	1	drop

In practice the TC 199 medium is prepared in lots of 500 ml, to which are added antibiotics (penicillin: 1,000,000 IU; streptomycin: 50 mg). They are stored at 4°C. The mixture of TC 199 medium, AB serum, and PHA is prepared immediately before use.

Place the carefully closed tubes in an inclined position into an incubator at 37°C for 3 days.

Accumulation of Mitoses

After approximately 70 hours of incubation, add three drops of a 0.04-percent solution of colchicine to 7 ml of culture medium.

Return the tubes to the incubator for 2 hours.

Hypotonic Treatment

Composition of the hypotonic solution:

Human serum	1 vol
Distilled water	5 vol

After centrifugation (800 rpm for 10 min) discard the supernatant and replace with approximately 8 ml of hypotonic solution. Resuspend the cells and replace the tubes at 37°C for 8 min.

Fixation

After centrifugation and removal of the supernatant, resuspend gently the cells in Carnoy's solution:

Ethanol	6 vol
Chloroform	3 vol
Acetic acid	1 vol

Let stand 30 min at ambient temperature. Centrifuge. Tip off the supernatant and resuspend gently in:

Ethanol	3 vol
Acetic acid	1 vol

Spreading

Spreading on slides can be done immediately or after a variable delay, even up to several days (the tubes being kept in the refrigerator).

Centrifuge. Discard the supernatant. Gently resuspend the cells in several drops of the second fixative.

The slides are kept in a beaker of distilled water at 4°C (see below). Take one slide. Tilt on filter paper to leave only a thin film of water. With a Pasteur pipette, let fall one or two drops of the cell suspension from a height of 10–20 cm.

Allow the slides to air-dry without heating.

Technical Variations

Other techniques differ principally by the hypotonic solution and the fixative.

The hypotonic solution, for instance, can be:

- A 0.075 M solution of KCl
- Distilled water, either pure or combined with a variable quantity of isotonic medium
- A 1 percent solution of sodium citrate

For fixation, ethanol can be replaced by methanol.

Cleaning of the Slides

This washing is essential, especially in cases where good-quality banding is desired (R-banding in particular).

The slides must remain about 24 hours in a sulfochromic mixture:

Potassium bichromate	100 g
Distilled water	1,000 ml
H_2SO_4	500 ml

Rinse under running water for one hour. Then rinse each slide individually in distilled water. They can then be kept in a beaker of distilled water and stored at 4°C.

BONE MARROW CULTURES

Bone marrow cells in spontaneous division theoretically constitute an excellent material for cytogenetic study. They were used for some time before the development of techniques of stimulation of blood lymphocytes. The necessity of a sternal puncture constitutes an obstacle to frequent use of this method. It is no longer utilized except for patients with malignant hemopathies. In such cases, chromosomal anomalies are acquired and are not necessarily present in the blood lymphocytes, while they are always present in the medullar cells. The Philadelphia chromosome in particular (Ph[1]), specific to chronic myeloid leukemia (CML), can be absent from peripheral blood at the onset of the illness or during periods of remission. For this reason, while the discovery of a Ph[1] in the blood is of great diagnostic value, its absence is not an argument against the diagnosis.

References

Nava, 1969; Hamerton, 1971.

Sampling

Samples are intended for direct study and for short-term cultures (24 and 48 h).

It is thus necessary to obtain at least 0.5 ml of bone marrow. This is distributed in equal aliquots into two tubes containing:

TC 199 medium	10 ml
Heparin 5,000 IU/ml	0.5 ml

And for one tube only, reserved for direct study:

Colchicine	0.4 μg/ml

Direct Study

The tube containing colchicine is placed at 37°C for 2 h.

The method of harvesting is the same as that for short-term cultures.

Short-term Cultures

Since cell concentration is variable, a preliminary count is necessary.

Inoculate 5–7 10^5 cells per ml of medium:

TC 199 medium	5 ml
Human AB serum	1.5 ml

It is preferable to set up two parallel cultures, one with PHA, the other without.

It is thus appropriate to prepare four tubes, two for 24-h incubation and two for 48-h incubation.

Accumulation of Mitoses

Add colchicine two hours before harvesting (0.04 percent solution, 3 drops per tube).

Replace in the incubator for two hours.

Hypotonic Treatment

Use a 1 percent sodium citrate solution at 37°C for 10–20 min.

Fixation

Two fixations are necessary, with the following solutions:
First fixative:

Absolute alcohol	3 vol
Acetic acid	1 vol
20 min at 0°C	

Second fixative:

45 percent acetic acid, cooled to 0°C

Spreading

The slides must be flame dried (air drying does not produce sufficient dispersion).

Blood Cultures (Hemopathies)

When it is not possible to obtain a marrow sample, it becomes necessary to base the examination on a blood sample. Short-term cultures (24 and 48 h) are then required, preferably without PHA stimulation and with a preliminary cell count as for bone marrow.

Technical Variants

Hypotonic solution:

KCl 0.075 M, 6–8 min

Fixative:

Acetic acid	1 vol
Methanol	3 vol

It is not necessary to flame the slides.

Banding Techniques

Banding of preparations made with pathological marrow cells is often difficult because of the lesser quality of the metaphases.
R-banding: the time of heat denaturation must be reduced. Optimum time is often of the order of several minutes.
G-banding: this technique seems to give better results than R-banding, especially after a KCl hypotonic treatment.

FIBROBLAST CULTURES

Reference

Grouchy et al., 1970.

Application

The technique is applicable to karyotype analysis of all cells in stationary culture (fibroblasts, "epithelioid" cells, hybrid cells).

Sampling of Skin or Fascia

A skin biopsy is a simple procedure. Clean the skin (at the inside of the arm or forearm or at the tip of the shoulderblade) with alcohol followed by ether. Pinch the skin with curved forceps in such a way that one millimeter projects into the curved portion of the forceps, along a length of 0.5 cm. Cut the fragment with a scalpel while pressing against the forceps.
A fascia biopsy is a surgical procedure. It should be asked for only if an operation is anticipated (celioscopy, ovarian biopsy, etc.).
Keep the sample in sterile physiological saline solution.

Setting Up the Culture

With two scalpels, cut the biopsy sample into smaller fragments in a drop of physiological saline solution.
With a curved-tip Pasteur pipette arrange the explants on the inner surface of a 30 ml sterile plastic flask.
Cut a piece of perforated cellophane in the form of the wall of the flask and slightly smaller. Soak for a few moments in absolute alcohol, then let dry; with two forceps roll the sheet into a cylinder, like a cigarette paper, insert it into the flask, and spread it onto the explants with forceps.
Add the culture medium (6 ml for a 30 ml flask). The following media can be used:

Human AB serum	2.5 ml
TC 199 medium	3 ml
Chick embryo extract	0.5 ml
Antibiotics	

or:

NCTC 109 medium	40 percent
RPMI 1640 medium	40 percent
Fetal calf serum	20 percent
Antibiotics	

Explant Transfer

Renew the medium every 5 or 6 days, depending on the pH. When the fibroblast discs have attained sufficient size (in 8–10 days), withdraw the cellophane and carefully remove the explants, which can then either be subcultured to maintain the culture, or else frozen. Renew the medium.

Transfer of the Primary Cell Culture

After 5 days discard the medium.

Rinse the flask with 1 ml of trypsin (2.5 per 1,000) to clean out residues.

Add again 0.5 ml of trypsin (2.5 per 1,000). Let activate 2 min at ambient temperature.

Break the fibroblast discs with a Pasteur pipette to obtain a homogeneous suspension.

Add 6 ml of medium. Replace in the incubator, flat.

The aim of the procedure is to obtain a homogeneous cell culture in the flask used for the primary culture.

After 4–5 days this culture can be transferred to two or more flasks in order to maintain the culture and obtain a chromosomal preparation.

Chromosomal Preparation

After 24 h add 3 drops of a 0.04 percent colchicine solution to the flask intended for karyotype analysis.

Relace for 4 h at 37°C.

Pour the supernatant into a centrifuge tube. Centrifuge for 5 min at 800 rpm. This is to recover the numerous dividing cells in suspension in the medium.

Add 1 ml of trypsin (2.5 per 1,000) to the flask and agitate gently.

Tip off the supernatant from the centrifuge tube. Add the cell suspension originating from the flask. Resuspend.

Add 10 ml H_2O and 2 drops of human or animal serum. Leave at ambient temperature for 20 min.

Resuspend the cells.

Add 4 ml of fixative:

Acetic acid	1 vol
Methanol	3 vol

Let sit 10 min.

Centrifuge. Discard the supernatant.

Add 4 ml of fixative.

Let activate 30 min.

Centrifuge 5 min at 800 rpm. Remove the supernatant.

Add 4 ml of fixative and resuspend the cells.

Centrifuge immediately.

Repeat the procedure once, if the sediment is large enough.

Resuspend the cells in 10 drops of fixative.

Rinse the slides in running water while rubbing with the thumb to clean them. Place the slides in cold distilled H_2O (about 0°C).

Remove one slide, leaving a film of water. Let 1 drop of the cell suspension fall onto the slide.

Stain. There is a choice of stains. We propose:

Giemsa	4 ml
Phosphate buffer (pH 6.5)	4 ml
Distilled H_2O	92 ml

Heteroploid Cells in Culture

In cases of heteroploid cells, the cultures are usually made in larger flasks, e.g., 250 ml. It is then normal to use 2 ml of trypsin to detach the cells.

For spreading and fixation a relatively deep layer of water can be left on the slide.

Let 2–3 drops of the cell suspension fall onto the slide. Flame immediately for a rather long while.

REFERENCES

DUTRILLAUX B, COUTURIER J: Techniques d'analyses chromosomiques. *Monographie annuelle de la Soc. Fr. Bio. Clin.*, 1972, p. 5.

GROUCHY J de, ROUBIN M, BILLARDON C: Etudes chromosomiques à partir de cultures cellulaires. Modifications techniques. *Ann Génét* **13**:141–143, 1970.

HAMERTON JL: *Human Cytogenetics*. Vol. I. *General Cytogenetics*. Academic Press, New York, 1971.

NAVA C de: Les anomalies chromosomiques au cours des hémopathies malignes et non malignes. Une étude de 171 cas. *Monographie des Annales de Génétique*, 1969, p. 89.

Staining and Banding Techniques

Until the early 70's only the "standard" stain was in use. It allowed a uniform staining of the chromosome arms, and it was impossible to distinguish among certain chromosomes. Since the development of banding techniques, no cytogenetic examination can be considered valid unless such techniques are applied. Nonetheless, the standard technique remains useful for rapid identification of numerical anomalies, such as mosaicism, and of chromosomal polymorphisms, such as variant satellites, or secondary constrictions.

Different banding techniques appeared almost simultaneously as of 1970. The most useful are:

- Q-banding (staining by fluorescent derivatives of quinacrine, followed by examination in UV light)
- R-banding (staining after heat denaturation)
- G-banding (staining after an ionic or enzymatic treatment)

Other techniques are used with specific goals:

- C-banding: staining of the centromeres
- T-banding: staining of the telomeres
- NORs: staining of nucleolar organizers
- Culture in the presence of BrdU: especially for the study of late X replication, and of chromatid exchanges

The very first banding technique, that of autoradiography after tritiated thymidine labeling, has rendered immense service. It is now practically abandoned and will not be considered here.

STANDARD STAINING

The most used stain is Giemsa's solution, which is presented in different forms depending on the manufacturer. For example, the following can be used:

Giemsa	4 ml
Phosphate buffer (pH 6.5)	4 ml
Distilled water	92 ml

Other stains used are orcein, toluidine blue, Unna blue, cresyl violet, etc.

Q-BANDING

References

Caspersson et al., 1970; Zech, 1973; Evans et al., 1972; Lin et al., 1971.

Aside from the tritiated thymidine labeling technique, Q-banding is the first banding technique to be described (Caspersson et al., 1968; Zech, 1972). Owing to the subsequent development of less involved methods, not requiring microscopic observation in UV light, this technique remains useful primarily for study of the Y chromosome (see also: buccal smear) and of certain bands such as the centromere of chromosome 3.

Staining

Rehydrate the slides in the ethanol series (95, 70, and 50 percent).
Rinse in distilled water.
Soak the slides in a phosphate buffer at pH 6.7 (Sørensen's or McIlvaine's buffer) for 10 min.
Stain 20 min in an aqueous solution of either quinacrine dihydrochloride (Q) (5 mg/ml) or quinacrine mustard (QM) (50 μg/ml).
Rinse three times in buffer (same pH).
Mount in the buffer and seal with a rubber solution.

Microscopic Observation

Examine in UV light; Zeiss photomicroscope, transmitted light; stop filters 53 and 47, and excitation filter I.

Microphotography
This requires a very long exposure time, of the order of 3 to 4 min for Kodak microfilm. Tri-X film can be used, but the grain is not as fine.

Technical Variations

The fluorescent drug varies depending on the authors: atebrine, acridine yellow, or acriflavine. The results are not as good as with QM.
The pH of the buffer varies (from 4 to 7) as well as the staining time.
The aqueous solution can be replaced by an alcohol solution.

R-BANDING

References

Dutrillaux and Lejeune, 1971; Carpentier et al., 1972.

Procedure

Immerse the slides in Earle's solution at pH 6.5 and 87°C for a length of time dependent on the age of the preparations: about 30–40 min for recent slides (2–3 days), 8–10 min for slides 15 days or more old. The optimum time being variable, a series of slides can be made, increasing the incubation time by increments of 5 min in the vicinity of the time thought to be best.
Rinse and stain with the usual Giemsa stain.

Microscopic Observation

View in normal light. If the stain is excessively pale, phase contrast with an orange filter can be used.

T-BANDING

Reference

Dutrillaux, 1973.

Procedure

This method is derived from a modification of the R-banding technique.
The pH of the denaturation solution must fall between 4.9 and 5.5.

The duration of heat treatment is variable as for R-banding.

Staining can be done either by Giemsa solution or by acridine orange (see below). The latter permits a better differentiation of R- and T-bands.

R-BANDING (BrdU + ACRIDINE ORANGE)

References

Bobrow et al., 1972; Dutrillaux et al., 1973; Couturier et al., 1973.

Acridine orange is a fluorescent substance usually observed in UV light. This stain allows R-banding without preliminary treatment or after diverse treatments, but with no clear advantage over Giemsa staining. By contrast, the combination of the method of despiralization by BrdU (*vide infra*) and staining with acridine orange gives very valuable results, particularly for chromosomes 4, 5, 13, and X.

The method of culture in the presence of BrdU will be described in detail further below.

Staining Procedure

Aqueous solution of acridine orange (1 mg/ml) (stock solution) diluted upon use to 5 percent in Sørensen's buffer pH 6.7 M/15. Soak the slides 5 min in this solution. Rinse with distilled water or buffer.

Mount in buffer.

View in UV light.

G-BANDING

G-banding is obtained by a variety of techniques. Most of these can be grouped into one of two families:

- either proteolytic enzyme action
- or ionic treatment: ASG techniques (Acid-Saline-Giemsa or Alkaline-Saline-Giemsa)

Proteolytic Enzymes

The first enzymes used were pronase (Dutrillaux et al., 1971), α-chymotrypsin (Finaz and Grouchy, 1971), and trypsin (Seabright, 1971). These techniques have the great advantage of simplicity.

Recommended Procedure

Prepare the following solutions.

PBS Solution:

NaCl	16 g
KCl	0.4 g
Na_2HPO_4	2.3 g
KH_2PO_4	0.4 g
H_2O	qs 2,000 ml

Trypsin Solution: 2g/l of PBS.

Giemsa Solution:

Giemsa	1.5 ml
Methanol	1.5 ml
Citric acid 0.1 M	2 ml
Na_2HPO_4 0.2M	4 ml
H_2O	qs 50 ml

Treat the slides with the trypsin solution for 10 sec to 2 min, depending on the age of the slide (inversely to R-banding, time is directly proportional to the age of the slide).

Rinse in two changes of PBS.

Stain with Giemsa solution for 2 to 5 min.

Note: The PBS solution can be prepared in advance and stored at 4°C. The trypsin solution is distributed in 50 ml flasks (corresponding to one use) and frozen. The Giemsa solution must be prepared just before use.

Technical Variations

The trypsin can be diluted at varying concentrations in an isotonic saline solution.

Ionic Treatments

These techniques, derived from the procedure described by Pardue and Gall (1970), were developed principally in three laboratories: Sumner et al. (1971), Drets and Shaw (1971), and Schnedl (1971). They are usually preceded by a KCl hypotonic treatment and a methanol-acetic acid fixation.

Technique of Sumner et al.

Incubate the slides in SSC X 2 at 60°C for 1 h.

Rinse with distilled water.

Stain in Giemsa's solution (2 percent, pH 6.8) for 90 min.

SSC X 2 Formula:

NaCl	17.53 g
Trisodic citrate $2H_2O$	8.82 g
H_2O	qs 1,000 ml
Adjust to pH 7	

Schnedl's Technique

Soak the slides in NaOH (0.07 N solution) for 90 sec at ambient temperature.

Wash in the ethanol series (70, 96, 100 percent).

Allow the preparation to dry.

Incubate for 24 h in Sørensen's buffer at pH 6.8 and at 59°C.

Stain 20 min in a 10 percent Giemsa solution (pH 7).

C-BANDING

Specific staining of heterochromatic material—centromeres, secondary constrictions of chromosomes 1, 9, 16, and the long arm of chromosome Y—is at the origin of all banding techniques, except Q-banding. The original technique is from Pardue and Gall (1970). It was adapted to the human karyotype by Arrighi and Hsu (1971) and Yunis et al. (1971).

Numerous procedures have been described. The G-banding techniques (which are derivatives thereof) entail either ionic or heat treatment.

Only the most commonly used techniques will be described.

Proposed Procedure (Craig-Holmes and Shaw, 1971)

Somewhat "aged" slides are preferred.

Flame the slides.

Treat the slides with 0.2 N HCl for 15 min.

Rinse in distilled water.

Cover the slides with 0.05 N NaOH for 15–30 sec, depending on their age.

Rinse in two changes of 70 percent ethanol and three changes of 95 percent ethanol.

Allow to dry.

Incubate 24 h in SSC X 2 at 65°C.

Rinse several times in 70 percent and 95 percent ethanol.

Stain in Giemsa.

Technical Variations

The original technique of Arrighi and Hsu is very similar but requires a preliminary treatment with RNAse.

Technique of Yunis et al. (1971): Incubate the slides for 10 min at 85–100°C in 0.06 M phosphate buffer, pH 6.8.

Cool to 0°C.

Incubate in the same buffer at 65°C for varying times.

Treatment with urea (Dutrillaux and Couturier, 1972): Soak the slides in a 10 percent urea solution at 89°C for 30 sec to 4 min.

Rinse.

Stain in Giemsa.

Staining with Giemsa 11 (Bobrow et al., 1972): Treat the slides with 2 percent Giemsa in 0.05 N NaOH (pH 11) for about 8 min.

Rinse in distilled water.

This technique stains heterochromatin red. It was proposed to distinguish human chromosomes in cellular hybrids, man-mouse for example.

STAINING OF NUCLEOLAR ORGANIZERS (NORs)

In humans, those regions corresponding to nucleolar organizers are normally located on the short arm of acrocentric chromosomes. They may be displayed on chromosomal preparations by silver staining (Goodpasture and Bloom, 1975; Howell et al., 1975; Denton et al., 1976). Several techniques exist that are derived from the same principle.

Recommended Procedure

On the slide, place 3 or 4 drops of extemporaneously prepared silver nitrate solution (1 g of $AgNO_3$ in 2 ml of bi-distilled water; filtration on GSWP 02,500 millipore).

- Cover with a slide; elute
- Leave standing one night at 37°C in a moist chamber
- Remove the slide: rinse very rapidly in distilled water; dry
- Simultaneously add 2 drops of each of the following solutions:

    formol, at 3% in CH_3COONa buffer, 3 H_2O at 3.2 g %

    ammoniac silver: dissolve 8 g of $AgNO_3$ in 10 ml of distilled water; add 15 ml of concentrated NH_4OH. Mix and filter on millipore.
- Immediately cover with a slide.
- A light yellow-brown staining appears in 10 to 60″. Remove slide and rinse thoroughly in distilled water.

BrdU TREATMENTS

5-bromodeoxyuridine (BrdU) is an analog of thymidine. Its incorporation into late S-phase lymphocyte cultures constitutes a simple and efficient method of revealing chromosomal segments with late replication. Much simpler and giving superior results, it has completely replaced labeling by tritiated thymidine. The mechanism by which BrdU induces a despiralization of segments with late replication is unclear, and is probably not due only to its incorporation in the place of thymidine.

Incorporation of BrdU during the entire period of culture (during several cellular cycles) allows discrimination between chromatids of different generations and reveals chromatid exchanges.

Study of the Late Replicating X Chromosome

References

Zakharov et al. (1971); Dutrillaux and Fosse (1974).

Recommended Procedure

Set up a culture with the usual technique.

On the third day, 7 hours before harvesting, add to each 7 ml culture tube 0.5 ml of a 0.26 percent BrdU solution in physiological medium. Final concentration is approximately 200 µg/ml.

Replace at 37°C. Two hours before harvesting add colchicine in the usual concentration.

For hypotonic treatment use a 0.075 M KCl solution.

Fix in acetic alcohol.

Stain the slides with acridine orange, 5 mg/100 ml pH 6.7 buffer, for 10 min.

Rinse carefully in the buffer.

Mount. Observe in UV light.

Note: Before staining, the slides can be rehydrated in successive 90, 70, and 50 percent ethanol baths.

Practical Use

X chromosomes with late replication are easily recognizable. They are submetacentric and show in the middle of the short arm and in the first third of the long arm, a rather long, faintly stained region. The distal portion of the long arm is generally divided by one, or more rarely two, clear bands, less pronounced than the proximal band. The early replicating X chromosome shows the usual R-banding.

In cases of numerical and structural anomalies, observations are similar to those made with tritiated thymidine labeling, thus confirming the value of the method.

Numerical Anomalies: In cases where more than two X chromosomes are present, inactivation is observed for as many X's as there are X chromosomes minus one.

Structural Anomalies: In every case the rearranged X is inactivated.

X-Autosome Translocations: In the majority of cases the cell appears to maintain as best as possible its genetic balance. When a translocation is balanced, the normal X is inactivated, apparently so as not to inactivate the translocated autosome fragment. For unbalanced translocations the X carrying an autosomal segment in the trisomic state is inactivated. Inactivation extends more or less to the autosomal segment. As a general rule, however, this inactivation is not sufficient to inhibit phenotypic expression of the trisomy.

There are many exceptions to this briefly outlined situation.

Chromatid Exchanges

After two cell cycles in the presence of BrdU, two sister chromatids of a given chromosome acquire different tinctorial properties that permit recognition of the first and second generation chromatids. This discrimination is made possible by the use of either a fluorescent substance (acridine orange or 33,258 Hoescht) or the combination 33,258 Hoescht-Giemsa.

References

Latt (1973); Perry and Wolff (1974); Dutrillaux (1975).

Techniques of Culture and Preparation of Slides

Set up cultures in the usual manner.

Add at the outset of the culture 0.1 ml of a 0.5 mg/ml BrdU solution in physiological saline.

Harvest the cells in the usual manner.

For hypotonic treatment use the KCl solution.

Fix with the methanol-acetic acid solution.

Acridine Orange Staining

Treat the slides with acridine orange (5 mg/100 ml of phosphate buffer pH 6.7) for 5 min.

Mount. View in UV light.

Staining with 33,258 Hoescht

Treat the slides with 33,258 Hoescht (5 mg/100 ml of distilled water) for 20 min.

Rinse in distilled water.

Mount the slides and expose them to UV light (Philips ultraphil sunlamp MLU 300 W) for 20 min at 15 cm.

Remove the coverslip.

Stain in Giemsa (2 percent solution in phosphate buffer pH 6.8).

425

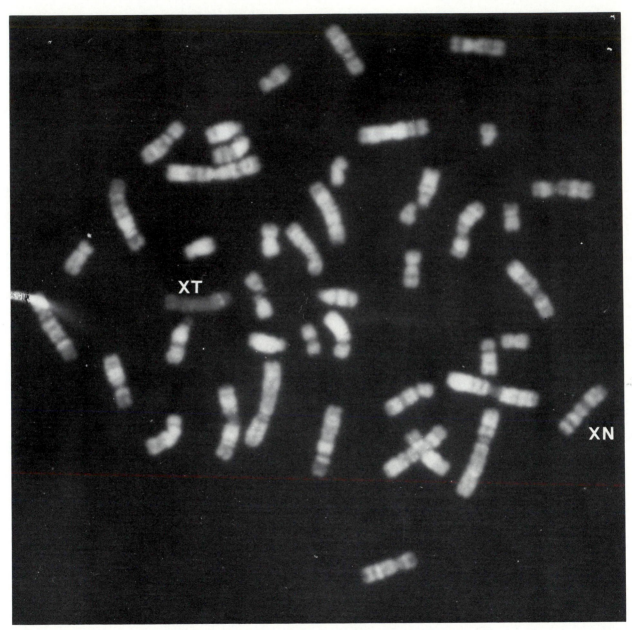

AI-1   Study of the late replicating X. XT, late replicating X; XN, normal replicating X.

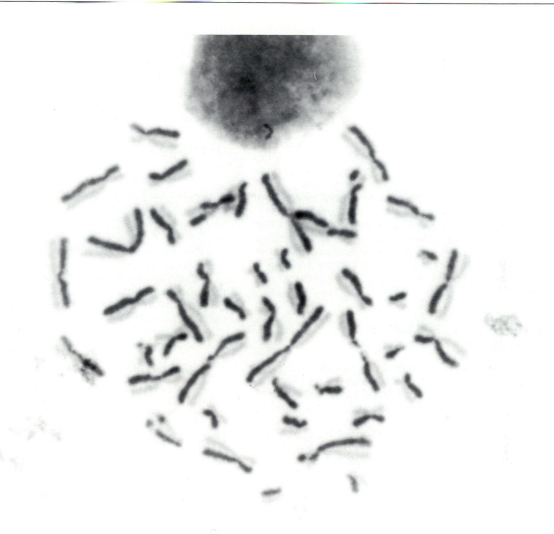

AI-2 Chromatid exchanges.

Technical Variations

BrdU can be added to the culture only after 48 h, the medium renewed on the third day, the dose of BrdU doubled 24 h later, and the culture harvested at the end of the fifth day (Dutrillaux, 1975).

Comments

The acridine orange technique is the most rapid and gives good results. That of Giemsa staining is more lengthy but has the advantage of allowing observation in normal light, and also preservation of the slides.

These techniques are theoretically most interesting: they confirm the semi-conservative chromosomal replication already indicated by tritiated thymidine labeling. In addition, they allow a detailed analysis of the number of chromatid exchanges, either spontaneous or after in vitro treatment with certain drugs, as well as in diseases with chromosomal breakage (Bloom's syndrome in particular).

HIGH RESOLUTION BANDING TECHNIQUES

Different methods have been perfected to obtain cells in late prophase or early prometaphase by synchronization of cultures. Under these conditions, the chromosomes are less condensed, and most of the bands differentiate into subbands, permitting a more sensitive analysis of the karyotype. The two most frequently employed agents for synchronization are amethopterine (methothrexate) and thymidine. The adding of colchicine is either suppressed or reduced to around 15 minutes.

Blocking by Thymidine

References

Dutrillaux, 1975; Viegas-Péquignot and Dutrillaux, 1978; Dutrillaux and Couturier, 1981.

Culture

Place whole blood in culture according to the standard method.

After 72 h in culture, add thymidine for a final concentration of 0.3 mg/ml.

Seventeen hours later (one night), rinse twice and return to culture, either in the normal medium or in a medium containing BrdU for a final concentration of 10 μg/ml.

The cells are harvested 5 to 6 h after rinsing.

Staining and Banding

Slides prepared using cultures without BrdU can be stained by Giemsa or undergo the usual banding methods: Q, C, G, or R banding.

Slides prepared using cultures with BrdU can be stained with acridine orange, or with Hoechst-Giemsa staining.

Blocking by Amethopterin

References

Yunis, 1976; Pai and Thomas, 1980; Camargo and Cervenka, 1980.

Culture

Place whole blood in culture according to the standard method.

After 72 h of culture, add amethopterin (methotrexate) for a final concentration of 10^{-7} M.

Seventeen hours later (one night), rinse twice and replace in culture in medium containing thymidine for a final concentration of 10^{-5}M.

Culture is stopped 5 to 6 h after rinsing.

Staining and Banding

Banding can be directly obtained by Wright's staining (3:1 mixture of phosphate buffer, pH 6.8, and a Wright alcoholic solution at 0.25% for 2'30 to 4').

Technical Variants

Other combinations have been proposed, including:

- Blocking by amethopterin (10^{-7}M; 17 h), then culture in the presence of BrdU (12 μg/ml; 5 1/2 hr), and action of actinomycin D (5.4 μg/ml; 1 h) (Yunis, 1981);
- Blocking by BrdU (200 μg/ml; 15–17 h), then culture in presence of thymidine 3 μg/ml; 6 to 7 h) (Dutrillaux and Viegas-Péquignot, 1981).

REFERENCES

General Reviews

BUCKTON KE: Identification of Human Mitotic Chromosomes. *Handbook of Genetics*, KING RC, Ed. Vol. 4. *Vertebrates of Genetic Interest*. Plenum Press, New York, 1975.

DUTRILLAUX B, COUTURIER J: La Pratique de l'Analyse Chromosomique. In *Techniques de Laboratoire*. Masson, Paris, New York, 1981, pp 86.

DUTRILLAUX B: Sur la nature et l'origine des chromosomes humains. *Monographies des Annales de Génétique*. Expansion Scientifique Ed., 1975, pp. 102.

HECHT F, WYANDT HE, HEATH MAGENIS E: The human cell nucleus: quinacrine and other differential stains in the study of chromatin and chromosomes. In *The Cell Nucleus*, Vol. II. BUSCH H, Ed. Academic Press, New York, 1974, pp. 33–121.

LUBS HA, McKENSIE WH, PATIL SR, MERRICK S: New staining methods for chromosomes. Chapter 12 in *Methods in Cell Biology*, Vol. VI. Academic Press, New York, 1973.

SCHWARZACHER HG, WOLF U: *Methods in Human Cytogenetics.* Springer-Verlag, New York, 1974.

Original Articles

ARRIGHI FE, HSU TC: Localization of heterochromatin in human chromosomes. *Cytogenetics* **10**:81–86. 1971.

BOBROW M, COLLACOTT H.E.A.L., MADAN K: Chromosome banding with acridine orange. *Lancet* **2**:1311, 1972.

BOBROW M, MADAN K, PEARSON PL: Staining of some specific regions of human chromosomes, particularly the secondary constriction region of no. 9. *Nature (New Biol)* **238**:122–124, 1972.

CARPENTIER S, DUTRILLAUX B, LEJEUNE J: Effet du milieu ionique sur la dénaturation thermique ménagée des chromosomes humains. *Ann Génét* **15**:203–205, 1972.

CASPERSSON T, ZECH L, JOHANSSON C: Differential binding of alkylating fluorochromes in human chromosomes. *Expt Cell Res* **60**:315–319, 1970.

COUTURIER J, DUTRILLAUX B, LEJEUNE J: Etude des fluorescences spécifiques des bandes R et des bandes Q des chromosomes humains. *C R Acad Sci* (Paris) **276**:339–342, 1973.

CRAIG-HOLMES AP, SHAW MW: Polymorphism of human constitutive heterochromatin. *Science* **174**:702–704, 1971.

DRETS ME, SHAW MW: Specific banding patterns of human chromosomes. *Proc Nat Acad Sc* **68**:2073–2077, 1971.

DUTRILLAUX B, LEJEUNE J: Sur une nouvelle technique d'analyse du caryotype humain. *C R Acad Sc* **272**:2638–2640, 1971.

DUTRILLAUX B, GROUCHY J de, FINAZ C, LEJEUNE J: Mise en évidence de la structure fine des chromosomes humains par digestion enzymatique (pronase en particulier). *C R Acad Sc* **273**:587–588, 1971.

DUTRILLAUX B: Nouveau système de marquage chromosomique: les bandes T. *Chromosoma* **41**:395–402, 1973.

DUTRILLAUX B, LAURENT C, COUTURIER J, LEJEUNE J: Coloration par l'acridine orange de chromosomes préalablement traités par le 5-bromo-déoxyuridine (BrdU). *C R Acad Sc* (Paris) **276**:3179–3181, 1973.

DUTRILLAUX B, FOSSE AM: Sur le mécanisme de la segmentation chromosomique induite par le BrdU (5-bromodéoxyuridine). *Ann Génét* **17**:207–211, 1974.

EVANS HJ, BUCKTON KG, SUMNER AT: Cytological mapping of human chromosomes: results obtained with quinacrine fluorescence and the acetic-saline-Giemsa techniques. *Chromosoma* **35**:310–325, 1971.

FINAZ C, GROUCHY J de: Le caryotype humain après traitement par l'α-chymotrypsine. *Ann Génét* **14**:309–311, 1971.

LATT SA: Microfluorometric detection of deoxyribonucleic acid replication in human metaphase chromosomes. *Proc Nat Acad Sci* **70**:3395–3399, 1973.

LIN CC, UCHIDA IA, BYRNES E: A suggestion for the nomenclature of the fluorescent patterns in human metaphase chromosomes. *Canad J Genet Cyto* **13**:361, 1971.

PARDUE ML, GALL JG: Chromosomal localization of mouse satellite DNA. *Science* **168**:1356–1358, 1970.

PERRY P, WOLFF S: New Giemsa method for the differential staining of sister chromatids. *Nature* **251**:156–158, 1974.

SCHNEDL W: Analysis of the human karyotype using a reassociation technique. *Chromosoma* **34**:448–454, 1971.

SEABRIGHT M: A rapid banding technique for human chromosomes. *Lancet* **2**:971–972, 1971.

SUMNER AT, EVANS HJ, BUCKLAND R: New techniques for distinguishing between human chromosomes. *Nature (New Biol)* **232**:31–32, 1971.

YUNIS JJ, ROLDAN L, YASMINEH WG, LEE JC: Staining of satellite DNA in metaphase chromosomes. *Nature* **231**:532–533, 1971.

ZAKHAROV AF, SELEZNEV JV, BENJUSCH VA, BARANOVSKAYA LI, DEMINTSCVA VT: Differentiation along human chromosomes in relation to their identification. *IVè Congrès International de Génétique Humaine, Excerpta Medica, Intern. Congress Series* **233**:193, 1971.

ZECH L: Fluorescence banding techniques. In *Nobel Symposia 23. Chromosome Identification*, CASPERSSON T, and ZECH L, Ed. Nobel Foundation, Stockholm, 1973, pp. 28–31.

Buccal Smears

In theory, examination of buccal smears permits the determination of numerical anomalies of the sex chromosomes X and Y. After standard staining, the number of Barr bodies reveals the number of X chromosomes carried by the individual. Quinacrine staining and observation in UV light reveals the number of Y chromosomes.

Smear Technique

With a metal spatula, gently scrape the buccal mucosa and spread, in several patches, on a clean slide.

Barr Bodies

Immediate observation: Immediate observation is recommended.

Stock orcein solution: 1 g of orcein (E. Gurr) in 45 ml of acetic acid. Heat just under boiling point for 20 min.

Let cool. Filter.

Store in darkness.

Upon use dilute to one-half in distilled water. Filter.

Let fall one drop on the smear, cover with a coverslip, and blot with an absorbent tissue so as to remove excess liquid.

Observe directly under the microscope.

Fixation: Fixation of several slides is recommended in order to provide a control in case of incoherent results.

Fix 20 min in Carnoy's solution:

Ethanol	6 vol
Chloroform	3 vol
Acetic acid	1 vol

Let dry.

Stain with a light stain, such as toluidine blue, methylene blue, orcein, etc.

In every case it is essential that only perfectly delimited nuclei be taken into account. The results must be given in percentages. On the average, about 15 to 30 percent of Barr bodies are present in a normal 46,XX female. In a normal 46,XY male or in a 45,X patient, no Barr bodies are present (0 percent).

Y Bodies

Fix 30 min in methanol.

Let dry.

Treat the slides with an aqueous solution of quinacrine dihydrochloride at 0.5 percent.

Rinse in distilled water.

Mount.

View in UV light.

The Y chromosome appears in the form of a fluorescent body, which, unlike the Barr body, is not necessarily attached to the nuclear membrane.

APPENDIX II
DERMATOGLYPHICS

The term *dermatoglyphics* was proposed by Cummins in 1926 to describe the fine ridged patterns of the epidermis of the palm and fingers (*glyphê* = engravement, *derma* = skin). By extension, study of dermatoglyphics also includes study of palmar and digital flexion creases.

Dermatoglyphics are situated not only on the palm and fingertips but also on the soles of the feet and the toes, but only the former are considered here, since plantar dermatoglyphics are still poorly known.

The nomenclature provided here is intended only to help the reader understand the descriptions given for each cytogenetic syndrome.

FLEXION CREASES

The Palm

From the fingers to the wrist the following are found in successive order:

- The distal palmar flexion crease, which extends from the ulnar margin of the hand to the second interdigital space
- The proximal flexion crease, which extends across the palm from the radial margin to the middle of the ulnar margin
- The thumb opposition crease, which delimits the thenar eminence

Coalescence of the proximal and distal flexion creases into a single crease is termed *unique transverse palmar crease* (UTPC). Incomplete fusion of the two creases gives an *equivalent* to UTPC.

The Fingers

Each finger is separated from the palm by the digito-palmar flexion crease and has two digital flexion creases, corresponding to the interphalangeal articulations. The thumb has only one flexion crease.

DERMATOGLYPHICS

The epidermal ridges alternate with furrows and are in parallel arrangement. The configuration of the ridges produces diverse patterns, which are the object of dermatoglyphic studies.

Palmar Dermatoglyphics

By convention, the margin of the palm has been divided into zones numbered 1 to 14 (see illustration on page 431).

A triradius is defined as the center of the junction of three regions of ridges. For each triradius, it is possible to demarcate the three regions with three lines.

The axial triradius and four digital triradii are distinguished.

Axial triradius: It is usually situated in the proximal region of the palm, at the convergence of thenar, hypothenar, and carpal ridges. It corresponds to fusion of the three embryonic primordia of the blade of the hand.

The normal position of the triradius is designated by t. It can be far distal, at the limit of the flexion creases—i.e., in t″—or else in an intermediate position, t′.

The line issuing from the axial triradius and extending into the palm, usually into 13, is called the main line or line T.

In some cases duplication of the axial triradius can be seen, for example in t and t′.

Digital triradii: Situated at the base of each of the four fingers, the digital triradii are termed, in order, a, b, c, and d. Of the three lines issuing from each triradius, two subtend the base of the corresponding finger, while the third radiant, or main line, is directed toward the center of the palm and terminates on the margin in one of the 14 regions referred to above. The main lines are identified by A, B, C, or D, corresponding to the digit subtended.

Main line index: The sum of the numbers corresponding to the different exit zones of each main line constitutes the number known as the main line index, which provides an estimate of the degree of transversality of these radiants.

The mean value of this index is 27 (see Table 1). It is known, for example, that in 21 trisomy the index is of the order of 31, thus reflecting the horizontalization of the dermal ridges.

AII-1 Palmar and digital dermatoglyphics. Diagram showing triradii, main lines, and creases. (Courtesy M.O. Rethoré.)

Axial triradius in t→13

No pattern on eminences

Digital triradii: a4; b5; c5; d7. Extra triradius in 9

Transversality index: 21

Interstitial pattern in 7 and 9

Finger patterns: I and II: whorl; III, IV, and V: ulnar loop

Normal palmar and digital flexion creases

Slight clinodactyly of V

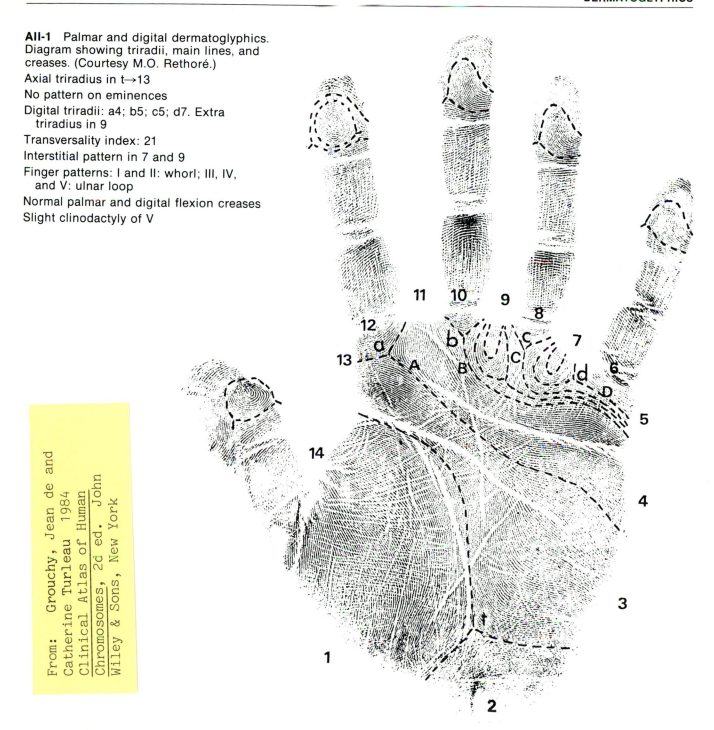

From: Grouchy, Jean de and Catherine Turleau 1984 Clinical Atlas of Human Chromosomes, 2d ed. John Wiley & Sons, New York

TABLE 1   **Palmar Dermatoglyphics**

	Right Hand		Left Hand		Right + Left Hands Males + Females
	Males	Females	Males	Females	
Transversality Index (mean)	27.54 ± 0.27	27.75 ± 0.27	25.71 ± 0.26	26.27 ± 0.26	26.9 ± 0.12
Axial triradius t	73.00	66.00	68.00	64.00	67.77
(100%) t'	19.00	27.00	27.50	28.50	25.50
t''	8.00	7.00	4.50	7.50	6.77
Hypothenar (Radial loops)	24.00	26.50	24.00	22.50	24.25
(100%) (Ulnar loops)	9.00	9.50	9.00	11.50	9.75
Single palmar crease (100%)	0.5	1.00	1.00	2.00	1.12
Patterns P_7		39.2		55.3	47.2
(100%) P_9		48.6		33.8	41.2
P_{11}		4.7		1.0	2.85

SOURCE: Turpin and Lejeune, 1953.

TABLE 2   **Frequency of Digital Figures**

	France (Turpin and Schutzenberger, 1949) %	Great Britain (Holt, 1964) %
Arches	9.02	4.98
Ulnar loops	58.17	63.54
Radial loops	4.1	5.36
Whorls	24.87	26.12
Complex patterns	3.84	—
	100.0	100.0

The atd angle: Formed by the lines joining the axial triradius t to the a and d digital triradii. Its value depends on the position of the axial triradius, for which it gives a numerical estimate.

"Pelotes" or interdigital patterns: The term "pelote" designates loop or whorl patterns, which can be situated between digital triradii. They are called P_7, P_9, or P_{11}, depending on their position.

Thenar and hypothenar patterns: Diverse patterns can be observed on the palmar eminences: loops, whorls, or complex patterns.

Digital Dermatoglyphics

The epidermal ridges situated on the fingertips can evince three kinds of patterns which have different frequencies in the general population (Table 2):

- The arch, the simplest pattern, without triradius
- The loop, adjacent to a single triradius, and having either a radial or ulnar opening
- The whorl, dependent on two triradii and having diverse forms, in rosette, double loop, racquet, etc.

For loops and whorls, the number of ridges from the center of the pattern to the triradius can be counted. For arches this number is equal to 0 (zero). The sum of the values thus obtained for the ten fingers gives the total ridge count. It is normally in the vicinity of 140–145 in males and 120–130 in females. It can vary in certain pathological conditions, sexual chromosome anomalies in particular.

REFERENCES

CUMINS H, MIDLO C: *Finger Prints, Palms and Soles: An Introduction to Dermatoglyphics.* Dover Publications, New York, 1961, p. 319.

HOLT SB: Current advances in our knowledge of the inheritance of variations in finger prints. *Proc Int Conf Hum Genetics,* Roma **3**:1450, 1964.

LAFOURCADE J, RETHORE MO: Les dermatoglyphes. In *Les Journées Parisiennes de Pédiatrie.* Ed. Médicales Flammarion, 1967.

PENROSE LS: Memorandum on dermatoglyphic nomenclature. *Birth Defects: Original Article Series,* Bergsma D, Ed Vol. IV, no. 3, 1968.

TURPIN R, SCHUTZENBERGER MP: L'étude des dermatoglyphes. *Sem Hôp* **25**:2553–2562, 1949.

TURPIN R, LEJEUNE J: Etude dermatoglyphique des paumes des mongoliens et de leurs parents et germains. *Sem Hôp* **29**:3955–3967, 1953.

APPENDIX III
TERMS USED IN CLINICAL DESCRIPTIONS

It is impossible to list all the terms used to describe the morphology of patients. We have simply noted a few that often are poorly understood. Most of them concern the ear, the morphology of which is highly significant in the different cytogenetic syndromes.

On the face profile we have indicated the landmarks defining the normal setting of the ear.

On the front view, the inner (A) and outer (B) canthal distances allow a rapid appreciation of the distance between the eyes. Normally, distance B equals three times distance A. In other words, the eyes must be separated by the length of one eye. Hyper- or hypotelorism is present if the inner canthal distance is superior or inferior to normal. The true definition of hypertelorism remains, however, radiological.

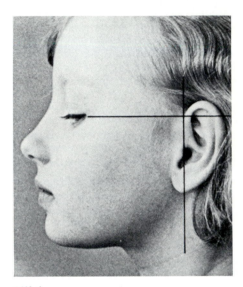

AIII-1

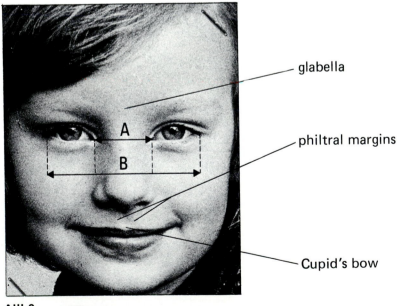

glabella

philtral margins

Cupid's bow

AIII-2

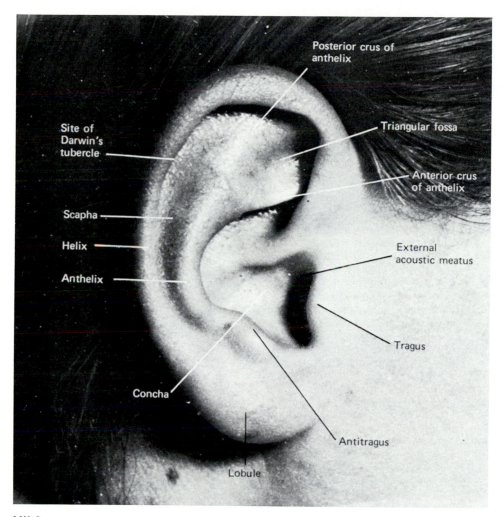

AIII-3

APPENDIX IV
TYPES AND MECHANISMS OF FORMATION OF CHROMOSOMAL ABERRATIONS

Chromosomal aberrations may be numerical or structural (with or without an anomaly in number).

NUMERICAL ABERRATIONS

By definition, numerical aberrations affect the number of chromosomes but not their structure, which remains normal. They may be homogeneous or in mosaic. When they are homogeneous, these aberrations result mainly from nondisjunctions during meiosis and may produce a trisomy (the presence of a supernumerary normal chromosome) or a monosomy (the loss of a chromosome). Trisomies are by far the most frequently encountered in infants born alive.

A nondisjunction is defined by the fact that two homologous chromatids migrate toward the same pole during anaphase, and "pass" into the same daughter cell together, rather than each of them migrating into a daughter cell. This nondisjunction may occur during the first or second maternal or paternal meiotic division. In certain cases, a study of chromosomal polymorphisms allows the distinction between the four possibilities. In addition, it has been established that maternal age is the principal etiological factor favoring nondisjunctions.

The most frequently occurring autosomal trisomies are 21, 18, 13, and 8. Trisomies of the sexual chromosomes are very frequent as well, and concern both X and Y: 47,XXX; 47,XXY; 47,XYY. More extensive anomalies in number also can be seen, such as 48,XXXX, etc.

Autosomal monosomies are probably produced at the same frequency, but are rarely observed at birth, no doubt due to the fact that they are eliminated at the earliest stages of embryonic life.

Insofar as sex chromosomes are concerned, monosomy 45,X is responsible for Turner's syndrome.

Numerical mosaicism is characterized by the presence of at least two different clones, and results from a postzygotic nondisjunction. Mosaicism is especially frequent in the case of sex chromosomes.

Mosaics are derived from a single zygote, whereas chimeras are derived from two zygotes, and may be due to accidents at fertilization, strictly speaking, such as a double fertilization.

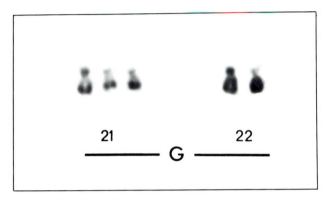

AIV-1 Free 21 trisomy.

Polyploids also are due to accidents of fertilization. The most frequently found is triploidy, characterized by the presence of three haploid sets of chromosomes: 69,XXX, or XXY, or XYY.

STRUCTURAL ABERRATIONS

Structural anomalies are the consequence of chromosomal breaks followed by one or more abnormal reunions. By definition, partial trisomies and monosomies result from structural rearrangements.

Structural aberrations may affect one or two homologous or nonhomologous chromosomes, and sometimes more.

They may be balanced or unbalanced. Balanced anomalies do not lead to an imbalance in chromosomal material, and generally do not have a phenotypical effect. However, during meiosis, they may cause the formation of unbalanced gametes and hence abnormal zygotes.

Unbalanced anomalies may occur *de novo* (*de novo* unbalanced deletions or translocations) or be the consequence of a balanced parental rearrangement.

Aberrations Affecting One Chromosome

Deletions: All deletions are probably intercalary, so as to respect the integrity of the telomere. They result

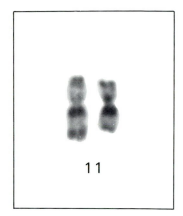

AIV-2   Terminal deletion of 11q.

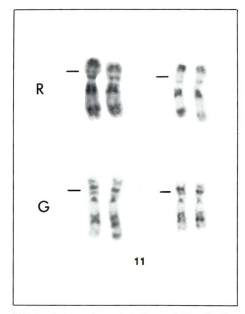

AIV-3   Intercalary deletion of 11p (indicated by a dash) responsible for the aniridia/Wilms tumor/ambiguous genitalia/mental retardation syndrome.

from two breaks occurring in the same chromosome arm and the loss of the intermediary segment. For convenience's sake, we will make the distinction between terminal deletions concerning the chromosomal extremity, and intercalary deletions, involving more proximal segments.

Deletions most often occur *de novo*. A minority of them (around 10–15%) result from malsegregation of a balanced parental rearrangement. *De novo* deletions may be considered as pure monosomies, whereas the other deletions are generally accompanied by a partial trisomy for another chromosome (duplication/deficiency).

Ring Chromosomes: These result from a break at the two extremities of a chromosome, followed by reunion and loss of the two distal segments. In practice, rings can be assimilated to a double deletion. Nonetheless, they cause complex mitotic phenomena, with varying duplication/deficiencies which make phenotypical interpretation difficult.

Inversions: These are due to two breaks on the same chromosome, followed by reunion after inversion of the intermediary segment. They are said to be pericentric if the centromere is included in the intermediary segment, and paracentric if the two breaks occur in the same arm.

These inversions are balanced rearrangements, but cause the formation of a pairing loop during meiosis, with the risk of producing abnormal gametes through duplication/deficiency.

These duplication/deficiencies involve the distal segments with respect to the breakpoints. The larger these segments, the higher the mortality, and the lower the risk of giving birth to a deformed child who is viable.

Isochromosomes: An isochromosome is an abnormal chromosome formed from two long or two short arms, with loss of the other arm. It may be mono- or dicentric, depending on the etiological mechanism. The most frequently observed isochromosome is that for the long arm of X, which constitutes a cytogenetic variant of Turner's syndrome.

Intrachromosomal Duplications: These are rearrangements leading to pure trisomy. Intrachromosomal duplications may occur either as tandem or as mirror duplications.

Aberrations Involving Two Chromosomes

These are mainly translocations. A translocation is characterized by two breaks on two different chromosomes, whether homologous or not, and abnormal reunion after an exchange of distal segments. A distinction is made between Robertsonian translocations, reciprocal translocations, and whole-arm translocations. These translocations may be balanced or unbalanced. They may arise *de novo,* or be transmitted.

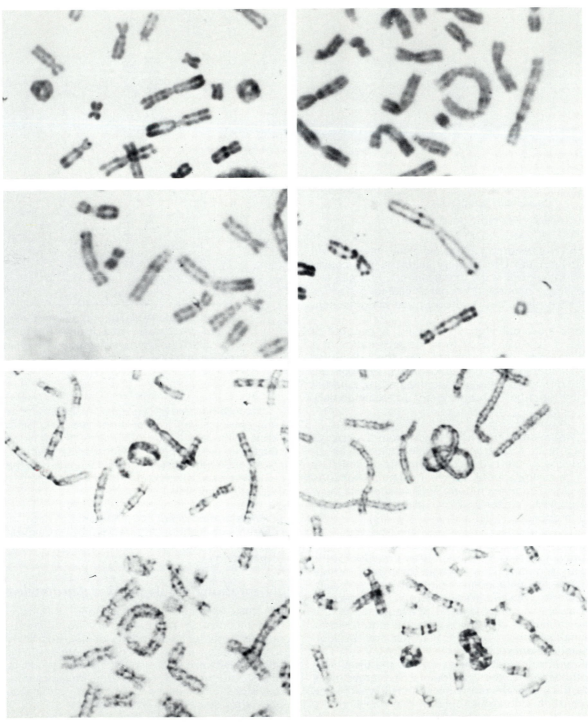

438 **AIV-4** Ring chromosome: r(4): various ring structures during mitosis.

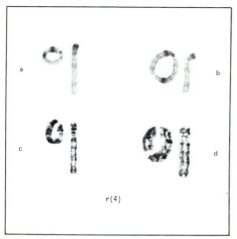

AIV-5 Ring chromosome: r(4): double ring.

Robertsonian Translocations: These occur between acrocentric chromosomes, either by centromeric fusion or by breaks in the juxtacentromeric regions. Balanced Robertsonian translocations do not have a phenotypical effect, although they do cause the loss of the short arm of the two translocated chromosomes. During meiosis, on the other hand, they may lead to formation of unbalanced gametes producing trisomic or monosomic zygotes for the totality of the chromosome. This risk is especially high for chromosomes 21 and 13, with the Robertsonian translocation responsible for the majority of familial forms of trisomies 21 and 13 (see chapters "21 Trisomy" and "13 Trisomy"). When the Robertsonian translocations occur between homologues, only unbalanced gametes can be formed.

Reciprocal Translocations: These translocations are due to exchanges of chromosomal segments between two chromosomes, with the breakpoints having been

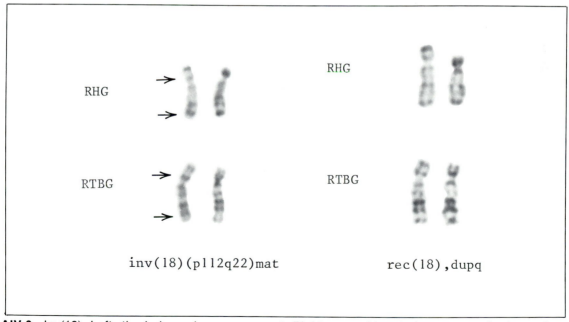

AIV-6 Inv(18). Left: the balanced rearrangement. Right: recombinant chromosome in the patient with duplication/deficiency: del (18p), dup(18q).

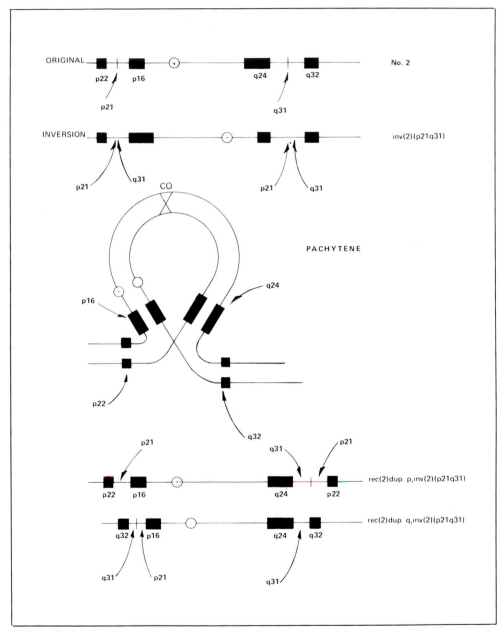

AIV-7 Pericentric inversion: diagram showing the recombinant chromosomes resulting from an inv(2) with breakpoints in 2p21 and 2q31 (in: ISCN, 1971).

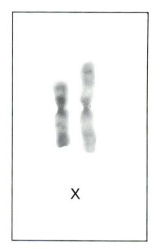

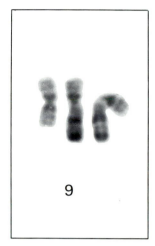

AIV-8 (left) Isochromosome X: i(Xq); (right) isochromosome 9p: i(9p) resulting in tetrasomy 9p.

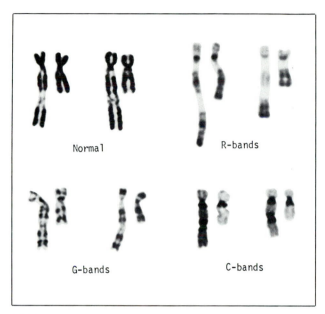

AIV-9 Intrachromosomal duplication of 9q (after applying different banding techniques).

produced elsewhere than in the juxtacentromeric regions.

Carriers of a balanced reciprocal translocation may form unbalanced gametes, depending on the following three types of meiotic segregation or disjunction:

Adjacent-1 type. One of the two rearranged chromosomes is transmitted with the normal homologue of the other chromosome. The result is a duplication of one of the translocated segments, and a monosomy of the other segment. Both reciprocal combinations are, of course, possible.

Adjacent-2 type. This is rare. One of the rearranged chromosomes is transmitted with its own homologue. There then exists a duplication/deficiency of the centric segments of the rearranged chromosomes (with no imbalance of the translocated portion).

These two types of segregation produce unbalanced zygotes having 46 chromosomes.

Type 3:1. This is rare, and the resulting zygotes have 45 or 47 chromosomes: or the two rearranged chromosomes are transmitted with a normal homologue; or else, only one rearranged chromosome is transmitted with the two normal homologues.

Factors that enhance adjacent-2 type segregation or

3:1 type disjunction are 1) involvement of an acrocentric at least; 2) the fact that the centric segments are short, or involve a secondary constriction, especially that of chromosome 9; and 3) the fact that the carrier of the balanced translocation is the mother.

Whole-arm Translocations: These translocations, at meiosis, produce disomic gametes for one arm, nullosomic gametes for the other. The zygotes are rarely viable. This type of translocation is observed most often in cases of sterility or repeated abortions.

Insertions: These are defined by the transfer of an intercalary chromosomal segment into another chromosomal arm. They result from a three-break mechanism, one on the recipient chromosome and the other two on the donor chromosome. The donor and recipient chromosomes moreover may be one and the same chromosome. The inserted segment may keep the same orientation with respect to the centromere, or show an inverse orientation.

During meiosis, the formation of nullosomic or disomic gametes for the inserted segment are observed.

441

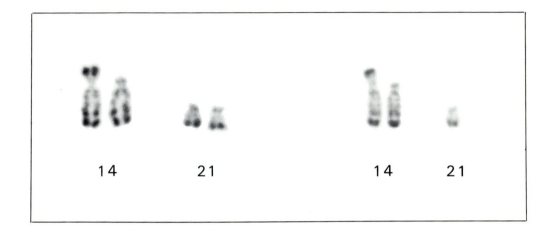

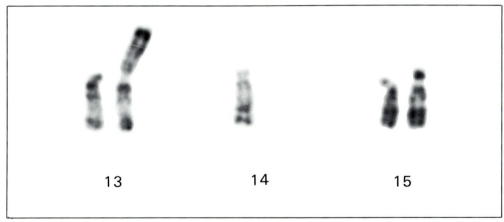

AIV-10   Robertsonian translocations: (top) t(14q21q). Right: balanced parental translocation. Left: translocation 21 trisomy. (bottom) Balanced t(13q14q).

AIV-11   Reciprocal translocation. Different possible segregations at meiosis of ▶ a t(2;5)(q21;q31). Letters A, B, C, and D designate chromosome segments between the break points and the telomere. The different types of unbalanced gametes are shown in the adjacent table (in ISCN, 1971).

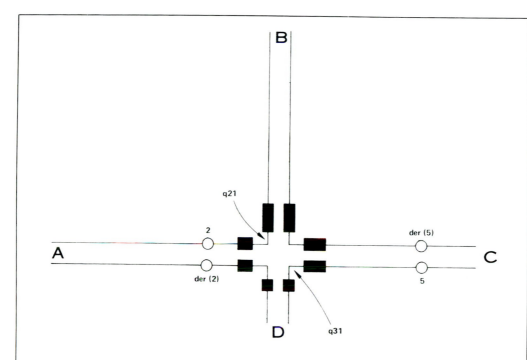

Designation of unbalanced karyotypes. Use of _der_ symbol to designate unbalanced karyotypes derived by segregation in a reciprocal translocation heterozygote. Based on the pachytene diagram above.

Disjunction (segregation)	Unbalanced gamete	Karyotype of zygote resulting from an unbalanced gamete fertilized by a normal gamete
Adjacent-1	AB CB	46,XX, − 5, + der(5),t(2;5)(q21;q31)mat
	AD CD	46,XY, − 2, + der(2),t(2;5)(q21;q31)mat
Adjacent-2[1]	AB AD	46,XY, − 5, + der(2),t(2;5)(q21;q31)mat
	CB CD	46,XY, − 2, + der(5),t(2;5)(q21;q31)mat
	AB AB	46,XX, + 2, − 5
	AD AD	46,XY, − 2, − 5, + der(2), + der(2),t(2;5)(q21;q31)mat
	CB CB	46,XY, − 2, − 5, + der(5), + der(5),t(2;5)(q21;q31)mat
	CD CD	46,XX, − 2, + 5
3:1[2]	AB CB CD	47,XX, + der(5),t(2;5)(q21;q31) mat
	AD	45,XY, − 2, − 5, + der(2),t(2;5)(q21;q31)mat
	CB CD AD	47,XX, − 2, + der(2), + der(5),t(2;5)(q21;q31)mat
	AB	45,XY, − 5
	CD AD AB	47,XY, + der(2),t(2;5)(q21;q31)mat
	CB	45,XX, − 2, − 5, + der(5),t(2;5)(q21;q31)mat
	AD AB CB	47,XX, − 5, + der(2), + der(5),t(2;5)(q21;q31)mat
	CD	45,XX, − 2

[1]Adjacent-2 disjunction minimally results in the first two unbalanced gametic types shown (AB AD, CB CD). Crossing-over in the interstitial segments between centromeres and points of exchange is necessary for the origin of the remaining four types.

[2]A further eight segregational types can occur at AII if there is crossing-over in the interstitial segments, making a total of 12 types of gametes with three chromosomes derived from the translocation quadrivalent.

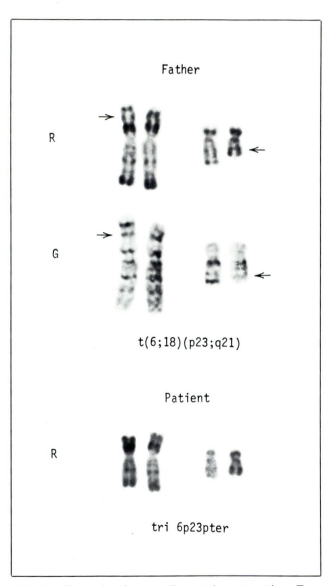

Father

R

G

t(6;18)(p23;q21)

Patient

R

tri 6p23pter

AIV-12 Example of type adjacent-1 segregation. Top: parental balanced translocation. Bottom: resulting partial trisomy.

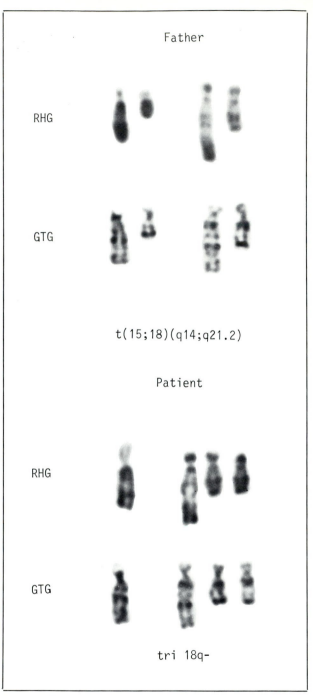

Father

RHG

GTG

t(15;18)(q14;q21.2)

Patient

RHG

GTG

tri 18q-

AIV-13 Example of a type adjacent-2 segregation.

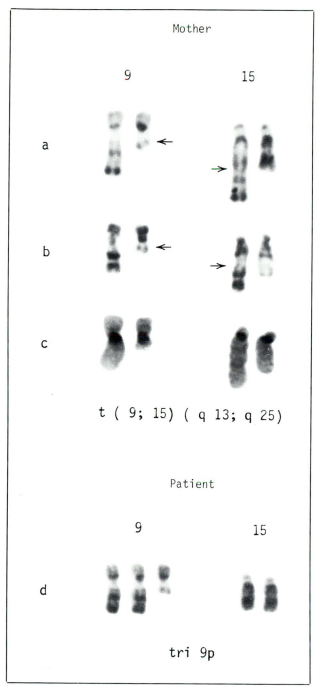

AIV-14 Example of a type 3:1 segregation.

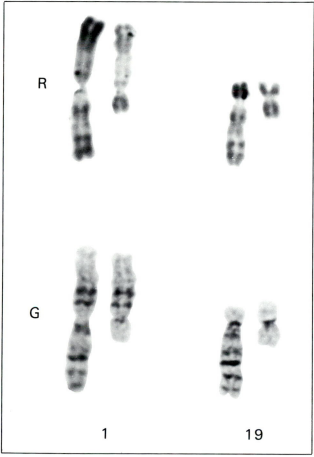

AIV-15 Whole-arm translocation between chromosomes 1 and 19.

445

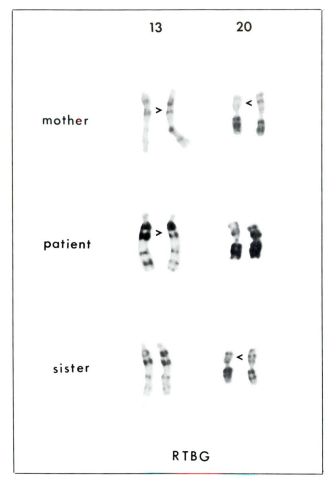

13   20

mother

patient

sister

RTBG

AIV-16 Balanced maternal insertion (20;13) resulting in two types of aneusomia: a monosomy 13q14 associated with retinoblastoma, or a trisomy 13q14 with no severe phenotypic correlation.

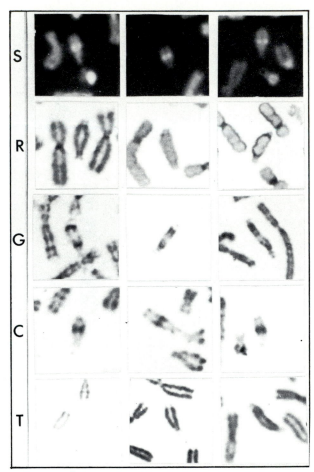

AIV-17 Pseudodicentric chromosome 21 observed in a patient with 21 trisomy.

Dicentric or Pseudodicentric Chromosomes: These chromosomes result from the fusion of two homologous or nonhomologous chromosomes, usually in the telomeric regions. When the two centromeres are sufficiently far from one another, one of them loses its function (pseudodicentric chromosome).

More Complex Rearrangements

Rearrangements exist involving more than two chromosomes and resulting in various configurations.

FRAGILE SITES

In certain individuals, a constitutional fragile site is observed which is localized on a given autosome. These fragile sites are transmitted as a dominant trait, and do not appear to have a phenotypical effect per se. A possible unfavorable role during meiosis remains open to question. Such fragile sites often are observed in couples who have experienced reproductive failure. Thus, they concern a given population, and only after systematic studies on normal populations will we be able to evaluate the genetic risk of these marker chromosomes. More-

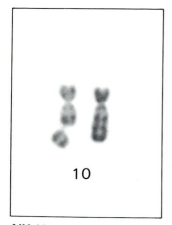

AIV-18 Fragile site 10q23.

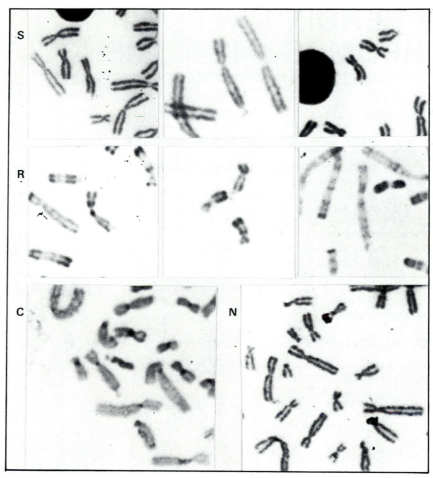

AIV-19 Fragile site Xq28. Standard staining, R- and C- banding. NORs (N) staining to show that there is no satellite.

over, it has not been excluded that there exists an ethnic variation of such polymorphisms.

The autosomes and the sites that are most often observed are 2q11, 10q23, 10q25, 11q13, 16p12, 16q22, and 20p11.

The case of chromosome X, with the site in Xq28, is very special due to its phenotypical effect, and a special chapter is devoted to it (p 410).

The expression of these fragile sites varies according to culture conditions and especially folic acid content in the medium.

REFERENCES

AURIAS A., PRIEUR M., DUTRILLAUX B., LEJEUNE J.: Systematic analysis of 95 reciprocal translocations of autosomes. *Hum. Genet.,* **45**:259–282, 1978.

GIRAUD F., AYME S., MATTEI J.F., MATTEI M.G.: Constitutional chromosomal breakage. *Hum. Genet.,* **34**:125–136, 1976.

HAMERTON J.L.: *Human Cytogenetics.* 2 vol., New York, Academic Press, 1971.

JALBERT P., SELE B.: Factors predisposing to adjacent 2 and 3:1 disjunctions: study of 161 human reciprocal translocations. *J. med. Genet.,* **16**:467–478, 1979.

SUTHERLAND G.R.: Heritable fragile sites on human chromosomes I. Factors affecting expression in lymphocyte culture. *Am. J. hum. Genet.,* **31**:125–135, 1979.

SUTHERLAND G.R.: Heritable fragile sites on human chromosomes II. Distribution, phenotypic effects, and cytogenetics. *Am. J. hum. Genet.,* **31**:136–148, 1979.

APPENDIX V
CHROMOSOME NOMENCLATURE

The nomenclature of human chromosomes (and of those of other species such as the hominoid primates) has been the subject of several conferences: in Denver (1960), London (1963), Chicago (1966), Paris (1971).

Each conference published a report completing that of the preceding conference.

The present nomenclature, called the Paris Nomenclature (1971), was detailed in a supplement published in 1975 by the Standing Committee, which was installed by the Paris Conference.

The nomenclature given below is an amalgamation of these different nomenclatures. It is limited to the description of human mitotic chromosomes and does not include the nomenclature of human meiotic chromosomes and of chromosomes from other species, such as the hominoid primates.

NORMAL KARYOTYPE

One writes down in sequence the total number of chromosomes (autosomes + sex chromosomes), a comma, and the sex chromosomes:

46,XY (normal male karyotype)
46,XX (normal female karyotype)

NUMERICAL ABERRATIONS

Sex Chromosomes

They are indicated very simply as follows:

45,X (Turner's syndrome)
47,XXY (Klinefelter's syndrome)

Autosomes

The + or − sign should be placed before the supernumerary or missing chromosome:

47,XY,+21 (21 trisomy in a boy)
47,XX,+18 (18 trisomy in a girl)
45,XX,−21 (21 monosomy in a girl)

Triploidy and Polyploidy

69,XXY (triploid cell)

Mosaicism

The different cell populations are indicated one after the other and separated by a single slash (/):

45,X/46,XY
46,XX/46,XY
46,XX/47,XX,+21
45,X/46,XX,/47,XXX

If it is necessary to distinguish between a *mosaic* (the different cell types derive from a single zygote) and a *chimera* (the different cell types derive from two or more zygotes) it is now recommended that the triplets *mos* and *chi* be used.

mos 45,X/46,XY   (mosaic due to the loss of the Y chromosome in a 46,XY zygote)
chi 46,XX/46,XY   (chimera resulting from a double fertilization or a fusion of two zygotes)

Structural Aberrations

The short arm of a chromosome is indicated by the letter *p*, the long arm by *q*, a satellite by *s*, the centromere by *cen*, a secondary constriction by *h*.

The plus sign indicates an increase in length (it is then placed after the symbol *p, q, s, h*).

The minus sign indicates a decrease in length (or deletion).

With Standard Staining (in the Absence of Banding)

46,XX,18p−   (deletion of the short arm of chromosome 18)
46,XX,Bq+   (presence of an additional chromosome fragment on the long arm of a B chromosome)
46,XY,9qh+   (increase in length of the secondary constriction of chromosome 9)
46,XY,Ds+   (large satellite on a D)

Identification of a Breakage Point After Banding

Each chromosome in the human somatic-cell complement is considered to consist of a continuous series of bands, with no unbanded areas. The bands are allocated to various regions along the chromosome arms and delimited by specific chromosome landmarks. The bands and the regions they belong to are identified by numbers, with the centromere serving as the point of reference for the numbering scheme (see Chart 1).

A given point on a chromosome is designated by:

- The chromosome number
- The arm symbol
- The region number
- The band number

These items are given in order without spacing or punctuation. For example, 3q24 corresponds to a point located approximately in the middle of the long arm of chromosome 3.

This nomenclature indicates the proximal edge of band 3q24. A numbering system is proposed to designate the location of breakpoints within bands on the basis of the relative distance of the breakpoint from the proximal margin of the band concerned. The proximal edge of a band, say, 3q24, is denoted as 3q2400. A point six-tenths of the distance from the proximal edge to the distal edge of this band would be denoted as 3q2406.

One other method consists of subdividing bands into subbands (Prieur et al., 1973; see Chart 2). A given point is then designated by three digits, for example, 12q13.2, and even four digits in the case of high resolution banding, for example, 12q13.12.

Rearrangements of Single Chromosomes

A single colon (:) indicates a chromosome break.
A double colon (::) indicates breakage and union.
An arrow (→) indicates "from to."
The triplet *ter* designates the distal end of a chromosome arm. A short system or a detailed system can be used. They are not mutually exclusive and can complement each other.

Deletions
A deletion is indicated by *del.*

Terminal Deletions
46,XX,del(1)(q21)

In the short system, the triplet *del* is followed by the chromosome number in parentheses and by the identification of the breakage points in parentheses.

46,XX,del(1)(pter→q21:)

In the detailed system applied to the same deletion, the single colon (:) indicates a break at band 1q21 and deletion of the long-arm segment distal to it. The remaining chromosome consists of the entire short arm of chromosome 1 and part of the long arm lying between the centromere and band 1q21.

Interstitial Deletions
46,XX,del(1)(q21q31)
46,XX,del(1)(pter→q21::q31→qter)

In the short system the two breakage points are not separated by a semicolon, since they are located on the same chromosome (see below: translocations).

The double colon (::) indicates breakage and union of bands 1q21 and 1q31 in the long arm of chromosome 1. The segment lying between these bands has been deleted.

Paracentric Inversions
Inversions are designated by *inv*:

46,XY,inv(2)(p13p24)
46,XY,inv(2)(pter→p24::p13→p24::p13→qter)

Breakage and union have occurred at bands 2p13 and 2p24 in the short arm of chromosome 2. The segment lying between these bands is still present but inverted, as indicated by the reverse order of the bands with respect to the centromere in this segment of the rearranged chromosome.

Pericentric Inversions
46,XY,inv(2)(p21q31)
46,XY,inv(2)(pter→p21::q31→p21::q31→qter)

Breakage and union have occurred at band 2p21 in the short arm and 2q31 in the long arm of chromosome 2. The segment lying between these bands is inverted.

Ring Chromosomes
Ring chromosomes are designated by the symbol *r*:

46,XY,r(2)(p21q31)
46,XY,r(2)(p21→q31)

Breakage has occurred at band 2p21 in the short arm and 2q31 in the long arm of chromosome 2. With deletion of the segments distal to these bands, the broken ends have joined to form a ring chromosome. Note the omission of the colon or double colon.

Isochromosomes
An isochromosome is designated by the letter *i*:

46,X,i(Xq)
46,X,i(X)(qter→cen→qter)

Breakpoints in this type of rearrangement are at or close to the centromere and cannot be specified. The designation indicates that both entire long arms of the X chromosome are present and separated by the centromere.

Dicentric Chromosomes
Dicentric chromosomes are designated by the triplet *dic*:

46,X,dic(Y)(q12)
46,X,dic(Y)(pter→q12::q12→pter)

Breakage and union have occurred at band Yq12 on sister chromatids to form a dicentric Y chromosome.

Direct Insertion Within a Chromosome
Insertions are indicated by the symbol *ins*:

46,XY,ins(2)(p13q21q31)
46,XY,ins(2)(pter→p13::q31→q21::p13→
q21::q31→pter)

Breakage and union have occurred at band 2p13 in the short arm and bands 2q21 and 2q31 in the long arm of chromosome 2. The long-arm segment between 2q21 and 2q31 has been inserted into the short arm at band 2p13. The original orientation of the inserted segment has been maintained in its new position.

Inverted Insertion Within a Chromosome
46,XY,inv ins(2)(p12q31q21)
46,XY,inv ins(2)(pter→p13::q21→q31::p13→
q21::q31→qter)

The orientation of the bands within the segment has been reversed with respect to the centromere.

Duplication of a Chromosome Segment
The symbol *dup* designates a duplication of a segment. It can be supplemented with the triplets *dir* or *inv* to indicate whether the duplication is direct or inverted.

46,XX,inv dup(2p)(p23→p14)
46,XX,inv dup(2p)(pter→p23::p14→p23::p23→qter)

The segment p23→p14 has been inverted and duplicated.

Rearrangements Occurring Between Two Chromosomes

A translocation is symbolized by the letter *t* followed by the chromosome numbers indicated in parentheses and separated by a semicolon.

In Standard Staining
46,XY,t(3q−;2p+)

Reciprocal translocation between a fragment of the long arm of a no. 3 chromosome and a fragment of the short arm of a no. 2.

45,XX,t(DqDq)

Robertsonian translocation or centromeric fusion between two D chromosomes.

After Banding
Both chromosomes are designated in a parenthesis; the chromosome having the lowest number is always specified first. If one of the rearranged chromosomes is a sex chromosome, then it should be listed first. The breakpoints are indicated in a second parenthesis.

If the type of translocation—Robertsonian, reciprocal, or tandem—is to be emphasized, *t* may be replaced with *rob, rcp,* or *tan,* respectively.

Reciprocal Translocations
46,XY,t(2;5)(p21;q31) or 46,XY,rcp(2;5)(p21;q31)
46,XY,t(2;5)(2pter→2q21::5q31→5qter;5pter→
5q31::2q21→2qter)

Breakage and union have occurred at bands 2q21 and 5q31 in the long arms of chromosomes 2 and 5, respectively. The segments distal to these bands have been exchanged between the two chromosomes.

Robertsonian Translocations
45,XX,t(13;14)(p11;q11) or 45,XX,rob(13;14)(p11;q11)
45,XX,t(13;14)(13qter→13p11::14q11→14qter)

This notation specifies in which arm of each chromosome breakage occurred; hence, which centromere has been retained.

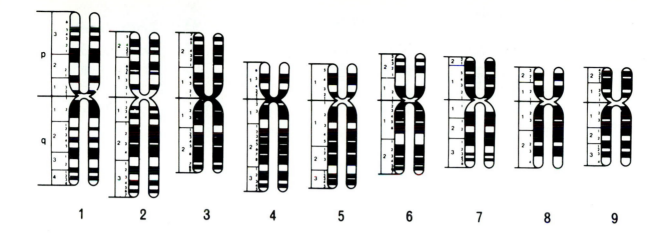

1   2   3   4   5   6   7   8   9

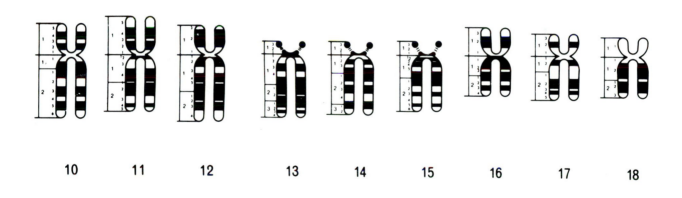

10   11   12   13   14   15   16   17   18

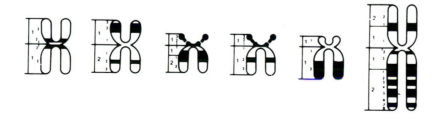

19   20   21   22   Y   X

AV-1   Paris nomenclature (1971).

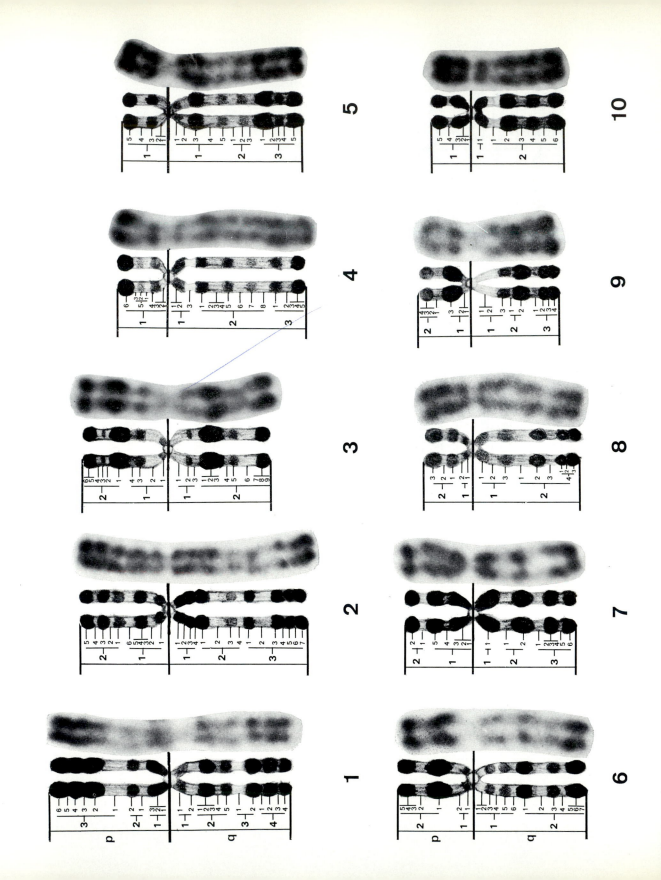

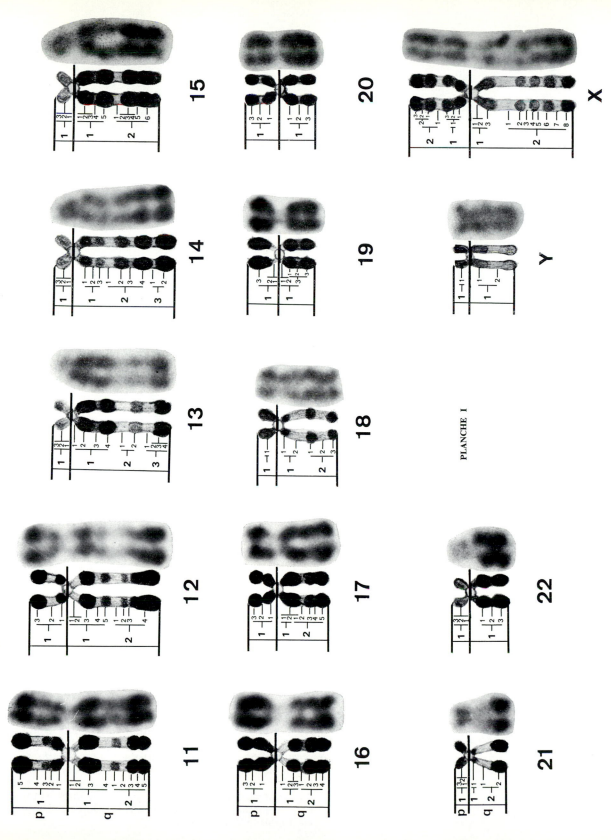

AV-2 and AV-3 PRIEUR M., DUTRILLAUX B., LEJEUNE J. (1973),—Planches descriptives des chromosomes humains. (Analyse en bandes R et nomenclature selon la conférence de Paris 1971). *Ann. Génét.*, **16**:39–46.

© Expansion Scientifique Française, Paris, 1973. Tous droits de reproduction strictement réservés.

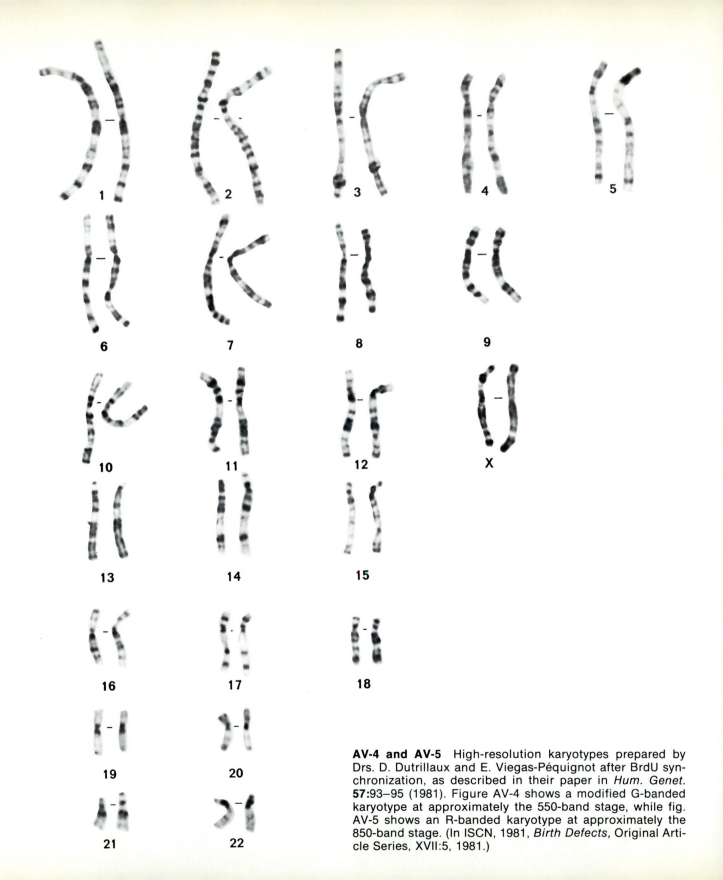

AV-4 and AV-5 High-resolution karyotypes prepared by Drs. D. Dutrillaux and E. Viegas-Péquignot after BrdU synchronization, as described in their paper in *Hum. Genet.* **57:**93–95 (1981). Figure AV-4 shows a modified G-banded karyotype at approximately the 550-band stage, while fig. AV-5 shows an R-banded karyotype at approximately the 850-band stage. (In ISCN, 1981, *Birth Defects,* Original Article Series, XVII:5, 1981.)

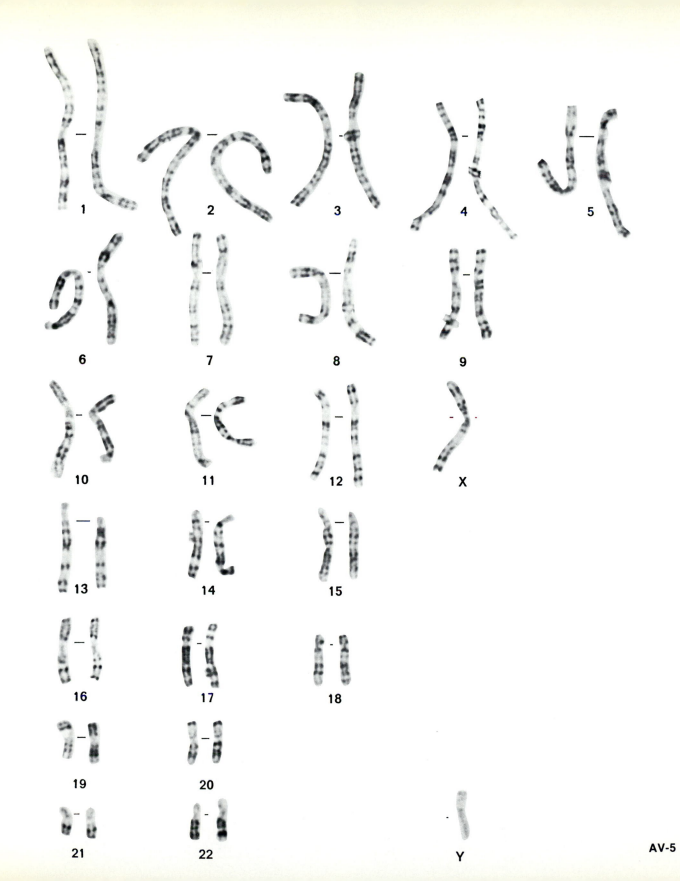

AV-5

If the origin of the centromere is not known, the following notation is used:

45,XX,t(13q14q)

or

45,XX,t(13;14)(13qter→cen→14qter)

Terminal Rearrangements

The double triplet *ter rea* can be used to describe this type of rearrangement. The expanded version of the Paris nomenclature and the triplet *cen* can be used to indicate the position of the primary constriction. For example:

45,XX,ter rea(12;14)(p13;p13)

45,XX,ter rea(12;14)(12qter→cen→12p13::14p13→14qter)

Breakage and union have occurred in band p13 of both chromosomes 12 and 14, and the centromere of chromosome 12 appears as the primary constriction. In this type of rearrangement *the chromosome carrying the primary constriction is always written first.*

Whole-arm Translocations

Whole-arm exchanges that involve nonacrocentric chromosomes where the position of the breakpoints relative to the centromeres is not known can be described, for example, as:

46,XY,t(2;3)(2p3p;2q3q)

46,XY,t(2;3)(2pter→cen→3pter;2qter→cen→3qter)

This exchange involves the whole arms of chromosomes 2 and 3 with the exchange of the respective short arms and long arms of both chromosomes. The derivative chromosomes would be der(2p3p) and der (2q3q). If it becomes possible to designate the centromere specifically as being derived from one or the other of the chromosomes involved, one can do so by preceding the symbol *cen* by the chromosome number, e.g., 2cen, if it is known that the centromere was derived from chromosome 2. For example:

46,XY,t(2;3)(2pter→2cen→3pter;2qter→3cen→3qter)

This indicates that the origin of the centromere in both derivative chromosomes is known.

Direct Insertion Between Two Chromosomes

46,XY,ins(5;2)(p14;q22q32)

46,XY,ins(5;2)(5pter→5p14::2q32→2q22::5p14→5qter;2pter→2q22::2q32→2qter)

Breakage and union have occurred at band 5p14 in the short arm and bands 2q22 and 2q32 in the long arm of chromosomes 5 and 2, respectively. The segment between 2q22 and 2q32 has been inserted into the short arm of chromosome 5 at band 5p14. The original orientation of the inserted segment has been maintained in its new position.

Inverted Insertion Between Two Chromosomes

46,XY,inv ins(5;2)(p14;q32q22)

46,XY,inv ins(5;2)(5pter→5p14::2q22→2q32::5p14→5qter;2pter→2q22::2q32→2qter)

Breakage and union have occurred at the same bands as in the previous example, and the insertion is the same except that the inserted segment has been inverted, i.e., 2q22 is now more distal to the centromere of the recipient chromosome than 2q32.

Derivative and Recombinant Chromosomes

Derivative chromosomes are products of a rearrangement and they segregate at meiosis without further change, whereas recombinant chromosomes arise *de novo* during gametogenesis in appropriate structural heterozygotes as predictable consequences of crossing-over in a displaced segment.

Derivative chromosomes are designated by *der* and recombinant chromosomes by *rec*.

The chromosome number used is that which indicates the origin of the centromere of the particular derivative or recombinant chromosome.

The symbol *pat* or *mat* indicates the paternal or maternal origin of the rearranged chromosome.

A maternal translocation 46,XX,t(2;5)(q21;q31) may produce two abnormal chromosomes der(2) and der(5). The unbalanced karyotype resulting from the malsegregation of this translocation may be 46,XX,der(2)mat or 46,XX,der(5)mat.

A parental pericentric inversion 46,XX,inv(2)(p21q31) may result, after crossing-over in the inversion loop (*aneusomie de recombinaison*), in duplication-deficiency. For instance:

46,XX,rec(2),dup p, inv(2)(p21q31)

or

46,XX,rec(2),dup q, inv(2)(p21q31)

Marker Chromosomes

A marker chromosome of completely unknown origin should be designated by the original Chicago Conference symbol *mar.*

CODE TO DESCRIBE BANDING TECHNIQUES

In this three-letter code, the first letter denotes the type of banding, the second letter the general technique, and the third letter the stain. For example:

Q- = Q-bands
QF- = Q-bands by fluorescence
QFQ = Q-bands by fluorescence using quina-crine
QFH = Q-bands by fluorescence using Hoechst 33,258
G- = G-bands
GT- = G-bands by trypsin
GTG = G-bands by trypsin using Giemsa
GTL = G-bands by trypsin using Leishman
GAG = G-bands by acetic saline using Giemsa
C- = C-bands
CB- = C-bands by barium hydroxide
CBG = C-bands by barium hydroxide using Giemsa
R- = R-bands
RF- = R-bands by fluorescence
RFA = R-bands by fluorescence using acridine orange
RH- = R-bands by heating
RHG = R-bands by heating using Giemsa
RB- = R-bands by BrdU
RBG = R-bands by BrdU using Giemsa
RBA = R-bands by BrdU using acridine orange
T- = T-bands
TH- = T-bands by heating
THG = T-bands by heating with Giemsa
THA = T-bands by heating with acridine orange

Any new triplet should be defined in the text of the publication in which it is first used.

This system may be extended by the use of a 4-letter code. For example:

RTBG: R bands, thymidine blocking, BrdU incorporation, Hoechst-Giemsa staining.

CHROMOSOME POLYMORPHISMS

The short terminology may be used, but where appropriate it could include a description of the technique used.

46,XY,1qh + (CBG)

Increase in length of the secondary constriction on the long arm of chromosome 1 after C-banding by barium hydroxide using Giemsa.

The detailed terminology is as follows. The triplet *var* is used before the chromosome number. Additional information regarding the variable region can then be conveyed by means of symbols set within brackets in the following order: 1) The location of the variable structure on the chromosome with either band numbers of code letters such as *cen, h, s,* etc. This is followed by a comma. 2) The banding technique used, denoted by the triplet code given above. 3) A numerical designation for the size and staining intensity of the variable region, higher numbers indicating greater size or staining intensity. Such numerical designations must be clearly defined. A zero indicates that size or intensity was not quantitated. The number of digits used to describe size must equal the number of digits used to describe intensity.

Size	Intensity
1 Very Small	1 Negative (no or almost no fluorescence)
2 Small	2 Pale (as on distal 1p)
3 Intermediate	3 Medium (as the two broad bands on 9q)
4 Large	4 Intense (as the distal half of 13q)
5 Very large	5 Brilliant (as on distal Yq)
0 No quantitation of size	0 No quantitation of intensity

When several techniques are used, each description should be separated by a comma. Their order is arbitrary. If more than one variable structure is present on the same homolog, each should be described in the same way in separate parentheses without separation by a comma.

If the same variant appears on more than one homolog, an asterisk followed by a number, e.g., *2, can be used to designate the number of chromosomes that conform to the initial *var* description. The parental origin of a chromosome can be indicated by inserting *pat* or *mat* after the last parenthesis but before the asterisk indicating the number of copies.

If more than one variable chromosome of a complement is to be described, these should be listed in descending order of chromosome size, the terms relating to each chromosome being separated by a comma. Bands on a given chromosome should be listed sequentially from the centromere outward, with those bands in the short arm listed first and those in the long arm last.

Examples:

46,XY,var(3)(cen,QFQ35)

Chromosome 3 with a centromeric region that, when Q-banded, is of intermediate size and fluoresces brilliantly.

46,XY,var(13)(p13,QFQ35)*2

Two chromosomes 13 with satellites (p13) that, when Q-banded, are of intermediate size and fluoresce brilliantly.

46,XY,var(13)(p13,QFQ55,CBG50)(q11.QFQ35,CBG20)

One chromosome 13 with very large satellites (p13) seen after both Q- and C-banding. These are brilliant after Q-banding, but C-banding intensity was not determined. In addition, band q11, when Q-banded, is of intermediate size and fluoresces brilliantly; when C-banded, it is likewise intermediate in size. C-banding intensity was not determined.

REFERENCES

Chicago Conference (1966): Standardization in Human Cytogenetics. The National Foundation. *Birth Defects: Original Article Series.* Vol. II, no. 2, 1966.

Denver Conference (1960): On the Nomenclature of Human Mitotic Chromosomes. *Ann. Hum. Genet.* **24**:319–325, 1960.

London Conference (1963): On the Normal Human Karyotype. *Cytogenetics* **2**:264–268, 1963.

Paris Conference (1971): Standardization in Human Cytogenetics. The National Foundation. *Birth Defects: Original Article Series.* Vol. VIII, no. 7, 1972.

Paris Conference (1971), Supplement (1975): Standardization in Human Cytogenetics. The National Foundation. *Birth Defects: Original Article Series.* Vol. XI, no. 9, 1975.

PRIEUR M., DUTRILLAUX B., LEJEUNE J.: Planches descriptives des chromosomes humains. Analyse en bandes R et nomenclature selon la Conférence de Paris 1971. *Ann. Génét.* **16**:39–46, 1973.

APPENDIX VI
THE HUMAN GENE MAP

All information concerning gene assignments in the index that follows was taken from Victor A. McKusick's letter, dated February 1, 1984. (This material was set in print at the very last moment in order to afford the most updated information possible. Some discrepancies will therefore exist with the Gene Maps in each chapter.) This letter is itself based on the seven-International Workshops on Human Gene Mapping (HGM) which have been held since 1973 under the auspices of the National Foundation March of Dimes (now called, "The March of Dimes Birth Defects Foundation"). The proceedings of the seven HGMs have been published in *Birth Defects: Original Article Series,* as well as in *Cytogenetics and Cell Genetics.*

			BD: OAS	Cytogenet. Cell Genet.	
New Haven	HGM-1	X(3)	: 1-216, 1974	13:	1-216, 1974
Rotterdam	HGM-2	XI(3)	: 1-310, 1975	14:	162-480, 1975
Baltimore	HGM-3	XII(7)	: 1-452, 1976	16:	1-452, 1976
Winnipeg	HGM-4	XIV(4)	: 1-730, 1978	22:	1-730, 1978
Edinburgh	HGM-5	XV(11)	: 1-236, 1979	25:	1-236, 1979
Oslo	HGM-6	XVIII(2)	: 1-343, 1982	32:	1-343, 1982
Los Angeles	HGM-7	XX(2)	: 1-666, 1984	37:	1-666, 1984

Key for Figure 1

The methods for mapping genes are symbolized as follows:

F   Linkage study in families, for example, the linkage of ABO blood group and nail-patella syndrome. (When a chromosomal heteromorphism or rearrangement is one trait, Fc is used, for example, the Duffy blood group locus on chromosome (chr.) 1. When a DNA polymorphism is one trait, Fd may be used, for example, Huntington disease on chr. 4.)

S   "Segregation" (cosegregation) of cellular traits and chromosomes (or segments of chromosomes), in particular clones from somatic cell hybrids, for example, thymidine kinase to chr. 17.

M   Microcell-mediated gene transfer (MMGT), for example, a collagen gene (COL1A1) to chr. 17.

C   Chromosome-mediated gene transfer (CMGT), for example, cotransfer of galactokinase and thymidine kinase. (In conjunction with this approach fluorescence-activated flow sorting can be used for transfer of specific chromosomes.)

R   Irradiation of cells followed by "rescue" through fusion with nonirradiated (nonhuman) cells (Goss-Harris method of radiation-induced gene segregation), for example, the order of genes on Xq. (Also called cotransference. The complement of cotransference = recombination.)

A   In situ DNA–DNA annealing ("hybridization"), for example, ribosomal RNA genes to acrocentric chromosomes, kappa light chain genes to chr. 2.

HS   DNA–cDNA molecular hybridization in solution ("Cot analysis"), for example, the assignment of Hb beta to chr. 11 in derivative hybrid cells.

RE   Restriction endonuclease techniques, for example, the fine structure map of the non-alpha-globin (NAG) region on (beta-globin cluster, HBBC) 11p, the physical linkage of three fibrinogen genes (on 4q) and APOA1 and APOC3 (on 11p).
Combined with somatic cell hybridization, for example, NAG (HBBC) to 11p.
Combined with chromosome sorting, for example, insulin to 11p.

D   Deletion mapping (concurrence of chromosomal deletion and phenotypic evidence of hemizygosity), trisomy mapping (presence of three alleles in the case of a highly polymorphic locus), or gene dosage effects (correlation of triplicate state of part or all of a chromosome with 50% more gene product). Examples: acid phosphatase-1 to chr. 2, glutathione reductase to chr. 8.

AAS   Deductions from the amino acid sequence of proteins, for example, linkage of delta- and beta-hemoglobin loci from the study of hemoglobin Lepore. (Includes deductions of hybrid protein structure by monoclonal antibodies, for example, close linkage of MN and Ss from the study of Lepore-like MNSs blood group antigen.)

LD Linkage disequilibrium, for example, beta- and delta-globin genes (HBB, HBD).

V Induction of microscopically evident chromosomal change by adenovirus (probably represents change comparable to "puffing" in insects; accompanied by activation of kinases), for example, adenovirus-12 changes on chr. 1 and 17.

Ch Chromosomal change associated with particular phenotype and not proved to represent linkage (Fc), deletion (D), or virus effect (V), for example, loss of 13q14 band in some cases of retinoblastoma. ("Fragile sites," observed in cultured cells with or without folate-deficient medium or BrdV treatment, fall into this class of method, for example, the fragile site at Xq27 in one form of X-linked mental retardation. Fragile sites are useful as markers in family linkage studies, for example, FS16q22 and haptoglobin.)

OT Ovarian teratoma (centromere mapping), for example, PGM3 and centromere of chr. 6.

EM Exclusion mapping, that is, narrowing the possible location of loci by exclusion of parts of the map by deletion mapping, extended to include negative lod scores from families with marker chromosomes and negative lod scores with other assigned loci, for example, support for assignment of MNSs to 4q.

H Based on presumed homology, for example, assignment of LDHC to 12p.

A12M1	Adenovirus-12 chromosome modification site-1C— 1q42-43 (V)
A12M2	Adenovirus-12 chromosome modification site-1A— 1p36 (V)
A12M3	Adenovirus-12 chromosome modification site-1B—1q21 (V)
A12M4	Adenovirus-12 chromosome modification site-17— 17q21-q22 (V)

AABT	Beta-amino acids, renal transport of—?chr. 21 (D)
ABL	Onc gene: Abelson strain of murine leukemia virus— 9q34 (S)
ABO	ABO blood group—9q34 (F)
ACMA	Actin, cardiac muscle alpha—15q11-qter (REa)
ACO1	Aconitase, soluble— 9p22-p13 (S)
ACO2	Aconitase, mitochrondrial— 22q11-q13 (S)
ACP1	Acid phosphatase-1—2p23 or 2p25 (D, S)
ACP2	Acid phosphatase-2— 11p12-cen (S)
ACTA	Actin, skeletal muscle alpha—1p12-qter (REa)
ACY1	Aminoacylase-1—3pter-q13 (S)
ADA	Adenosine deaminase— 20q13-qter (S, D)
ADCP1	Adenosine deaminase complexing protein-1—?chr. 6 (S)
ADCP2	Adenosine deaminase complexing protein-2— 2p23-2q32 (S)
ADH	Alcohol dehydrogenase, class 1—chr. 4 (REa)
ADH2	Alcohol dehydrogenase-2— 1pter-1p23 (S)
ADK	Adenosine kinase—10q11-q24 (S, D, EM)
ADRBR	Beta-adrenergic receptor— chr. 5 (S)
AF8T	Temperature-sensitive (AF8) complement—chr. 3 (S)
AFP	Alpha-fetoprotein—4q11-q13 (H, A)
AG(HBAC)	ALPHA GLOBIN GENE CLUSTER—16p12-pter (S, HS, A, D)
AHCY	SAHH (q.v.)
AHH	Aryl hydrocarbon hydroxylase—2p (S)
AIC	Aicardi syndrome—?Xp22 (Ch)
AK1	Adenylate kinase-1 (soluble)—9q34 (F, S, D)
AK2	Adenylate kinase-2 (mitochondrial)—1p34 (S, F, R)
AK3	Adenylate kinase-3 (mitochondrial)—9p24-p13 (S)

AKE	Acrokeratoelastoidosis—?2p (F)
ALAD	Delta-aminolevulinate dehydratase—(9q34; linked to ABO) (F)
ALB	Albumin—4q11-q13 (F, A, REa)
ALD	Adrenoleukodystrophy—(Xq28; linked to G6PD) (F)
AMY1	Amylase, salivary—1p21 (F, A, REa)
AMY2	Amylase, pancreatic—1p21 (F, A, REa)
AN1	Aniridia, type 1—(chr. 2; ?linked to ACP1) (F)
AN2	Aniridia-2—?11p13 (?separate from WAGR) (Ch)
APOA1	Apolipoprotein A-I—11p11-q13 (REa, RE, F)
APOC2	Apolipoprotein C-II—?11p11-q13 (F)
APOC3	Apolipoprotein C-III—11p11-q13 (REa, RE, F)
APOA4	Apolipoprotein A-IV—?11p11-p13 (F)
APOE	Apolipoprotein E—chr. 19 (F; linked to C3)
APRT	Adenine phosphoribosyltransferase—16q12-q22 (S, D)
ARSA	Arylsulfatase A—22q1331-qter (S)
ARSB	Arylsulfatase B—chr. 5 (S)
ASD2	Atrial septal defect, secundum type (chr. 6; linked to HLA) (F)
ASG	Aspermiogenesis factor—?1p13 or 1q25 (Ch)
ASL	Argininosuccinate lyase—7p21-q22 (S)
ASMD	Anterior segment mesenchymal dysgenesis (chr. 4; linked to MNSs) (F)
ASNRS (NARS)	Asparaginyl-tRNA synthetase—chr. 18 (REa)
ASNS	Asparagine synthetase—7p11-q11 (S)
ASS	Argininosuccinate synthetase—9q34 (S, D)
ASSP2	Argininosuccinate synthetase pseudogene-2—chr. 6 (REa)
AT3	Antithrombin III—1q23.1-q23.9 (F, D, A, REa)

ATN	Tyrosinase negative albinism—(?11p; linked to NAG in mouse) (H)
ATPM	OMR (q.v.)
AVR(IFR)	Antiviral state regulator—chr. 16 (D)
AVRR	Antiviral state repressor regulator—?5p (S)
B2M	Beta-2-microglobulin—15q21-q22 (S, H)
BA2R	BALB/c 3T3 ts2 temperature, sensitivity complementing—Xq13-Xq27 (S)
BAS	Beta-adrenergic stimulation, response to—?chr. 21 (D)
BCT1	Branched chain amino acid transferase-1—12pter-q12 (S)
BCT2	Branched chain amino acid transferase-2—chr. 19 (S)
BEVI	Baboon M7 virus infection—chr. 6 (S)
BF	Properdin factor B—6p213 (in MHC) (F)
BLVR	Biliverdin reductase—7p14-cen (S)
BLYM1	Oncogene BLYM1: chicken bursal lymphoma—1p32 (A)
BR1A, BR1B, etc.	Blastogenic response to specific synthetic polypeptides (Ir homologs)—6p21.3 (F)
BVIN	BALB virus induction, N-tropic—chr. 15 (S)
BVIX	BALB virus induction, xenotropic—chr. 11 (S)
BWS	Beckwith-Wiedemann syndrome—?11p13-p15 (Ch)
C2	Complement component-2—6p21.3 (in MHC) (F)
C3	Complement component-3—19pter-q13.2 (S)
C3BR	Receptor for C3b—6p21.3 (in MHC) (S)
C3DR	Receptor for C3d—6p21.3 (in MHC) (S)
C4BP	Complement component-4 binding protein—?chr. 6 (H)

C4F(C4A)	Complement component-4 fast—chr. 6 (in MHC) (F)
C4S(C4B)	Complement component-4 slow—chr. 6 (in MHC) (F)
C8	Complement component-8—1p (F)
CA2	Carbonic anhydrase II—chr. 8 (S)
CAE	Cataract, zonular pulverulent (1cen-q21; linked to Fy) (F)
CAH1(CA21H)	Congenital adrenal hyperplasia-1 (21-hydroxylase deficiency)—(6p21) (F)
CAT	Catalase—11p13 (S)
CB3S	Coxsackie B3 virus susceptibility—chr. 19 (S)
CBBM	Colorblindness, blue monochromatic—?Xq28 (F)
CBD	Colorblindness, deutan (Xq28) (F)
CBP	Colorblindness, protan (Xq28) (F)
CBS	Cystathionine beta-synthase—chr. 21 (S)
CBT	Carotid body tumor—?13q34 (F)
CC	Congenital cataract—(?16q; ?linked to HP) (F)
CDPX	Chondrodystrophia punctata, X-linked—?Xp22.32 (Ch)
CES	Cat eye syndrome—?22pter-q11 (D)
CG	Chorionic gonadotropin (?nonstructural locus) (?chr. 10 and 18) (S, REb)
CGA	Chorionic gonadotropin, alpha chain—6q11-q21 (?18p11) (REa)
CGB	Chorionic gonadotropin, beta chain—chr. 19 (REa)
CGD	Chronic granulomatous disease—(Xp22; linked to Xg) (F)
CHE1	Pseudocholinesterase-1—(chr. 3; linked to TF) (F)
CHR	Chromate resistance—5q35 (S)
CKBB	Creatine kinase, brain type—14q32-qter (?also 17) (S)
CLGN(CLG)	Collagenase (recessive dystrophic epidermolysis bullosa)—chr. 11 (S)
CML	Chronic myeloid leukemia—22q11.3 (Ch)
CMT1	Charcot-Marie-Tooth disease—(1q; linked to Fy) (F)
Co	Colton blood group (chr. 2; linked to Jk) (D, F)
COI	Coloboma of iris—?2p25.1-pter (Ch)
COL1A1	Collagen I alpha-1 chain—17q210-q220 (S, M, A, REa)
COL1A2	Collagen I alpha-2 chain—7q21-q22 (S, REa, D)
COL3A1	Collagen III alpha-1 chain—chr. 7 (S)
COL4A1	Collagen IV alpha-1 chain—?chr. 17 (REa)
COLM	Collagen, alpha-1(I)-like—chr. 12 (REa)
CP	Ceruloplasmin—(chr. 3; linked to TF) F
CPA	Carboxypeptidase A—7q22-qter (REa)
CPRO	Coproporphyrinogen oxidase—chr. 9 (S)
CPSD	Cathepsin D—11pter-q12 (S)
CRP	C-reactive protein—chr. 1 (REa)
CS	Citrate synthase, mitochondrial—12p11-qter (S)
CSCI	Corticosterone side-chain isomerase—(?6p; ?linked to MHC) (H)
CSA,CSB(CSH1)	Chorionic somatomammotropin (see PL)—17q210-q220 (S, REa, A)
CSL	Chorionic somatomammotropin-like—17q210-q220 (RE)
CTH	Cystathionase—chr. 16 (S)
CTRB	Chymotrypsinogen B—chr. 16 (REa)
CVS	Coronavirus 229E sensitivity—15q11-qter (S)
D1S3	Previously anonymous DNA segment found identical to SK (q.v.)
D4S10	DNA segment G8 (linked to Huntington disease)—chr. 4 (REa)
D14S1	DNA segment—14q32 (S)
DBH	Dopamine-beta-hydroxylase—(?chr. 9; ?linked to ABO) (F)

DC1	Ia (immune response) antigens to DC1 specificity—6p2105-p23 (F)	
DCE	Desmosterol-to-cholesterol enzyme—chr. 20 (F)	
DGI1	Dentinogenesis imperfecta-1—(chr. 4; linked to GC) (F)	
DGS	DiGeorge syndrome—22q11 (Ch)	
DHFR	Dihydrofolate reductase—5q11-q22 (S, REa, H)	
DHPR(QDPR)	Quinoid dihydropteridine reductase—chr. 4 (S)	
DHTR	Dihydrotestosterone receptor = TFM (q.v.)	
DIA1	NADH-diaphorase—22q1331-qter (S)	
DIA2	Diaphorase-2—?chr. 7 (S)	
DIA4	Diaphorase-4—16q12-q21 (S)	
DIPI(VDI)	Defective interfering particle induction, control of—chr. 16 (S)	
DJS	Dubin-Johnson syndrome—chr. 13 (LD)	
DLX1	Dyslexia-1—chr. 15 (Fc)	
DMD	Duchenne muscular dystrophy—?Xp12-p21 (Ch)	
DM	Myotonic dystrophy—(chr. 19; in group linked to C3) (F)	
DNCM	Cytoplasmic membrane DNA—9qh (H, A)	
DNL	Lysosomal DNAase—chr. 19 (S)	
Do	Dombrock blood group—(1p; linked to PGD) (F)	
DTS	Diphtheria toxin sensitivity—5q15-qter (S)	
E11S	Echo 11 sensitivity—19q (S)	
EBR1	CLGN (q.v.)	
EBS1	Epidermolysis bullosa, Ogna type (?chr. 16; linked to GPT1) (F)	
EF2	Elongation factor-2—chr. 19 (S)	
EGFR	Epidermal growth factor, receptor for—7p13-p22 (S)	
EL1	Elliptocytosis-1—(1p; linked to Rh) (F)	
EL2	Elliptocytosis-2—(?linkage to	

	chr. 1 markers other than Rh) (F)
ELA1	Elastase-1—chr. 12 (REa)
EMTB(RPS14)	Emetine resistance (ribosomal protein S14)—5q31-q35 (S)
ENO1	Enolase-1—1p36-1pter (S, F, R)
ENO2	Enolase-2—12p11-p12 (S)
ERBA	Oncogene: avian erythroblastic leukemia virus—chr. 17 (S)
ERBB	Oncogene ERBB—7pter-q22 (S)
ERV1	Endogenous retrovirus-1—chr. 18 (S)
ESA4	Esterase-A4—11q13-q22 (S)
ESAT	Esterase activator—chr. 14 (S)
ESB3	Esterase-B3—chr. 16 (S)
ESD	Esterase D—13q14.1 (S, F, D)
EXT	Multiple exostosis—(?9q3; ?linked to ABO) (F)
F7(CF7)	Clotting factor VII—13q34 (D)
F7E	Clotting factor VII expression (chr. 8) (D)
F8C	Clotting factor VIII, procoagulant component—HEMA (q.v.)
F9	Clotting factor IX = HEMB (q.v.)
F10(CF10)	Clotting factor X—13q34 (D)
F12(HAF)	Clotting factor XII (Hageman factor)—6p23-pter (D)
F13A	Factor XIII, component A—(6p, linked to MHC) (F)
FCP	F-cell production—(11p) (F)
FDH	Formaldehyde dehydrogenase—4p14-qter (S)
FEA	F9 embryonic antigen—(?6p; ?linked to HLA) (H)
FES	Onc gene: feline sarcoma virus—15q25-q26 (S)
FGA	Fibrinogen, alpha chain—4q21-q31 (REa, H)
FGB	Fibrinogen, beta chain—4q21-q31 (REa)
FGG	Fibrinogen, gamma chain—(chr. 4; linked to MNSs) (F, REa)

FH	Fumarate hydratase—1q42.1 (S, R, D)
FHC(HC)	Familial hypercholesterol-emia—19pter-q13 F, S
FMS	Oncogene FMS (McDonough feline sarcoma virus)—5q34 (S)
FN1	Fibronectin—chr. 2 (S) (see also chr. 8 and 11)
FOS	Oncogene FOS: FBJ osteo-sarcoma virus—2q22-q34 (REa, A)
FPGS	Folylpolyglutamate synthe-tase—9cen-9q34 (S)
FS	FRAGILE SITE, observed in cultured cells, with or without folate-deficient medium, or BrdU—2q11.2; 2q13; 3p14.2; 6p23; 6q26; 7p11.2; 8q22.3; 9p21.1; 9q31; 10q23.3; 10q25.2; 11q13.3; 11q23.3; 12q13.1; 16p12.3; 16q22.1; 16q23; 17p12; 20p11.23; Xq26; Xq27.3
FTH	Ferritin heavy chain—chr. 19 (S)
FTL	Ferritin light chain—chr. 19 (S)
FUCA1	Alpha-L-fucosidase-1—1p32-p34 (S, F, R)
FUSA2	Alpha-L-fucosidase-2—(?chr. 4; linked to PLG) F
FUSE	Polykaryocytosis inducer—chr. 10 (S)
Fy	Duffy blood group—1q12-q21 (F, Fc)
GAA	Acid alpha-glucosidase—chr. 17 (S)
GALB	Alpha-galactosidase B—22q13-qter (S)
GALE	Galactose-4-epimerase—1p32-pter (S, LD)
GALK	Galactokinase—17q21-q22 (S, C, R)
GALT	Galactose-1-phosphate uridyltransferase—9p13-p21 (S, D, F)
GANAB	Neutral alpha-glucosidase AB—11q13-qter (S)
GANC	Neutral alpha-glucosidase C—chr. 15 (S)

GAPD	Glyceraldehyde-3-phosphate dehydrogenase—12p13 (S, D)
GARS	Glycinamide ribonucleotide synthetase—chr. 21 (S)
GBA	Acid beta-glucosidase—1q (S)
GC	Group-specific com-ponent—4q11-q13 (F, Fc)
GCF1	Growth rate controlling fac-tor-1—chr. 7 (S)
GCF2	Growth rate controlling fac-tor-2—chr. 16 (S)
GCG	Glucagon—2p36-p37 (REa,A)
GCPS	Greig craniopolysyndactyly syndrome—?3p21.1 or 7p13 (Ch)
GCTG	Gamma-glutamylcyclotrans-ferase—7p14pter (S)
GDH	Glucose dehydrogenase—1pter-p36.13 (S)
GH-PL	GROWTH HORMONE/PLA-CENTAL LACTOGEN GENE FAMILY—17q210-q220 (S, REa, A)
GHN	Growth hormone, normal—17q210-q220 (S, REa, A)
GHS	Goldenhar syndrome—?7p (Ch)
GHV	Growth hormone, variant—17q210-q220 (S, REa, A)
GLA	Alpha-galactosidase A—Xq22-q24 (F, S, R)
GLAT	Galactose + activator—chr. 2 (S)
GLAU1	Congenital glaucoma-1—?chr. 11 (Ch)
GLB1	Beta-galactosidase-1—3p21-cen (S)
GLB2	Beta-galactosidase-2—22q13-qter (S)
GLO1	Glyoxalase I—6p21.3-p21.2 (F, S)
GLYB	Glycine auxotroph B, com-plementation of hamster—8q21.1-qter (S)
GOT1	Glutamate oxaloacetate transaminase, soluble—10q25.3-q26.1 (S, D)
GOT2	Glutamate oxaloacetate transaminase, mitochon-drial—16p12-q22 (S)

G6PD	Glucose-6-phosphate dehydrogenase—Xq28 (F, S, R)	HBBC	NAG (q.v.)
GPD1	Alpha-glycerophosphate dehydrogenase (glycerol-3-phosphate dehydrogenase)—chr. 12 (S)	HBD	Hemoglobin delta chain—11p15 (LD, AAS, F, RE, S)
		HBG1	Hemoglobin gamma chain, ala as AA 136—11p15 (AAS, RE)
GPHB	GLYCOPEPTIDE HORMONE BETA CLUSTER—chr. 19	HBG2	Hemoglobin gamma chain, gly as AA 136—11p15 (AAS, RE)
GPI	Glucosephosphate isomerase—19p13-q13 (S, D)	HBGR	Hemoglobin gamma regulator—11p15 (RE)
GPT1	Glutamate pyruvate transaminase, ?soluble red cell—?16pter-p11 (S)	HBHR	Hemoglobin-H-related mental retardation—?16p (F)
GPT2	Glutamate pyruvate transaminase, ?soluble liver—?8q13-qter (S, EM)	HBE	Hemoglobin epsilon chain—11p1205-p1208 (AAS, RE)
GPX1	Glutathione peroxidase-1—3p13-q12 (S)	HBZ	Hemoglobin zeta chain—16p12-pter (REa, A)
GRL	Glucocorticoid receptor, lymphocyte—chr. 5 (S)	HBZP	Hemoglobin zeta pseudogene—16p12-pter (REa, A)
GRP78	Glucose-regulated protein, GRP78—9cen-9qter (REa)	HC	FHC (q.v.)
		HCVS	CVS (q.v.)
GRS	Gardner syndrome—?2q14.3-q21.3 (Ch)	HD	Huntington disease—chr. 4 (Fd)
GSAS	Glutamate-gamma-semialdehyde synthetase—chr. 10 (S)	HEMA(F8C)	Classic hemophilia (hemophilia A)—(Xq28) (F)
GSR	Glutathione reductase—8p21.1 (S, D)	HEMB(F9)	Hemophilia B—(Xq26-q27) REa, A, F
GST3	Glutathione S-transferase-3—11q13-q22 (S)	HEXA	Hexosaminidase A—15q22-152q25.1 (S)
GUK1 & 2	Guanylate kinase-1 and 2—1q32-q42 (S)	HEXB	Hexosaminidase B—5q13 (S)
GUSB	Beta-glucuronidase—7cen-q22 (S)	HFE	Hemochromatosis (chr. 6; linked to HLA) (LD, F)
GUSM	Beta-glucuronidase modifier—chr. 19 (S)	Hh	Bombay phenotype—(chr. 19; linked to Lutheran) (F)
H	HISTONE GENE FAMILY (H1, H2A, H2B, H3, H4)—7q32-q36 (A)	HK1	Hexokinase-1—10pter-p11 (S)
		HLA-A,B,C	Human leukocyte antigens—6p21.3 (F, S, A)
HADH	Hydroxyacyl-CoA dehydrogenase—chr. 7 (S)	HLA-D/DR	Human leukocyte antigen, D-related—6p21.3 (F, S, A)
HAF	F12 (q.v.)	HLA-DC	Human leukocyte antigen, DC type—6p21.3 (F, A)
HAGH	Glyoxalase II (hydroxyacyl glutathione hydrolase)—chr. 16 (S)	HLA-SB	Human leukocyte antigen, SB type—6p21.3 (F, S, A)
HBA	Hemoglobin alpha chain—16p12-pter (HS, REa, A, D).	HP	Haptoglobin—16q22 (Fc)
		HPA1	Hpa I restriction endonuclease polymorphism—11p1205-p1208 (RE)
HBAC	AG (q.v.)	HPRT	Hypoxanthine-guanine phosphoribosyltransferase—Xq26-q27 (F, S, R)
HBB	Hemoglobin beta chain—11p15 (LD, AAS, F, RE, S)	HRAS1(RASH1)	Harvey rat sarcoma-1 proto-oncogene—11p15.5-p15.1 (S)

HRAS2(RASH2) — Harvey rat sarcoma-2 proto-oncogene—X chr. (REa)—pseudogene

HTOR — 5-Hydroxytryptamine oxygenase regulator—chr. 21 (D)

HV1S — Herpesvirus-sensitivity (?chr. 3 and/or 11) (S)

HYA — Y histocompatibility antigen, locus A—Y chr. (F)

HYB — Y histocompatibility antigen, locus B—X chr. (Ch)—?regulator

HYC — Y histocompatibility antigen, locus C—?X chr. (Ch.)—?receptor

IDA(IDUA) — Alpha-L-iduronidase—22pter-q11 (S)

IDDM — Insulin-dependent diabetes mellitus—(?chr.6; ?linked to HLA) (F)

IDH1 — Isocitrate dehydrogenase, soluble—2q32-qter (S)

IDH2 — Isocitrate dehydrogenase, mitochondrial—15q21-qter (S)

IF1, IF2 — Interferon 1 and 2—?chr. 2 and 5, respectively (S)

IFF(IFB) — Interferon, fibroblast—9p24-p13 (REa)

IFI(IFG) — Interferon, immune—12q24.1 (A)

IFL(IFA) — INTERFERON, LEUKOCYTE, GENE CLUSTER—9qter-p13 (REa)

IFR — AVR (q.v.)

IFRC — Interferon receptor (antiviral protein)—21q21-qter (S, D)

IGAS — Immunoglobulin heavy chains attachment site—chr. 2 (S)

IGEP — IgE pseudogene—chr. 9 (A)

IGH — IMMUNOGLOBULIN HEAVY CHAIN GENE CLUSTER—14q32.3 (S)

IGHA1 — Gene for constant region of heavy chain of IgA1—14q32 (S, RE)

IGHA2 — Gene for constant region of heavy chain of IgA2—14q32 (S, RE)

IGHD — Gene for constant region of heavy chain of IgD—14q32 (S, RE)

IGHE — Gene for constant region of heavy chain of IgE—14q32 (S, RE)

IGHG1 — Gene for constant region of heavy chain of IgG1—14q32 (S, RE)

IGHG2 — Gene for constant region of heavy chain of IgG2—14q32 (S, RE)

IGHG3 — Gene for constant region of heavy chain of IgG3—14q32 (S, RE)

IGHG4 — Gene for constant region of heavy chain of IgG4—14q32 (S, RE)

IGHJ — Gene (multiple) for J (joining) region of heavy chain—chr. 14 (S, RE)

IGHM — Gene for constant region of heavy chain of IgM—chr. 14 (S, RE)

IGHV — Gene (multiple) for variable region of heavy chain—chr. 14 (S, RE)

IGK — IMMUNOGLOBULIN KAPPA LIGHT CHAIN GENE CLUSTER—2p (A)

IGKC — Gene for constant region of kappa light chain—2p (A)

IGKJ — Gene (multiple) for J (joining) region of kappa light chain—2p (A)

IGKV — Gene (multiple) for variable region of kappa light chain—2p (A)

IGL — IMMUNOGLOBULIN LAMBDA LIGHT CHAIN GENE CLUSTER—chr. 22 (REa)

IGLC — Gene for constant region of lambda light chain—chr. 22 (REa, A)

IGLJ — Gene (multiple) for J (joining) region of lambda light chain—chr. 22 (REa)

IGLV — Gene (multiple) for variable region of lambda light chain—chr. 22 (REa)

IHG — Immune response to syn-

	thetic polypeptide— HGAL—6p21.3 (F)
IL2	TCGF (q.v.)
INS	Insulin—11p15-p15.5 (S, A, REb)
INSL	Insulin-like DNA sequence— 6p23-q12 (REa)
INT1	Oncogene INT: putative murine mammary cancer oncogene—12pter-q14 (REa)
IRDN	Insulin-related DNA polymor- phism—11p13-p15.5 (RE)
IR	BR1A, BR1B, etc. (q.v.) ?homologs of Ir gene in mouse
IS	Immune suppression— 6p2105-6p2300 (F)
ITG(BR1A, BR1B)	Immune response to syn- thetic polypeptide— TGAL—6p21.3 (F)
ITPA	Inosine triphosphatase—20p (S)
Jk	Kidd blood group—(chr. 2; linked to IGKC) (F)
KAR	Aromatic alpha-keto acid re- ductase—12p (S)
KRAS1(RASK1)	Kirsten rat sarcoma proto- oncogene-1—6p23-q12 (S)—?pseudogene
KRAS2(RASK2)	Kirsten rat sarcoma proto- oncogene-2—chr. 12 (S)
LAG5	Leukocyte antigen group five—chr. 4 (S)
LAP	Laryngeal adductor paraly- sis—(?chr. 6; ?linked to HLA) (F)
LARS(RNTLS)	Leucyl-tRNA synthetase— chr. 5 (S)
LCAT	Lecithin-cholesterol acyl- transferase—(16q22; linked to HP) (F, LD)
LCH	Lentil agglutinin binding— chr. 14 (S)
LDHA	Lactate dehydrogenase A— 11p1203-p1208 (S)
LDHB	Lactate dehydrogenase B— 12p121-p122 (S, D)
LDHC	Lactate dehydrogenase C—

	(?12p; linked to LDHB in pigeon) (H)
Le	Lewis blood group (chr. 19; linked to C3) (F)
LEU7	Leu-7 membrane antigen of natural killer lympho- cytes—chr. 11 (S)
LGS	Langer-Giedion syndrome— 8q23.3 (Ch)
LHB	Luteinizing hormone beta subunit—chr. 19 (RE)
LIPA	Lysosomal acid lipase-A— chr. 10 (S)
LIPB	Lysosomal acid lipase-B— chr. 16 (S)
LSD	Letterer-Siwe disease— ?13q14-q31 (Ch)
Lu	Lutheran blood group—chr. 19 (F)
M130	External membrane protein- 130—chr. 10 (S)
M195	External membrane protein- 195—chr. 14 (S)
M7VS1	Baboon M7 virus sensitivity- 1—chr. 19 (S)
MANA	Cytoplasmic alpha-D-man- nosidase—15q11-qter (S)
MANB	Lysosomal alpha-D-man- nosidase—19pter-q13 (S)
MAP97(MFJ1)	Melanoma-associated anti- gen p97—chr. 3 (S)
MAR	Macrocytic anemia, refrac- tory—5q (Ch)
MARS	MTRNS (q.v.)
MDH1	Malate dehydrogenase, solu- ble—2p23 (S)
MDH2	Malate dehydrogenase, mito- chondrial—7p22-q22 (S)
MDI	Manic-depressive illness (?chr. 6; ?linked to HLA) (F)
MDLS	Miller-Dieker lissencephaly syndrome—?17p13 (Ch)
ME1	Malic enzyme, soluble—6q12 (S)
ME2	Malic enzyme, mitochon- drial—?chr. 11 (H)
MEN2	Multiple endocrine neopla- sia, type II (Sipple syn- drome)—20p12.2 (Ch)

MHC	MAJOR HISTOCOMPATIBIL-ITY COMPLEX—6p2105-p23 (F, S)		NARS	ASNRS (q.v.)
MIC7	Attached cell antigen 28.3.7 identified by Imperial Cancer monoclonal—chr. 15 (S)		NB	Neuroblastoma—1p32-1pter (?1p34) (Ch)
			NDF	Neutrophil differentiation factor (chr. 6) (LD)
MLRW	Mixed lymphocyte reaction, weak (chr. 6) (F)		NEU1	Neuraminidase-1—?6p2105-6p2300 (?linked to HLA) (H)
MMC	Malignant melanoma, cutaneous—(?1p; ?linked to Rh) (F)		NGF	Nerve growth factor—1p21-1pter (REa)
MN	MN blood group—4q28-q31 (F, Fc, AAS)		NHCP1	Nonhistone chromosomal protein-1—chr. 7 (S)
MOS	Onc gene: Moloney murine sarcoma virus—8q22 (S)		NHCP2	Nonhistone chromosomal protein-2—chr. 16 (S)
MPI	Mannosephosphate isomerase—15q22-qter (S)		NF1	Neurofibromatosis—(?chr. 19; linked to DM) (F)
MRBC	Monkey red blood cell receptor—chr. 6 (S)		NF2	Neurofibromatosis—(?chr. 4; linked to GC) (F)
MS	Menkes syndrome—Xp11-Xp21 (Fd)		NF3	Familial intestinal neurofibromatosis—?12q21-q24.2 (Ch)
MTR	5-Methyltetrahydrofolate: L-homocysteine S-methyltransferase (tetrahydropteroyl-glutamate methyltransferase)—chr. 1 (S)		NLP1	Neoplastic lymphoproliferation-1—?8q24.3 (Ch)
			NLP2	Neoplastic lymphoproliferation-2—?18q24 (Ch)
MTRNS (MARS)	Methionyl-tRNA synthetase—chr. 12 (S)		NM	Neutrophil migration—7q22-qter (D)
MYB	Onc gene: avian myeloblastosis virus—6p21-q22 (S)		NMYC	Neuroblastoma MYC oncogene—2p (S,A)
MYC	Onc gene: myelocytomatosis virus—8q24 (A)		NP	Nucleoside phosphorylase—14q13 (S, D)
MYHS	MYOSIN, SKELETAL, HEAVY CHAIN GENE FAMILY—17p11-pter (REa)		NPS1	Nail-patella syndrome—(9q3; linked to AK-1) (F)
MYHSA1	Myosin, skeletal, heavy chain, adult-1—17p11-pter (REa)		NRAS	Oncogene NRAS—1p (REa)
MYHSA2	Myosin, skeletal, heavy chain, adult-2—17p11-pter (REa)		OA1	Ocular albinism-1 (Nettleship-Falls type)—(Xp22; linked to Xg) (F)
MYHSE1	Myosin, skeletal, heavy chain, embryonic-1—17p11-pter (REa)		OA2	Ocular albinism-2 (Forsius-Eriksson type)—(Xp22; linked to Xg) (F)
			OAK	Optic atrophy, Kjer type—(?chr. 2; ?linked to Jk) (F)
NAG(HBBC)	NON-ALPHA-GLOBIN CLUSTER (HEMOGLOBIN BETA CLUSTER)—11p1205-1208 (S, RE)		OIAS	2′,5′-oligoisoadenylate synthetase—chr. 11 (S)
			OMPD	Orotidylmonophosphate decarboxylase (with OPRT, q.v.)
			OMR(ATPM)	Oligomycin resistance (mitochondrial ATPase)—chr. 10 (S)
NAGA	N-acetyl-alpha-D-galactosaminidase—22q13 (S)		OPRT	Orotate phosphoribosyltransferase—OMP decarboxylase—3cen-q21 (S)

ORM	Orosomucoid—(9q34; linked to ABO) (F)	PKM2	Pyruvate kinase-3—15q22-qter (S,D)
OTC	Ornithine transcarbamoy-lase—Xp21 (Ch)	PKU	PAH (q.v.)
		PL	Placental lactogen (same as CSA, CSB, q.v.)—17q210-17q220 (S, REa, A)
PAH(PKU)	Phenylalanine hydroxy-lase—chr. 12 (REa)	PLA	Plasminogen activator—chr. 6 (S)
PAIS	Phosphoribosylamino-imidazole synthetase—chr. 21 (S)	PLG	Plasminogen—chr. 4 (S)
		PLT1	Primed lymphocyte test-1—6p2105-p23 (F)
PBGD(UPS)	Porphobilinogen deamin-ase—11q23-qter (S)	POMC	Proopiomelanocortin—2p23 (REa)
PDB	Paget disease of bone—(?chr. 6; ?linked to HLA) (F)	PP	Inorganic pyrophospha-tase—10q11.1-q24 (S)
PDGF	Platelet-derived growth fac-tor—SIS (q.v.)	PPAT	Phosphoribosylpyrophos-phate amidotrans-ferase—4pter-q21 (S)
PEPA	Peptidase A—18q23 (S, D)	PRGS	Phosphoribosylglycinamide synthetase—21q22 (S,H)
PEPB	Peptidase B—12q21 (S)		
PEPC	Peptidase C—1q (S, R)	PRL	Prolactin—6p23-q12 (S)
PEPD	Peptidase D—19pter-q13 (S)	PRPS	Phosphoribosylpyrophos-phate synthetase—X chr. (F, S)
PEPS	Peptidase S—4p12-q12 (S)		
PFGS	Phosphoribosyl formylgly-cinamidine synthetase—chr. 14 (S)	PSP	Phosphoserine phospha-tase—7qter-q22 (S)
PFKL	Phosphofructokinase, liver—21q22 (S)	PTH	Parathyroid hormone—11pter-p11 (REa)
PFKM	Phosphofructokinase, mus-cle—1cen-1q32 (S)	PVS	Polio virus sensitivity—19q (S)
PFKP(PFKF)	Phosphofructokinase, platelet—10pter-p11.1 (S)	PWS	Prader-Willi syndrome—15q11-q12 (Ch)
PG	Pepsinogen—?chr. 6 (F,H)		
PGAM1	Phosphoglycerate mutase A—10q25.3-q26.1 (D, H)	1qh	Centric heterochromatic seg-ment, long arm, chr. 1 (for-merly "uncoiler")
PGD	6-Phosphogluconate dehy-drogenase—1p36.13-pter (F-S)	9qh	Centric heterochromatic seg-ment, long arm, chr. 9
PGFT	Phosphoribosylglycinamide formyltransferase—14q22-qter (S)	QDPR	DHPR (q.v.)
PGK	Phosphoglycerate kinase—Xq13 (F, S)	RACH	Regulator of acetylcholines-terase—chr. 2 (D)
PGM1	Phosphoglucomutase-1—1p221 (F, S, R)	RAF1	Oncogene RAF1—3p25 (S)
PGM2	Phosphoglucomutase-2—4p14-q12 (S)	RAF2	Oncogene RAF2—chr. 4 (pseudogene) (S)
PGM3	Phosphoglucomutase-3—6q12 (S, F, OT)	RB1	Retinoblastoma-1—13q14.1 (Ch)
PGP	Phosphoglycolate phospha-tase—16p13-p12 (S)	RCC	Renal cell carcinoma—?3p21 (Ch)
PI	Alpha-1-antitrypsin—14q (F,S)	Rd	Radin blood group—(1p32-p34; linked to Rh) F
		REN	Renin—1p21-qter (REa)

469

Rh	Rhesus blood group (1p32-p36.11) (F-S,D)
RGS	Rieger syndrome—?4q23-q27 (Ch)
RN5S	5S RNA gene(s)—1q42-q43 (A)
RNTMI(TRM1,2)	Initiator methionine tRNA—6p23-q12 (REa)
RNR	RIBOSOMAL RNA—13p12, 14p12, 15p12, 21p12, 22p12 (A)
RNU1	RNA, U1 small nuclear—1p36 (S)
RP1	Retinitis pigmentosa-1 (?chr. 1) (F)
RPE	Ribulose 5-phosphate 3-epimerase—2q32-qter (S)
RPS14	Ribosomal protein S14 = EMTB (q.v.)
RS	Retinoschisis—(Xp22; linked to Xg) (F)
RWS	Ragweed sensitivity—(chr. 6; ?linked to HLA) (F)
S7	Surface antigen 7 = EGFR (q.v.)
SAHH(AHCY)	S-adenosylhomocysteine hydrolase—20cen-20q13.1- (S)
Sc	Scianna blood group—(1p32-p34) (F)
SCA1	Spinocerebellar ataxia I—(chr. 6; linked to HLA) (F)
SCCL	Small-cell cancer lung—3p14-p23 (Ch)
SDH	Succinate dehydrogenase—1p22.1-qter (S)
Se	Secretor—(chr. 19; in group linked to C3) F
Sf	Stoltzfus blood group—(4q; linked to MNSs) (F)
SGP75	Surface antigen, glycoprotein 75,000—chr. 11 (S)
SHMT	Serine hydroxymethyltransferase—12q12-q14 (S)
SIS(PDGF)	Oncogene: simian sarcoma virus—chr. 22 (S)
SK(D1S3)	Oncogene Sloan-Kettering chicken virus—1q12-1qter (REa)
SOD1	Superoxide dismutase, soluble—21q211 (S,D)

SOD2	Superoxide dismutase, mitochondrial—6q21 (S)
SORD	Sorbitol dehydrogenase—15pter-q21 (S, H)
SP3	Spermatogenesis factor-3 (azoospermia third factor)—?Yq11 (D)
SPA2, SPA5	Surface polypeptide, anonymous—chr. 2, 5, respectively (S)
SPC	SALIVARY PROTEIN COMPLEX—?6p (F)
SPH	Spherocytosis—8p11, ?14q (F, Ch)
SRC	Oncogene SRC (Rous sarcoma)—chr. 20 (REa)
Ss	Ss blood group—4q28-q31 (F, Fc, AAS)
SST	Somatostatin—3q27-3q28 (REa, A)
STA	Stature—Y chr. (D)
STS	Steroid sulfatase—Xp22.3 (F, S)
TCGF(IL2)	T-cell growth factor (interleukin-2)—chr. 4 (REa)
TCN2	Transcobalamin II—?chr. 16 or 17 (H)
TDF	Testis determining factor—Y chr. (?same as HYA) (F)
TF	Transferrin—chr. 3 (S, H)
TFM(DHTR)	Testicular feminization syndrome (dihydrotestosterone receptor)—Xp11-q13 (S)
TFR	Transferrin receptor—chr. 3 (S, H, REa)
TG	Thyroglobulin—chr. 8 (S)
THC	Thrombocytosis, primary—21q (Ch)
TK1	Thymidine kinase, soluble—17q21-q22 (S, C, R)
TK2	Thymidine kinase, mitochondrial—chr. 16 (S)
TKCR	TKCR syndrome of Goeminne—?Xq28 (Ch)
TPI1 & 2	Triosephosphate isomerase-1 and 2—TPI-1 on 12p13 (S, D)
TRC	T-cell receptor for MHC antigens—?chr. 14 (H)

TRM1,2	RNTMI (q.v.)	VDI	DIPI (q.v.)
TRY1	Trypsin-1—7q22-qter (REa)	VMD1	Macular dystrophy, atypical
TS	Tooth size—Yq11 (D)		vitelliform—?16p (F)
TYS	Sclerotylosis—(4q; linked to		
	MNSs) (F)	WAGR	Wilms' tumor/aniridia/
			gonadoblastoma/retarda-
			tion—11p13 (Ch)
UGP1	Uridyl diphosphate glucose	WARS	Tryptophanyl-tRNA synthe-
	pyrophosphorylase-1—		tase—14q21-qter (S)
	1q21-q23 (S, R)	WS1	Waardenburg syndrome-1—
UGP2	Uridyl diphosphate glucose		(?9q34; ?linked to ABO)
	pyrophosphorylase-2—chr.		(F)
	2 (S)		
UMPK	Uridine monophosphate	Xg	Xg blood group (Xp22.3;
	kinase—1p32 (S, R)		linked to STS) (F)
UP	Uridine phosphorylase—chr.	Xk	Kell blood group precursor
	7 (S)		(Xp22.3; linked to Xg) (F)
UPS	Uroporphyrinogen I synthase	XPA	Xeroderma pigmentosum A
	= PBGD (q.v.)		complementation—1q (S)
UVDR	Ultraviolet damage repair—	XRS	X-ray sensitivity—?13q14
	chr. 13 (S)		(Ch)

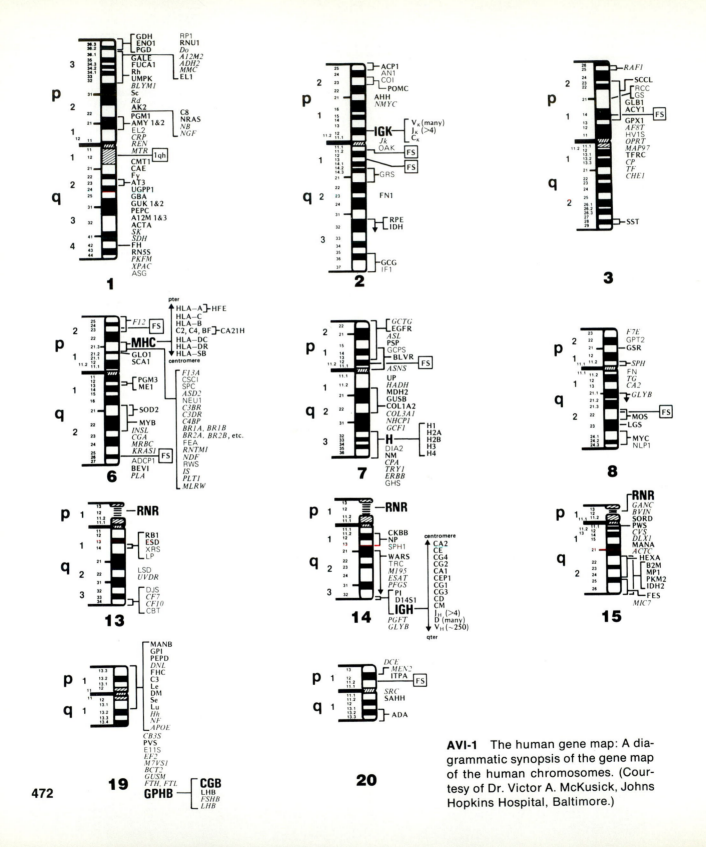

AVI-1   The human gene map: A diagrammatic synopsis of the gene map of the human chromosomes. (Courtesy of Dr. Victor A. McKusick, Johns Hopkins Hospital, Baltimore.)

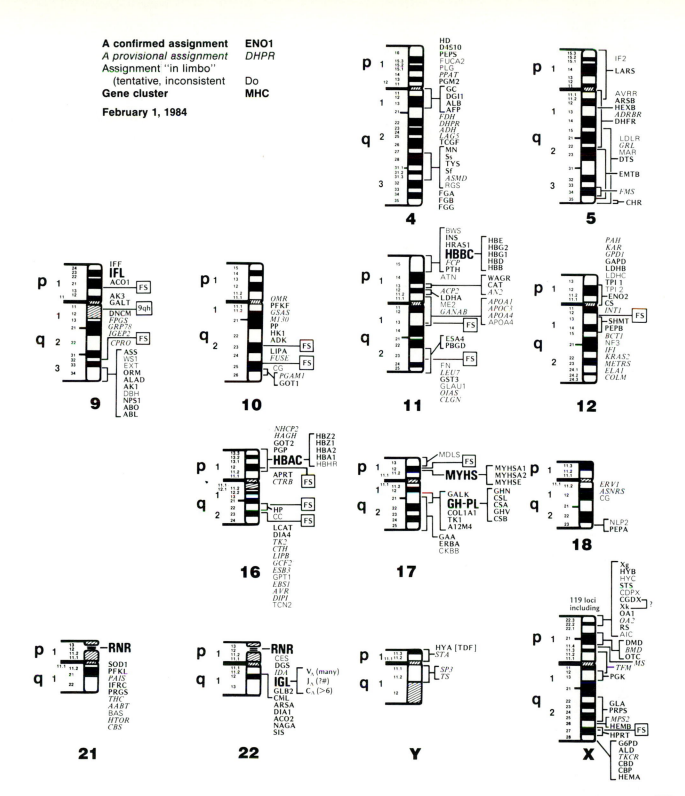

A confirmed assignment   ENO1
A provisional assignment   *DHPR*
Assignment "in limbo"
    (tentative, inconsistent   Do
Gene cluster   **MHC**

February 1, 1984

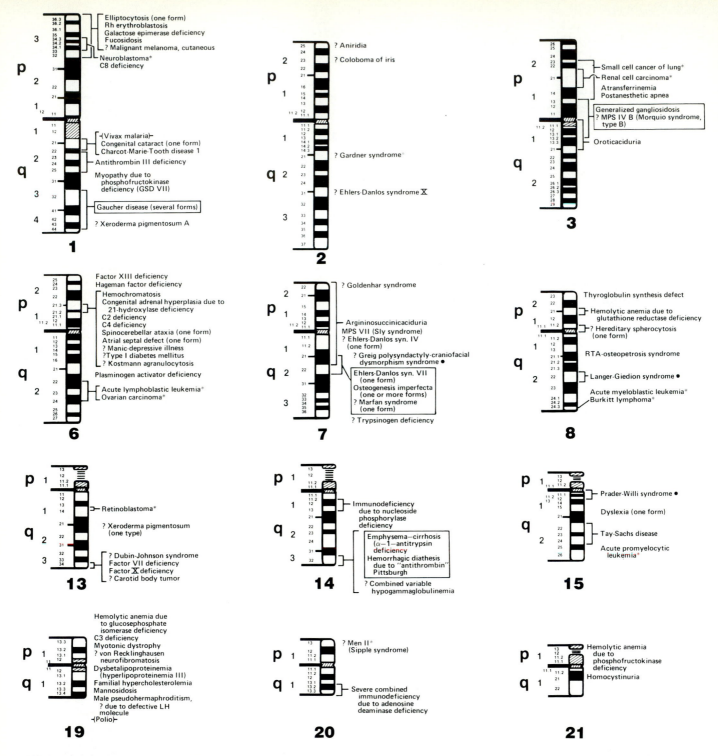

AVI-2 Diagrammatic synopsis of the morbid anatomy of the human genome. The location of the particular disease mutations, as indicated here, has been determined either by study of the disease itself or (more often) by mapping of the "normal" (wild-type) gene. Diphtheria, polio, and vivax malaria are specific infections susceptibility to which is determined by genes at the sites indicated. Several deficiency states that apparently are not accompanied by "disease" are specially labeled. Malformation syndromes and neoplasms associated with specific restricted chromosomal changes are included, as well as some neoplasms that have been related to oncogenes. (Courtesy of Dr. Victor A. McKusick, Johns Hopkins Hospital, Baltimore.)

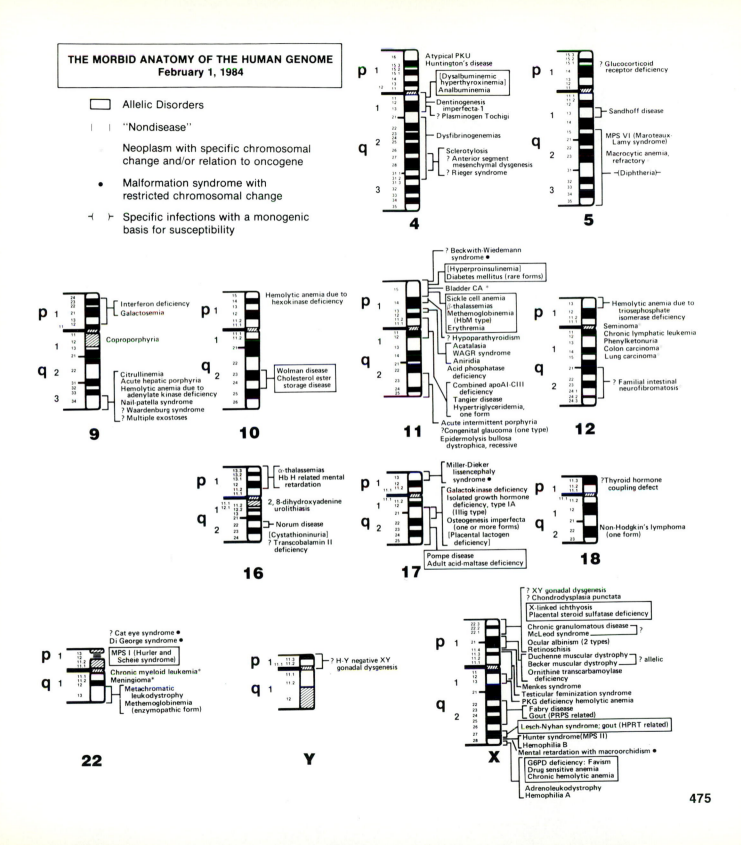

THE MORBID ANATOMY OF THE HUMAN GENOME
February 1, 1984

☐ Allelic Disorders

| | "Nondisease"

Neoplasm with specific chromosomal change and/or relation to oncogene

● Malformation syndrome with restricted chromosomal change

⊣ ⊢ Specific infections with a monogenic basis for susceptibility

4

p1
- Atypical PKU
- Huntington's disease
- [Dysalbuminemic hyperthyroxinemia]
- Analbuminemia
- Dentinogenesis imperfecta-1
- ? Plasminogen Tochigi

q2 q3
- Dysfibrinogenemias
- Sclerotylosis
- ? Anterior segment mesenchymal dysgenesis
- ? Rieger syndrome

5

p1
- ? Glucocorticoid receptor deficiency

q1 q2
- Sandhoff disease
- MPS VI (Maroteaux-Lamy syndrome)
- Macrocytic anemia, refractory
- ⊣(Diphtheria)⊢

9

p1
- Interferon deficiency
- Galactosemia

1
- Coproporphyria

q2 q3
- Citrullinemia
- Acute hepatic porphyria
- Hemolytic anemia due to adenylate kinase deficiency
- Nail-patella syndrome
- ? Waardenburg syndrome
- ? Multiple exostoses

10

p1
- Hemolytic anemia due to hexokinase deficiency

q2
- Wolman disease
- Cholesterol ester storage disease

11

p1
- ? Beckwith-Wiedemann syndrome ●
- [Hyperproinsulinemia]
- Diabetes mellitus (rare forms)
- Bladder CA *
- Sickle cell anemia
- β-thalassemias
- Methemoglobinemia (HbM type)
- Erythremia
- ? Hypoparathyroidism
- Acatalasia
- WAGR syndrome
- Aniridia

q2
- Acid phosphatase deficiency
- Combined apoAI-CIII deficiency
- Tangier disease
- Hypertriglyceridemia, one form
- Acute intermittent porphyria
- ?Congenital glaucoma (one type)
- Epidermolysis bullosa dystrophica, recessive

12

p1
- Hemolytic anemia due to triosephosphate isomerase deficiency
- Seminoma
- Chronic lymphatic leukemia
- Phenylketonuria
- Colon carcinoma
- Lung carcinoma *

q2
- ? Familial intestinal neurofibromatosis

16

p1
- α-thalassemias
- Hb H related mental retardation
- 2, 8-dihydroxyadenine urolithiasis

q2
- Norum disease
- [Cystathioninuria]
- ? Transcobalamin II deficiency

17

p1
- Miller-Dieker lissencephaly syndrome ●
- Galactokinase deficiency
- Isolated growth hormone deficiency, type IA (Illig type)
- Osteogenesis imperfecta (one or more forms)
- [Placental lactogen deficiency]
- Pompe disease
- Adult acid-maltase deficiency

18

p1
- ?Thyroid hormone coupling defect

q2
- Non-Hodgkin's lymphoma (one form)

22

p1
- ? Cat eye syndrome ●
- Di George syndrome ●
- MPS I (Hurler and Scheie syndrome)
- Chronic myeloid leukemia*
- Meningioma*

q1
- Metachromatic leukodystrophy
- Methemoglobinemia (enzymopathic form)

Y

p1
- ? H-Y negative XY gonadal dysgenesis

X

p1
- ? XY gonadal dysgenesis
- ? Chondrodysplasia punctata
- X-linked ichthyosis
- Placental steroid sulfatase deficiency
- Chronic granulomatous disease ?
- McLeod syndrome
- Ocular albinism (2 types)
- Retinoschisis
- Duchenne muscular dystrophy ? allelic
- Becker muscular dystrophy
- Ornithine transcarbamoylase deficiency
- Menkes syndrome

q2
- Testicular feminization syndrome
- PKG deficiency hemolytic anemia
- Fabry disease
- Gout (PRPS related)
- Lesch-Nyhan syndrome; gout (HPRT related)
- Hunter syndrome(MPS II)
- Hemophilia B
- Mental retardation with macroorchidism ●
- G6PD deficiency: Favism
- Drug sensitive anemia
- Chronic hemolytic anemia
- Adrenoleukodystrophy
- Hemophilia A

475

APPENDIX VII
SYNDROME FINDER

An index, *stricto sensu,* did not seem very appropriate to us in the case of the present *Atlas.* Indeed, since many terms, such as chromosome, deletion, trisomy, cranium, etc., appear with great frequency, it would have been rather meaningless to index them practically at each page of the book. We therefore endeavored to construct "syndrome finder" by classifying the most important symptoms according to the plan adopted in each chapter and indicating in which syndrome they appear more specifically. We thought that this "syndrome finder" would be more helpful for the reader than a classical index.

However, its preparation raised several difficulties. A great number of symptoms are common to many different chromosomal aberrations. The most striking example is mental retardation, which is present in all the autosomal disorders and a fraction of the sex chromosome anomalies. This is also the case, though to a lesser degree, for growth retardation. We have therefore decided not to include mental retardation in the list of symptoms. As for growth retardation we have considered this only when its severity suggests a very precise syndrome, e.g., Turner's syndrome.

This rule is, however, difficult to apply to other symptoms that do not have great diagnostic value in chromosomal diseases, such as microcephaly or epicanthus. Any list of syndromes corresponding to these symptoms would certainly vary from one author to another. Conversely, some symptoms have great diagnostic value, such as trigonocephaly or hexadactyly. This is immediately apparent from the small number of syndromes corresponding to certain symptoms.

The symbolism we have used may not always be consistent with the headings of the different chapters. It is in agreement with the Paris Nomenclature and was found convenient in the present instance. For example, del 5p was preferred to monosomy 5p.

GENERALIZATIONS

Mean birth weight:	
• <2,500 g	r(1), del 4p, tri 6p, del 7q, tri 7q, tri 9, tri 9q, tri 11q, del 13q, tri 16p, tri 16q
• 2,500–3,000 g	tri 1q, del 1q, tri 2q, tri 2p, tri 4p, tri 4q, del 4q, del 5p, tri 5q, tri 6q, del 7p, tri 7p, tri 8p, del 8p, tri 8q, tri 9p, tri 10p, tri 10q, del 11q, del 12p, tri 12q, tri 13, tri 14q1, tri 17q, tri 18, tri 18q2, del 18p, del 18q, tri 21, r(21), 45,X
• normal	tri 3p, tri 3q, tri 5p, tri 8, del 9p, tri 12p, tri 13qter, tri 14, tri 15q1, tri 20p, r(22)
Small crown-heel length at birth (46 cm)	del 4p, tri 6p, del 8p, tri 14, tri 16p, tri 17q, tri 19q, 45,X
Hypotonia	del 1q, del 7q, tri 10p, tri 11p, tri 11q, tri 12p, del 18q, tri 21
Hypertonia	del 9p, tri 16p, r(21)
Severe hypotrophy	r(1), del 4p, tri 9q, tri 10p, tri 11p, tri 16p, tri 18, del 18q
Corpulent infants	tri 12p
Short stature	45,X
High stature	47,XYY, 48,XXYY
Dolichomorphism	tri 2p, del 4p, tri 8, tri 18
Peculiar cry	del 5p
Laryngeal hypotonia	del 4q
Recurrent infections	del 7q, del 11q
Survival in adulthood	tri 2q, tri 5p, del 5p, tri 8, tri 8q, del 9p, tri 18q2
Subnormal growth	tri 2q, tri 5p, tri 8, tri 18q2, tri 20p

CRANIUM

Microcephaly   del 1q, del 4p, tri 4q, del 5p, tri 5q, r(6), del 7q, tri 9p, tri 9, tri 9q, tri 10q, tri 11q, del 12p, del 13q, tri 13qter, tri 15q2, tri 16p, tri 17q, tri 19q, tri 21, r(21), r(22)

Macrocephaly   tri 12q

Dolichocephaly   del 4p, tri 7p, tri 9q, tri 10p, tri 15q2, tri 18, tri 18q2, r(21)

Brachycephaly   tri 1q, del 1q, tri 4q, tri 6q, del 7q, tri 9p, tri 10q, tri 19q, tri 21

Trigonocephaly   del 7p, del 9p, del 11q, tri 13qter

Turricephaly   tri 12p

Craniostenosis   del 7p, tri 13

Wide sutures   tri 1q, tri 7p, tri 9p, tri 19q

FOREHEAD

• high and bossing   tri 2p, tri 3p, tri 7q, tri 8, tri 8p, tri 8q, tri 9, tri 9q, tri 10q, tri 11p, tri 12p, tri 14, tri 17q, r(21)

• receding   tri 4q, tri 13, tri 13q1

• low   tri 4p, tri 16q, tri 17q

• narrow   del 12p, tri 13

Temporal indentation   tri 3p

Abnormal hairline on forehead   tri 10p, tri 11p, tri 17q, tri 18q2

FACE

Round and flat   del 1q, tri 3p, tri 4p, del 5p, tri 6q, del 7q, tri 11q, tri 16p, del 18p, tri 19q, tri 20p, tri 21

Oval   tri 15q1, tri 18q2, 48,XXXX, 49,XXXXX, 49,XXXXY

Long   tri 5p, tri 8, tri 15q2

Triangular   tri 1q, del 5p, 45,X

Depressed midface   del 18q

High cheekbones   tri 15q1, tri 20p

Elfin face   tri 1q

Round cheeks   tri 3p, tri 7p, del 7q, tri 10q, tri 11p, tri 12q, tri 15q1, tri 15q2

EYES

Palpebral fissures slanted:

• upwards/outwards   tri 3q, del 7q, tri 8q, del 9p, tri 21, 48,XXXX, 49,XXXXX, 49,XXXXY

• downwards/outwards   tri 1q, tri 6q, del 7p, tri 10q, tri 16q, tri 19q, r(21), 45,X

Hypertelorism   tri 2q, tri 2p, tri 4p, tri 7p, tri 8q, tri 11p, del 13q, tri 16p, tri 16q, tri 19q, r(22), 49,XXXXY

Hypotelorism   tri 5p, del 7p, tri 13qter, tri 14q1

Epicanthus   del 1q, del 5p, del 9p, tri 11p, del 13q, tri 21, 45,X, 48,XXXX

Exophthalmos   del 4p, del 7p, del 9p

Deeply set eyes   tri 9p, tri 9, tri 9q, tri 14, tri 15q1, del 18q

Ptosis   tri 6p, del 11q, del 13q, tri 14, del 18p, tri 19q

Blepharophymosis   tri 6p, tri 10q

Arched eyebrows   tri 6q, tri 10q

Synophris   tri 3q, tri 4p, tri 13qter

Large glabella   del 4p

Protruding glabella   tri 4p

Brushfield spots   tri 21

Small eyes   tri 4p, tri 10q, tri 14q1

Large eyes   tri 5q

Squinty eyes	tri 17q
Doe's eyes	r(22)
Minor ocular anomalies	tri 2p, tri 3q, del 4p, tri 4q, tri 6p, del 13q, tri 13q1, tri 14, del 18q, 45,X
Severe microphthalmia/ anophthalmia	tri 13
Retinoblastoma	del 13q
Aniridia	del 11p

NOSE

Nosebridge	
• protruding	tri 4q, tri 5p, del 8p, tri 8q, tri 10p, tri 18, tri 18q2, r(21)
• flat	del 4q, del 5p, del 7p, del 9p, del 11q, tri 11q, tri 12q, tri 16q, tri 21, 49,XXXXX, 49,XXXXY
• wide	tri 3q, del 4p, tri 8q, del 9p, tri 10p, tri 11p, del 11q, tri 12q, tri 14q1
Aplastic bones (boxer's nose)	tri 4p
Greek profile	del 4p, del 13
Short nose	del 1q, tri 2q, tri 2p, del 4q, tri 6p, del 7p, tri 7p, tri 7q, tri 12p tri 14, tri 16p
Long nose	del 12p
Bulbous nose	del 7q, tri 8p, tri 9p, del 11q, tri 14, tri 16p, tri 18q2
Beaked nose	tri 7p, tri 9q, tri 11p
Broad nose	del 4p, tri 6q, del 7q, tri 11p, tri 13, tri 17q
Pointed nose	tri 7q, tri 14q
Nares oriented:	
• upwards	del 4q, del 7p, tri 11q, tri 16p, tri 21
• downwards	del 4p, tri 9p

UPPER LIP

Long upper lip	tri 4p, tri 5q, tri 6p, del 9p, tri 11q, tri 13qter, r(21)
Short upper lip	del 4p, tri 4q, tri 9p, tri 12q, del 18q, tri 19q
Deep philtrum	del 4p, tri 4q, tri 15q2, tri 19q, r(21)
Flat philtrum	tri 4p, tri 5q, tri 6q, del 9p, del 11q, tri 14q1, del 18q
Unilateral grin	tri 9p
Harelip	tri 3p, del 4p, del 4q, del 7q, tri 10p, tri 11p, tri 13

MOUTH

Cleft palate	tri 3p, del 4q, del 7p, tri 7p, tri 7q, tri 8q, tri 9, tri 9q, tri 10p, tri 11p, tri 13, tri 14q1, tri 17q, tri 19q
High arched palate	tri 4p, del 7p, tri 7p, del 8p, tri 8q, del 9p, tri 9, tri 9q, tri 10q, tri 13qter, tri 14, tri 14q1, tri 18q2, 45,X
Small mouth	tri 2p, del 4q, tri 5q, tri 6p, tri 6q, del 7p, del 8p, del 9p, tri 9q, tri 21, r(21)
Wide mouth	tri 2q, tri 3p, tri 4p, del 11q, tri 17q, del 18p
Down-turned mouth	tri 3p, tri 3q, del 4p, tri 7p, del 11q, tri 12q, tri 17q, tri 19q, 45,X
Thin upper lip	tri 2q
Thin lips	tri 5q, tri 6p, tri 6q, tri 10p, del 11q, tri 12q, tri 13qter, tri 16q, tri 17
Everted lower lip	tri 2q, del 4q, tri 7q, tri 8, tri 8p
Particular shapes:	
• oval	tri 14q1
• turtle beak	tri 10p
• carplike	tri 5q, tri 6q, del 18q, tri 19q
• pursed	tri 4q
• rabbit teeth	tri 4p, del 13q
Dental anomalies	tri 4p, del 5p, tri 6q, del 8p, tri 13qter, tri 18q2, del 18p, 45,X
Macroglossia	tri 21

CHIN

Micrognathia	tri 1q, del 1q, tri 2q, tri 2p, tri 3p, tri 3q, tri 4q, del 4q, del 5p, tri 6p, r(6), tri 7q, tri 8, tri 8q, tri 9, tri 9q, tri 11q, tri 14, tri 15q2, tri 16p, tri 16q, tri 17q, tri 18, tri 18q2, r(21), 45,X
Protruding chin	tri 10p, del 18q
Large mandible	tri 5p
Effaced angles of the mandible	tri 4p, tri 6q, tri 18q2, r(22)

EARS

Low set	tri 2p, tri 2q, tri 3q, tri 4p, tri 4q, tri 5q, tri 6p, del 7p, tri 7q, del 9p, tri 9, tri 10p, tri 10q, del 11q, del 12p, tri 12q, tri 14, tri 15q2, tri 16p, tri 16q, tri 18, del 18p, tri 19q, 45,X
Posteriorly rotated	tri 4q, del 7p, tri 7q, tri 8p, tri 10p, tri 10q, tri 12q, tri 14, tri 17q, del 18p, tri 19q
Small	tri 1q, tri 21
Large	tri 2p, tri 2q, tri 4q, tri 7q, tri 8, tri 8p, tri 9q, tri 10p, del 13q, del 18p, r(21), r(22)
Detached	tri 5q, tri 8, tri 8q, tri 14, del 18p
Faunlike	del 4q, tri 18
Flat helix	tri 6p, tri 9q, tri 13
Folded helix	tri 4p, tri 4q, tri 14, del 18q
Protruding anthelix	tri 4q, tri 9, del 11q, tri 12p, tri 13qter, tri 15q2, tri 16p, del 18q, r(21)
Aplastic anthelix	del 18p
Aplastic or adherent lobe	del 4p, tri 4q, tri 6p, del 9p, del 11q, tri 12q, tri 13qter
Protruding tragus	del 18q
Protruding antitragus	del 4p, del 18q
Preauricular dimples, fistula, tags	del 4p, tri 11q

NECK, THORAX, ABDOMEN

Short neck with redundant skin	tri 3p, tri 3q, tri 4q, tri 8p, del 9p, tri 9p, tri 9q, tri 12p, tri 12q, del 13q, tri 14q1, tri 17q, tri 18q2
Long, thin neck	del 4p
Flat nape	tri 21
Pterygium colli	tri 6q, del 9p, 45,X
Low hairline	tri 3q, tri 4p, tri 6q, 45,X
Short sternum	tri 18
Large thorax	del 8p, tri 9p, tri 19q, 45,X
Narrow pelvis	tri 9q, tri 18
Scoliosis	tri 2p, del 4p, tri 4p, tri 6q, tri 8p, tri 10p, tri 10q, tri 19p
Decreased acetabular angles	tri 21
Hernias	tri 4q, tri 5q, del 7q, tri 9p, tri 11q, tri 13qter, tri 16q, tri 18, tri 21
Gynecomastia	47,XXY, ♂46,XX, 48,XXYY
Wide-spaced nipples	tri 4p, tri 4q, tri 6q, del 7q, del 8p, del 9p, tri 9p, tri 11q, tri 13qter, tri 19q, 45,X
Supernumerary nipples	tri 12p

LIMBS

Slender limbs	tri 2p, del 4p
Stocky limbs	tri 3q, tri 12q, tri 17q

479

Hyperflexed limbs	tri 9q, tri 10p, tri 11q, tri 16q
Pleading posture	tri 11q, tri 18
Cubitus valgus	45,X
Radioulnar synostosis	tri 11q, 49,XXXXY, 49,XXXXX
Syndactyly	del 13q, r(22)
Froglike position	del 18q
Absent knee cap	tri 8
Genu valgum	45,X
Ligamental hyperlaxity	tri 10q, tri 17q
Deep furrows	tri 8
Hexadactyly	tri 4p, tri 13, r(13), tri 13qter, tri 17q
Collar bone agenesis	tri 11q

Hands

Long hands	del 12p
Short stubby hands	tri 3p, tri 3q, tri 12p, tri 18q2, del 18p, tri 19q, tri 21
Long palms	tri 9p
Long fingers	tri 1q, del 4p, del 7p, tri 9q, del 12p, del 18q
Arachnodactylia	tri 2p
Brachymesophalangia	tri 9p
Elongated 2nd phalange	del 9p
Brachymetacarpia	45,X
Tapered fingers	tri 9q, tri 10q, del 18q
Overlapping fingers	tri 9q, tri 13, tri 18
Thumb anomalies	tri 2q, tri 4p, tri 4q, tri 5q, del 13q, tri 13qter, tri 16p, tri 16q
Anomalies of the radial axis	r(4)
Dysplastic nails	tri 3q, tri 8, tri 9p, tri 13, tri 18, 45,X
Proximally implanted thumbs	tri 3q, del 4p, del 4q, del 18p
Abnormal fingers	del 4q, tri 6q, del 7q, tri 8q, tri 9p, tri 10p, del 11q, tri 14q1

Feet

Club feet	tri 4p, del 7p
Rocker-bottom feet	tri 7p, tri 13, tri 16p, tri 18
Dorsiflexion of big toe	tri 4p, tri 9q, tri 10q, tri 16p, tri 16q, tri 18
Abnormal toes	tri 9p, tri 10q, del 11q

GENITALIA

Minor anomalies in male (crypt-orchidism, micropenis)	tri 1q, del 1q, tri 2p, tri 3p, tri 3q, tri 4p, tri 4q, tri 6q, tri 7p, del 7q, del 8p, tri 8q, del 9p, tri 9, tri 11q, tri 13, del 13q, tri 13qter, tri 14, tri 15q2, tri 17q, tri 18, tri 18q2, del 18q, tri 19q, r(21)
Testicular atrophy	47,XXY, ♂46,XX, 48,XXYY, 49,XXXXY
Macroorchidism	fra(X)
Hypospadias	del 4p, tri 6q, del 9p, del 13q, del 18q, tri 19q
Hyperplasia of labia minora	del 9p
Hypoplasia of labia minora	del 18q
Streak gonads	45,X
Abnormal uterus	tri 13, del 13q
Anal atresia and perineal malformations	tri 11q, del 13q, tri 16q, tri 18
Primary amenorrhea	45,X

Absence of puberty	45,X
Ambiguous genitalia	del 11p, tri 16q

SKIN

Lymphoedema	del 8p, 45,X
Hemangioma	tri 13, tri 13qter
Marbelized skin	tri 21
Nevi	45,X
Dimples	del 4p, tri 9p, tri 9q, del 18q
Ulceration of the scalp	del 4p, tri 13
Dry skin and scarce, thin hair	tri 6p, tri 16q
Thick subcutaneous tissue	del 7q
Hirsutism	tri 3q

MALFORMATIONS

Cardiac	tri 1q, tri 2p, tri 3p, tri 3q, del 4p, tri 4q, del 4q, tri 5q, tri 7p, tri 8p, del 8p, tri 8q, tri 9, tri 9q, del 11q, tri 11q, tri 13, del 13q, tri 14, tri 15q2, tri 18, tri 21, r(21), 45,X
Renal and urinary tract	tri 3q, tri 4q, del 4q, tri 7p, tri 9, tri 10p, del 11q, tri 11q, tri 13, del 13q, tri 18, r(21), 45,X
Digestive	tri 1q, tri 3p, tri 11q, tri 13, tri 18, tri 21, r(21)
Cerebral (arhinencephaly, cebocephaly, etc.)	del 7q, tri 9, tri 13, del 13q, tri 13q1, del 18p
Corpus callosum agenesis	tri 8p, tri 11q
Severe ocular anomalies	tri 3q, del 4p, tri 13
Osteoarticular, minor	tri 2p, tri 3p, del 4p, tri 4p, tri 6q, del 7p, tri 8q, tri 10p, tri 12q, tri 13, del 13q, tri 15q2, tri 16q, del 18q, tri 21, r(21), 45,X, 48,XXXX, 49,XXXXX, 49,XXXXY
Osteoarticular, severe	tri 6q, tri 8, tri 8p, tri 9

DERMATOGLYPHICS

Abnormal palmar creases	tri 3q, tri 4p, del 4q, del 5p, del 7p, del 7q, tri 8, tri 8p, tri 9p, tri 9, tri 10p, tri 12p, tri 13, tri 18, tri 21
Immature ridges	tri 10p, tri 21
Hypermature ridges	r(21)
Axial triradius t'	del 5p, tri 9p
t''	del 9p, tri 10p, tri 18, tri 21
t'''	tri 13
Excess of arches	del 4p, tri 8, tri 8p, tri 9p, tri 13, tri 18, tri 18q2, tri 20p
Excess of whorls	tri 3p, tri 4p, del 9p, tri 10p, del 18q
Excess of ulnar loops	tri 21
Supernumerary flexion creases on fingers	del 9p, tri 9q
Absence of triradii b and c	tri 9p

SIMILARITIES WITH KNOWN SYNDROMES

Aarskog syndrome	tri 2p
Cornelia de Lange syndrome	tri 3q, tri 13qter
Pierre Robin syndrome	tri 11q

FIGURE CREDITS

1.1 Yunis E., Egel H., Zuniga R., Ramirez E. *Hum. Genet.*, **36**:113–116 (1977).

1.2 Lungarotti M.S., Falorni A., Calabro A., Passalacqua F., Dallapiccola B. *J. med. Genet.*, **17**:398–401 (1980).

1.3 Taysi K., Sekhon G.S. *Hum. Genet.*, **44**:277–285 (1978).

1.4 Fryns J.P., Meelenaere A. de, Pedersen J., Van den Berghe H. *Ann. Génét.* **23**:181–182 (1980).

1.5 Neu R.L., Gardner L.I. *Clin. Genet.*, **4**:474–479 (1973).

1.6 Pan S.F., Fatora S.R., Sorg R., Garver K.L., Steele M.W. *Clin. Genet.*, **12**:303–313 (1977).

1.7 Schinzel A. *Hum. Genet.*, **49**:167–173 (1979).

1.8 Andrie M., Erlach A., Mayr W.R., Rett A. *Hum. Genet.*, **41**:115–120 (1978).

1.9 Personal collection.

1.10 Mankinen C.B., Sears J.W., Alvarez V.R. *Birth Defects, Orig. Art. Ser.*, **XII-5**:131–136 (1976).

1.11 Personal collection.

1.12 J. Melnick collection.

1.13 Bobrow M., Emerson P.M., Spriggs A.I., Ellis H.L. *Amer. J. Dis. Child.*, **126**:257–260 (1973).

1.14 Gordon R.R., Cooke P. *Lancet*, **2**:1212–1213 (1964).

1.15 C.B. Wolf collection.

2.1 G. Schwanitz collection.

2.2 Sekhon G.S., Taysi K., Rath R. *Hum. Genet.*, **44**:99–103.

2.3 Mayer U., Schwanitz G., Grosse K.P., Etzold R. *Ann. Génét.*, **21**:172–176 (1978).

2.4 Armendares S., Salamanca-Gomez F. *Clin. Genet.*, **13**:17–24 (1978).

2.5 to 2.8 Laurent C., Biémont M.-Cl., Guibaud P., Guillot J., Noël B., Quack B., Geneviève M., Cressens M.-L. *Ann. Génét.*, **21**:13–18 (1978).

2.9 Zabel B., Hansen S., Hartmann W. *Hum. Genet.*, **32**:101–104 (1976).

2.10 to 2.13 Laurent C., Biémont M.-Cl., Guibaud P., Guillot J., Noël B., Quack B., Geneviève M., Cressens M.-L. *Ann. Génét.*, **21**:13–18 (1978).

2.14 Dennis N.R., Neu R.L., Bannerman R.M. *Am. J. med. Genet.*, **1**:271–277 (1978)

3.1 Mulcahy M.T., Pemberton P.J., Sprague P. *Ann. Génét.*, **22**:217–220 (1979).

3.2 Chiyo H.A., Kuroki Y., Matsui I., Niitsu N., Nakagome Y. *J. med. Genet.*, **13**:525–527.

3.3 Schwanitz G., Schmid R.D., Grosse G., Grahn-Liebe E. *J. Génét. hum.*, **25**:141–150 (1977).

3.4 and 3.5 Wilson G.N., Hieber V.C., Schmickel R.D. *J. Ped.*, **93**:783–788 (1978).

3.6 Fineman R.M., Hecht F., Ablow R.C. et al. *Pediatrics* **61**:611–618 (1978).

3.7 Schwanitz G., Schmid R.D., Grosse G., Grahn-Liebe E. *J. Génét. hum.*, **25**:141–150 (1977)

3.8 and 3.16 Fineman R.M., Hecht F., Ablow R.C. et al. *Pediatrics* **61**:611–618 (1978).

3.9, 3.10, and 3.13 Fineman R.M., Hecht F., Albow R.C. et al. *Pediatrics*, **61**:611–618 (1978).

3.11 Mulcahy M.T., Pemberton P.J., Sprague P. *Ann. Génét.*, **22**:217–220 (1979).

3.12 and 3.14 Allderdice P.W., Browne N., Murphy D.P. *Amer. J. Hum. Genet.* **37**:699–718 (1975).

3.15 Boué J., Hirshhorn K., Lucas M., Gautier M., Moszer M., Bach Ch. *Ann. Pédiat.* **21**:567–573 (1974).

3.17 Verjaal M., Nef J. de. *Am. J. Dis. Child.*, **132**:43–45 (1978).

3.18 Gonzales J., Lesourd S., Braconnier A. *Ann. Génét.*, **23**:119–122 (1980).

3.19 and 3.20 Rethoré M.O., Lejeune J., Carpentier S., Prieur M., Dutrillaux B., Seringe P., Rossier A., Job J.C. *Ann. Génét.*, **15**:159–165 (1972).

3.21 Ballesta F., Vehi L. *Ann. Génét.*, **17**:287–290 (1974).

3.22 Yunis J.J. *Am. J. Dis. Child.*, **132**:30–33 (1978).

3.23 Charrow J., Cohen M.M., Meeker D. *Am. J. med. Genet.*, 431–436 (1981).

3.24 Say B., Barber J., Bobrow M., Jones K., Coldwell J.G. *J. Pediat.*, **88**:447–450 (1976).

3.25 Parloir C., Fryns J.P., Van den Berghe H. *Hum. Genet.*, **47**:239–244 (1979).

4.1 Wolf U., Reinwein H., Porsch R., Schroter R., Baitsch H. *Humangenetik*, **1**:397–413 (1965).

4.2 J.P. Fryns collection.

4.3 Arias D., Passarge E., Engle M.A., German J. Human chromosomal deletion: Two patients with the 4p– syndrome. *J. Pediat.*, **76**:82–88 (1970).

4.4 Institut de Progénèse (J. Lejeune) collection.

4.5 Personal collection.

4.6 Personal collection.

4.7, 4.8, and 4.10 Institut de Progénèse (J. Lejeune) collection.

4.9 Wolf U., Reinwein H., Porsch R., Schroter R., Baitsch H. *Humangenetik*, **1**:397–413 (1965).

4.11 Institut de Progénèse (J. Lejeune) collection.

4.12 Greek warrior's helmet.

4.13 Institut de Progénèse (J. Lejeune) collection.

4.14 Schinzel A., Schmid W. *Humangenetik*, **15**:163–171 (1972).

4.15 Rethoré M.O., Dutrillaux B., Job J.C., Lejeune J. *Ann. Génét.*, **17**:109–114 (1974).

4.16 Giraud F., Mattei J.F., Mattei M.G., Ayme S., Bernard R. *Humangenetik*, **30**:89–108 (1975).

4.17 Hustinx T.W.J., Gabreels F.J.M., Kirkels V.G.H.J., Korten J.J., Scheres J.M.J.C., Joosten E.M.G., Rutten F.J. *Ann. Génét.*, **18**:13–19 (1975).

4.18 Giovanelli G., Forabosco A., Dutrillaux B. *Ann. Génét.*, **17**:119–124 (1974).

4.19 Schrocksnadel H., Feichtinger C., Scheminsky C. *Humangenetik*, **29**:329–335 (1975).

4.20 and 4.21 Crane J., Sujansky E., Smith A. *Am. J. med. Genet.*, **4**:219–229 (1979).

4.22 Giovanelli G., Forabosco A., Dutrillaux B. *Ann. Génét.*, **17**:119–124 (1974).

4.23 La Chapelle A. de, Koivisto M., Schroder J. *J. med. Genet.*, **10**:384–389 (1973).

4.24 Dutrillaux B., Laurent C., Forabosco A., Noël B., Suerinc E., Biémont M.C., Cotton J.B. *Ann. Génét.*, **18**:21–27 (1975).

4.25 Canki N., Debevec M., Rainer S., Rethoré M.O. *Ann. Génét.*, **20**:195–198 (1977).

4.26 and 4.27 Dutrillaux B., Laurent C., Forabosco A., Noël B., Suerinc E. *Ann. Génét.*, **18**:21–27 (1975).

4.28 and 4.30 Andrle M., Erlach A., Rett A. *Hum. Genet.*, **49**:179–183 (1979).

4.29 Cervenka J., Reza Djavadi G., Gorlin R.J. *Hum. Genet.*, **34**:1–7 (1976).

4.31 and 4.32 Vogel W., Siebers J.W., Gunkel J., Haas B., Knorr-Gartner H., Niethammer D.G., Noël B. *Humangenetik*, **28**:103–112 (1975).

4.33 and 4.34 Mitchell J.A., Packman S., Loughman W.D., Fineman R.M., Zackai E., Patil S.R., Emanuel B., Bartley J.A., Hanson J.W. *Am. J. med. Genet.*, **8**:73–89 (1981).

4.35 Rethoré M.O., Couturier J., Mselati J.C., Cochois D., Lavaud J., Lejeune J. *Ann. Génét.*, **22**:214–216 (1979).

4.36 Mitchell J.A., Packman S., Loughman W.D., Fineman R.M., Zackai E., Patil S.R., Emanuel B., Bartley J.A., Hanson J.W. *Am. J. med. Genet.*, **8**:73–89 (1981).

4.37 Golbus M.S., Conte F.A., Daentl D.L. *J. med. Genet.*, **10**:83–85 (1973).

5.1, 5.2, 5.3, and 5.4 Personal collection.

5.5 and 5.6 Personal collection.

5.7 Institut de Progénèse (J. Lejeune) collection.

5.8 and 5.9 Personal collection.

5.10 Opitz J.M., Patau K. New Chromosomal and Malformation Syndromes. *Birth Defects, Orig. Art. Ser.*, **XI-5**:191–200 (1975).

5.11 Leschot N.J., Lim K.S. *Hum. Genet.*, **46**:271–278 (1979).

5.12 Brimblecombe F.S.W., Lewis F.J., Vowles M. *J. med. Genet.*, **14**:271–274 (1977).

5.13 Zabel B., Baumann W., Gehler J., Conrad G. *J. med. Genet.*, **15**:143–147 (1978).

5.14 and 5.15 Stoll C., Rethoré M.O., Laurent C., Lejeune J. *Arch. franç. Pédiat.*, **32**:551–561 (1975).

5.16 Lejeune J., Lafourcade J., Berger R., Rethoré M.O., *Ann. Génét.*, **8**:11–15 (1965).

5.17 Noël B., Quack B., Thiriet M. *Ann. Génét.*, **11**:247–252 (1968).

5.18 Osztovics M., Kiss P. *Clin. Genet.*, **8**:112–116 (1975).

5.19 Seabright M., Gregson N.M., Johnson M. *J. med. Genet.*, **17**:444–446 (1980).

5.20 Rodewald A., Zankl M., Gley E.O., Zang K.D. *Hum. Genet.*, **55**:191–198 (1980).

5.21 Bartsch-Sandhoff M., Liersch R. *Ann. Génét.*, **20**:281–284 (1977).

5.22 and 5.23 Curry C.J.R., Loughman W.D., Francke U., Hall B.D., Golbus M.S., Derstine J., Epstein C.J. *Clin. Genet.*, **15**:454–461 (1979).

5.24 Rodewald A., Zankl M., Gley E.O., Zang K.D. *Hum. Genet.*, **55**:191–198 (1980).

5.25 Bartsch-Sandhoff M., Liersch R. *Ann. Génét.*, **20**:281–284 (1977).

6.1 Turleau C., Chavin-Colin F., Grouchy J. de. *Ann. Génét.*, **21**:88–91 (1978).

6.2 Bernheim A., Berger R., Vaugier G., Thieffry J.C., Matet Y. *Hum. Genet.*, **48**:13–16 (1979).

6.3 Côté G.B., Papadakou-Lagoyanni S., Sbyrakis S. *J. med. Genet.*, **15**:479–481 (1978).

6.4 Rosi G., Venti G., Migliorini-Bruschelli G., Donti E., Bocchini V., Armellini R. *Hum. Genet.*, **51**:67–72 (1979).

6.5 and 6.6 Therkelsen A.J., Klinge T., Henningsen K., Mikkelsen M., Schmidt G. *Ann. Génét.*, **14**:13–21 (1971).

6.7 Chiyo H., Kuroki Y., Matsui I., Yanagida K., Nakagome Y. *Humangenetik*, **30**:63–67 (1975).

6.8 Neu R.L., Gallien J.U., Steinberg-Warren N., Wynn R.J., Bannerman R.M. *Ann. Génét.*, **24**:167–169 (1981).

6.9 Schroer R.J., Culp D.M., Stevenson R.E., Potts W.E., Taylor H.A., Simensen R.J. *Clin. Genet.*, **18**:83–87 (1980).

6.10 Schmid W., D'Apuzzo V., Rossi E. *Hum. Genet.*, **46**:279–284 (1979).

6.11 and 6.12 Clark C.E., Cowell H.R., Telfer M.A., Casey P.A. *Am. J. med. Genet.*, **5**:171–178 (1980).

6.13 Personal collection.

6.14 Van Den Berghe H., Fryns J.P., Cassiman J.J., David G. *Ann. Génét.*, **17**:29–35 (1974).

6.15 Fried K., Rosenblatt M., Mundel G., Krikler R. *Clin. Genet.*, **7**:192–196 (1975).

6.16 Moore C.M., Heller R.H., Thomas G.H. *J. med. Genet.*, **10**:299–303 (1973).

7.1 and 7.2 Moric-Pettrovic S., Laca Z., Krajgher A. *Ann. Génét.*, **19**:133–136 (1976).

7.3 and 7.4 Schmid M., Wolf J., Nestler H., Krone W. *Hum. Genet.*, **49**:283–289 (1979).

7.5 Alfi O.S., Donnell G.N., Kramer S.L. *J. med. Genet.*, **10**:187–189 (1973).

7.6 Vogel W., Siebers J.W., Reinwein H. *Ann. Génét.*, **16**:277–280 (1973).

7.7 Bass H.N., Crandall B.F., Marcy S.M. *J. Pediat.*, **83**:1034–1038 (1973).

7.8 Turleau C., Rossier A., Montis G. de, Roubin M., Chavin-Colin F., Grouchy J. de. *Ann. Génét.*, **19**:37–42 (1976).

7.9 Grace E., Sutherland G.R., Stark G.D., Bain A.D. *Ann. Génét.*, **16**:51–54 (1973).

7.10 Serville F., Broustet A., Sandler B., Bourdeau M.J., Leloup M. *Ann. Génét.*, **18**:67–70 (1975).

7.11 R. Berger collection.

7.12 Bernstein R., Dawson B., Morcom G., Wagner J., Jenkins T. *Clin. Genet.*, **17**:228–237 (1980).

7.13 Friedrich U., Osterballe O., Stenbjerg S., Jorgensen J. *Hum. Genet.*, **51**:231–235 (1979).

7.14 Nielsen K.B., Egede F., Mouridsen I., Mohr J. *J. med. Genet.*, **16**:461–466 (1979).

7.15 Turleau C., Rossier A., Montis G. de, Roubin M., Chavin-Colin F., Grouchy J. de. *Ann. Génét.*, **19**:37–42 (1976).

7.16 to 7.19 Harris E.L., Wappner R.S., Palmer C.G., Dinno N., Seashore M.R., Breg W.R. *Clin. Genet.*, **12**:233–238 (1977).

7.20 Nielsen K.B., Egede F., Mouridsen I., Mohr J. *J. med. Genet.*, **16**:461–466 (1979).

7.21 Bernstein R., Dawson B., Morcom G., Wagner J., Jenkins T. *Clin. Genet.*, **17**:228 237 (1980).

7.22 Ayraud N., Rovinski J., Lambert J.C., Galiana A. *Ann. Génét.*, **19**:265–268 (1976).

7.23 Higginson G., Weaver D.D., Magenis R.E., Prescott G.H., Haag C., Hepburn D.J. *Clin. Genet.*, **10**:307–312 (1976).

7.24 Klep-de Pater J.M., Bijlsma J.B., Bleeker-Wagemakers E.M., de France H.F., de Vries-Ekkers C.M.A.M. *J. med. Genet.*, **16**:151–154 (1979).

7.25 Valentine H., Sergovich F. *Birth Defects, Orig. Art. Ser.* **XIII, 3B**:261–262 (1977).

7.26 and 7.28 Seabright M., Lewis G.M. *Hum. Genet.*, **42**:223–226 (1978).

7.27 and 7.29 Crawfurd M. d'A., Kessel I., Liberman M., McKeown J.A., Mandalia P.Y., Ridler M.A.C. *J. med. Genet.*, **16**:453–460 (1979).

7.30 Klep-de Pater J.M., Bijlsma J.B., Bleeker-Wagemakers E.M., de France H.F., de Vries-Ekkers C.M.A.M. *J. med. Genet.*, **16**:151–154 (1979).

7.31 Carnevale A., Frias S., Del Castillo V. *Clin. Genet.*, **14**:202–206 (1978).

7.32 Friedrich U., Lyngbye T., Oster J. *Humangenetik*, **26**:161–165 (1975).

7.33 McPherson E., Hall J.G., Hickman R. *Hum. Genet.*, **35**:117–123 (1976).

7.34 Miller M., Kaufman G., Reed G., Bilenker R., Schinzel A. *Am. J. med. Genet.*, **4**:323–332 (1979).

7.35 Institut de Progénèse (J. Lejeune) collection.

8.1 Biljsma J.B., Wijffels J.C., Tegelaers W.H.N. *Helv. paediat. Acta*, **27**:281–298 (1972).

8.2 Fineman R.M., Ablow R.C., Howard R.O., Albright J., Breg W.R. *Pediatrics*, **56**:762–767 (1975).

8.3 Tuncbilek E., Halicioglu C., Say B. *Humangenetik*, **23**:23–29 (1974).

8.4 Institut de Progénèse (J. Lejeune) collection.

8.5 Kondo I., Tomisawa T., Matsuura A., Yoshinori I., Yamashita A., Hara Y. *Ann. Paediat. Japon.*, **21**:20–28 (1975).

8.6 Dallapicola B., Gallenga P.E., Pinca A., Capra L., Proc. 4th Int. Congr. Neurogenet. Neuroophthalm. 1973. *Acta Genet. Med. Gemel.* **23**:253–258 (1974).

8.7 Jacobsen P., Mikkelsen M., Rosleff F. *Ann. Génét.*, **17**:87–94 (1974).

8.8 Crandall B.F., Bass H.N., Marcy S.M., Glovsky M., Fish C.H. *J. med. Genet.*, **11**:393–398 (1974) (Mental Retardation Research Center U.C.L.A.).

8.9 Jaibert P., Jobert J., Patet J., Mouriquand C., Roget J. *Ann. Génét.*, **9**:109–112 (1966).

8.10 Grouchy J. de, Turleau C., Léonard C. *Ann. Génét.*, **14**:69–72 (1971).

8.11 Institut de Progénèse (J. Lejeune) collection.

8.12 and 8.14 Malpuech G., Dutrillaux B., Fonck Y., Gaulme J., Bouche B. *Arch. franç. Pédiat.*, **29**:853–859 (1972).

8.13 and 8.15 Institut de Progénèse (J. Lejeune) collection.

8.16 Crandall B.F., Bass H.N., Marcy S.M., Slovsky M., Fish C.H. *J. med. Genet.*, **11**:393–398 (1974) (Mental Retardation Research Center, U.C.L.A.).

8.17 Ballesta F., Fernandez E., Mila M. *J. Génét. hum.*, **28**:361–366 (1980).

8.18 Abuelo D., Perl D.P., Henkle C., Richardson A. *J. med. Genet.*, **14**:463–466 (1977).

8.19 Fineman R.M., Ablow R.C., Breg W.R., Wing S.D., Rose J.S., Rothman S.L.G., Warpinski J. *Clin. Genet.*, **16**:390–398 (1979).

8.20 and 8.21 Schinzel A. *Hum. Genet.*, **37**:17–26 (1977).

8.22 Institut de Progénèse (J. Lejeune) collection.

8.23 Lejeune J., Rethoré M.O. In: Caspersson T., Zech L. *Chromosome Identification* p 214. New York, Academic Press, 1973.

8.24 Sanchez O., Yunis J.J. *Humangenetik*, **23**:297–303 (1974).

8.25 Laurent C., Biémont M.C., Midenet M., Couturier J., Dutrillaux B. *Lyon med.*, **232**:609–615 (1974).

8.26 Funderburk S.J., Barrett C.T., Klisak I. *Ann Génét.*, **21**:219–222 (1978).

8.27 Jones L.A., Dengler D.R., Taysi K., Shackelford G.D., Hartmann A.F. *J. med. Genet.*, **17**:232–235 (1980).

8.28 Hongell K., Knuutila S., Westermarck T. *Clin. Genet.*, **13**:237–240 (1978).

8.29 Fineman R.M., Ablow R.C., Breg W.R., Wing D., Rose J.S., Rothman L.G., Warpinski J. *Clin Genet.*, **16**:390–398 (1979).

8.30 Chiyo H.A., Nakagome Y., Matsui I., Kuroki Y., Kobayashi H., Ono K. *Clin. Genet.*, **7**:328–333 (1975).

8.31 Hongell K., Knuutila S., Westermarck T. *Clin. Genet.*, **13**:237–240 (1978).

8.32 Rosenthal M., Krmpotic E., Bocian M., Szego K. *Clin. Genet.*, **4**:507–516 (1973).

8.33 Trisomy 8q–. Institut de Progénèse (J. Lejeune) collection.

8.34 Reiss J.A., Brenes P.M., Chamberlin J., Magenis R.E., Lovrlen E.W. *Hum. Genet.*, **47**:135–140 (1979).

8.35 Orye E., Craen M. *Clin. Genet.*, **9**:289–301 (1976).

8.36 Rodewald A., Stengel-Rutkowski S., Schulz P., Cleve H. *Europ. J. Pediat.*, **125**:45–57 (1977).

8.37 Taillemite J.L., Channarond J., Tinel H., Mulliez N., Roux C. *Ann. Génét.*, **18**:251–255 (1975).

9.1 Turleau C., Grouchy J. de, Chavin-Colin F., Roubin M., Langmaid H. *Ann. Génét.*, **17**:167–174 (1974).

9.2 Rethoré M.O., Ferrand J., Dutrillaux B., Lejeune J. *Ann. Génét.*, **17**:157–161 (1974).

9.3 Schwanitz G., Schamberger U., Rott H.D., Wieczorek V. *Ann. Génét.*, **17**:163–166 (1974).

9.4 Orye E., Verhaaren H., Van Egmond H., Devloo-Blancquaert A. *Clin. Genet.,* **7**:134–143 (1975).

9.5 Centerwall W.R., Mayeski C.A., Cha C.C. *Humangenetik,* **29**:91–98 (1975).

9.6 Weber F., Muller H., Sparkes R. New Chromosomal and Malformation Syndromes. *Birth Defects, Orig. Art. Ser.* **XI-5**:201–205 (1975).

9.7 Rethoré M.O., Lafourcade J. In: *Journées de Pédiatrie,* Paris, Flammarion (1974).

9.8 Mulcahy M.T., Jenkyn J. *Clin. Genet.,* **8**:199–204 (1975).

9.9 Podruch P.E., Weisskopf B. Trisomy for the short arm of chromosome 9 in two generations, with balanced translocations t(15p + ;9q –) in three generations. *J. Pediat.,* **85**:92–95 (1974).

9.10 Turleau C., Grouchy J. de, Roubin M., Chavin-Colin F., Cachin O. *Ann. Génét.,* **18**:125–129 (1975).

9.11 Jacobsen P., Hobolth N., Mikkelsen M. *Clin. Genet.,* **7**:317–324 (1975).

9.12 and 9.13 Lewandowski R.C., Yunis J.J., Lehrke R., O'Leary J., Swaiman K.F., Sanchez O. *Amer. J. Dis. Child.,* **130**:663–667 (1976).

9.14 Institut de Progénèse (J. Lejeune) collection.

9.15 Blanck C.E., Colver D.C.B., Potter A.M., McHugh J., Lorber J. *Clin. Genet.,* **7**:261–273 (1975).

9.16 Turleau C., Grouchy J. de, Roubin M., Chavin-Colin F., Cachin O. *Ann. Génét.,* **18**:125–129 (1975).

9.17 Moedjono S.J., Crandall B.F., Sparkes R.S. *J. med. Genet.,* **17**:227–242 (1980).

9.18 Personal collection. Mosaic.

9.19 Abe T., Morita M., Kawai K., Misawa S., Takino T., Hashimoto H., Nakagome Y. *Ann. Génét.,* **20**:111–114 (1977).

9.20 and 9.21 Mulcahy M.T. *Ann. Génét.,* **21**:47–49 (1978).

9.22 Kuroki Y., Yokota S., Nakai H., Yamamoto Y., Matsui I. *Hum. Genet.,* **38**:107–111 (1977).

9.23 to 9.26 Alfi O., Donnell G.N., Allderdice P.W., Derencsenyi A. *Ann. Génét.,* **19**:11–16 (1976).

9.27 to 9.29 Institut de Progénèse (J. Lejeune, M. Prieur) collection.

9.30 F. Serville collection.

9.31 and 9.32 Alfi O., Donnell G.N., Allderdice P.W., Derencsenyi A. *Ann. Génét.,* **19**:11–16 (1976).

9.33 Rutten F.J., Hustinx T.W.J., Dunk-Tillemans A.A.W., Scheres J.M.J.C., Tjon Y.S.T. *Ann. Génét.,* **21**:51–55 (1978).

9.34 Fraisse J., Lauras B., Ooghe M.J., Freycon F., Rethoré M.O. *Ann. Génét.,* **17**:175–180 (1974).

9.35 Fryns J.P., Lambrechts A., Jansseune H., Van den Berghe H. *Hum. Genet.,* **50**:29–32 (1979).

9.36 Jacobsen P., Mikkelsen M., Rosleff F. *Clin. Genet.,* **4**:434–441 (1973).

9.37 and 9.38 Turleau C., Grouchy J. de, Chavin-Colin F., Roubin M., Brissaud P.E., Repéssé G., Safar A., Borniche P. *Humangenetik,* **29**:233–241 (1975).

9.39 Pescia G., Jotterand-Bellomo M. Crousaz H. de, Payot M., Martin D. *Ann. Génét.,* **22**:558–585 (1979).

9.40 Mattei J.F., Mattei M.G., Ardissone J.P., Taramasco H., Giraud F. *Clin. Genet.,* **17**:129–136 (1980).

9.41 Faed M., Robertson J., Brown S., Smail P.J., Muckhart R.D. *J. med. Genet.,* **13**:239–242 (1976).

9.42 Sutherland G.R., Carter R.F., Morris L.L. *Hum. Genet.,* **32**:133–140 (1976).

9.43 Bowen P., Ying K.L., Chung G.S.H. Trisomy 9 mosaicism in a newborn infant with multiple malformations. *J. Pediat.,* **85**:95–97 (1974).

9.44 Mace S.E., Macintyre M.N., Turk K.B., Johnson W.E. *J. Pediatr.,* **92**:446–448 (1978).

9.45 Sutherland G.R., Carter R.F., Morris L.L. *Hum. Genet.,* **32**:133–140 (1976). 9q – trisomy.

9.46 Lewandowski R.C., Yunis J.J. *Clin. Genet.,* **11**:306–310 (1977).

9.47 Tropp M.R., Currie M. *Hum. Genet.,* **38**:131–135 (1977).

9.48 Haslam R.H.A., Broske S.P., Moore C.M., Thomas G.H., Neill C.A. *J. med. Genet.,* **10**:180–184 (1973).

10.1 and 10.2 Prieur M., Forabosco A., Dutrillaux B., Laurent C., Bernasconi S., Lejeune J. *Ann. Génét.,* **18**:217–222 (1975).

10.3 Roux C., Taillemite J.L., Baheux-Morlier G. *Ann. Génét.,* **17**:59–62 (1974).

10.4 Kroyer S., Niebuhr E. *Ann. Génét.,* **18**:50–55 (1975).

10.5 Talvik T., Mikelsaar A.V., Mikelsaar R., Käossar M., Tüür S. *Humangenetik,* **19**:215–226 (1973).

10.6 Yunis J.J., Sanchez O. A new syndrome resulting from partial trisomy for the distal third of the long arm of chromosome 10. *J. Pediat.,* **84**:567–570 (1974).

10.7 Laurent C., Bovier-Lapierre M., Dutrillaux B. *Humangenetik,* **18**:321–327 (1973).

10.8 Kroyer S., Niebuhr E. *Ann. Génét.,* **18**:50–55 (1975).

10.9 to 10.14 Klep-de Pater J.M., Bijlsma J.B., France H.F. de, Leschot N.J., Duijndan-Van den Berge M., Van Hemel J.O. *Hum. Genet.,* **46**:29–40 (1979).

10.15 and 10.16 Back E., Kosmutzky J., Schuwald A., Hameister H. *Ann. Génét.,* **22**:195–198 (1979).

10.17 and 10.18 Cantu J.M., Salamanca F., Buentello L., Carnevale A., Armendares S. *Ann. Génét.,* **18**:5–11 (1975).

10.19 Hustinx T.H.W.J., Ter Haar B.G.A., Scheres J.M.J.C., Rutten F.J. *Clin. Genet.,* **6**:408–415 (1974).

10.20 Turleau C., Doussau de Bazignan M., Roubin M., Grouchy J. de. *Ann Génét.,* **19**:61–64 (1976).

10.21 Stengel-Rutkowski S., Murken J.D., Frenkenberger R., Riechert M., Spiess H., Rodewald A., Stene J. *Europ. J. Pediatr.,* **126**:109–125 (1977).

10.22 Grosse K.P., Schwanitz G., Singer H., Wieczorek V. *Humangenetik,* **29**:141–144 (1975).

10.23 and 10.24 Schleiermacher E., Schliebitz U., Steffens C. *Humangenetik,* **23**:163–172 (1974).

10.25 Stengel-Rutkowski S., Murken J.D., Frenkenberger R., Riechert M., Spiess H., Rodewald A., Stene J. *Europ. J. Pediatr.,* **126**:109–125 (1977).

10.26 and 10.27 Chieri P. de, Spatuzza E., Bonich J.M. *Hum. Genet.,* **45**:71–75 (1978).

10.28 Aller V., Abrisqueta J.A., Pérez-Castillo A., Del Mazo J., Martin-Lucas M.A. de, Torres M.L. *Hum. Genet.,* **46**:129–134 (1979).

10.29 Shokeir M.H.K., Ray M., Hamerton J.L., Bauder F., O'Brien H. *J. med. Genet.,* **12**:99–103 (1975).

10.30 Francke U., Kernahan C., Bradshaw C. *Humangenetik,* **26**:343–351 (1975).

10.31 Bourrouillou G., Colombies P., Gallegos D., Manelfe C., Rochiccioli P. *Ann. Génét.,* **24**:61–64 (1981).

10.32 Turleau C., Grouchy J. de, Ponsot G., Bouygues D. *Hum. Genet.,* **47**:233–237 (1979).

11.1 and 11.2 Rott H.D., Schwanitz G., Grosse K.P., Alexandrow G. *Humangenetik,* **14**:300–305 (1972).

11.3 Giraud F., Mattei J.F., Mattei M.G., Bernard R. *Humangenetik,* **28**:343–347 (1975).

11.4 Laurent C., Biémont M.C., Bethenod M., Cret L., David M. *Ann. Génét.,* **18**:179–184 (1975).

11.5 Aurias A., Turc C., Michiels Y., Sinet P.M., Graveleau D., Lejeune J. *Ann. Génét.,* **18**:185–188 (1975).

11.6 Tusques J., Grislain J.R., André M.J., Mainard R., Rival J.M., Cadudal J.L. *Ann. Génét.,* **15**:167–172 (1972).

11.7 Jacobsen P., Hauge M., Henningsen K., Hobolth N., Mikkelsen M., Philip J. *Hum. Hered.,* **23**:568–585 (1973).

11.8 to 11.11 Fraccaro et al. *Hum. Genet.,* **56**:21–51 (1980).

11.12 to 11.17 Fraccaro et al. *Hum. Genet.,* **56**:21–51 (1980).

11.18 Turleau C., Chavin-Colin F., Roubin M., Thomas D., Grouchy J. de. *Ann. Génét.,* **18**:257–260 (1975).

11.19 and 11.20 Jacobsen P., Hauge M., Henningsen K., Hobolth N., Mikkelsen M., Philip J. *Hum. Hered.,* **23**:568–585 (1973), Basle, S. Karger

11.21 Léonard C., Courpotin C., Labrune B., Lepercq G., Kachaner J., Caut P. *Ann. Génét.,* **22**:115–120 (1979).

11.22 Engel E., Hirshberg C.S., Cassidy S.B., McGee B.J. *Am. J. ment. Defic.,* **80**:473–475 (1976).

11.23 Laurent C., Biémont M.C., Veyron M., Guilhot J., Guibaud P. *Ann. Génét.,* **22**:239–241 (1979).

11.24 Faust J., Vogel W., Loning B. *Clin. Genet.,* **6**:90–97 (1974).

11.25 Palmer C.G., Poland C., Reed T., Kojetin J. *Hum. Genet.,* **31**:219–225 (1976).

11.26 Sanchez O., Yunis J.J., Escobar J.L. *Humangenetik,* **22**:59–65 (1974).

11.27 Falk R.E., Carrel R.E., Valente M., Grandall B.F., Sparkes R.S. *Am. J. ment. Def.,* **77**:383–388 (1973).

11.28 Rethoré M.O., Junien C., Aurias A., Couturier J., Dutrillaux B., Kaplan J.C., Lejeune J. *Ann. Génét.,* **23**:35–39 (1980).

12.1 Rethoré M.O., Kaplan J.C., Junien C., Cruveiller J., Dutrillaux B., Aurias A., Carpentier S., Lafourcade J., Lejeune J. *Ann. Génét.* **18**:81–87 (1975).

12.2 and 12.3 Armendares S., Salamanca F., Nava S., Ramirez S., Cantu J.M. *Ann. Génét.*, **18**:89–94 (1975).

12.4 Uchida I.A., Lin C.C. Identification of partial 12 trisomy by quinacrine fluorescence. *J. Pediat.*, **82**:269–272 (1973).

12.5 Parslow M., Chambers D., Drummond M., Hunter W. *Hum. Genet.*, **47**:253–260 (1979).

12.6 Biederman B., Bowen P., Robertson C., Schiff D. *Hum. Genet.*, **36**:35–41 (1977).

12.7 and 12.8 Hansteen I.L., Schirmer L., Hestetun S. *Clin. Genet.*, **13**:339–349 (1978).

12.9 Kondo I., Hamaguchi H., Haneda T. *Hum. Genet.*, **46**:135–140 (1979).

12.10 Tenconi R., Baccichetti C., Anglani F., Pellegrino P.A., Kaplan J.C., Junien C. *Ann. Génét.*, **18**:95–98 (1975).

12.11 and 12.12 Orye E., Craen M. *Humangenetik*, **28**:335–342 (1975).

12.13 Malpuech G., Rethoré M.O., Junien C., Geneix A. *Lyon med.*, **233**:275–279 (1975).

12.14 Magnelli N.C., Therman E. *J. med. Genet.*, **12**:105–108 (1975).

12.15 Harrod M.J.E., Byrne J.B., Dev V.G., Francke U. *Am. J. med. Genet.*, **7**:123–129 (1980).

13.1 Personal collection.

13.2 Institut de Progénèse (J. Lejeune) collection.

13.3 Personal collection.

13.4 J. Boué collection.

13.5 Institut de Progénèse (J. Lejeune) collection.

13.6 and 13.7 Masterson J.G., Law E.M., Donovan D.E., O'Brien N.G. *J. Irish med. Ass.*, **61**:195–199 (1968).

13.8 Grouchy J. de, Turleau C., Danis F., Kohout G., Briard M.L. *Ann. Génét.*, **21**:247–251 (1978).

13.9 Habedank M. *Hum. Genet.*, **52**:91–99 (1979).

13.10 and 13.11 Wilroy R.S., Summitt R.L., Martens P.R. New Chromosomal and Malformation Syndromes. *Birth Defects Orig. Art. Ser.* **XI-5**:217–222 (1975).

13.12 Hornstein L., Soukup S. *Clin. Genet.*, **19**:81–86 (1981).

13.13 Jotterand M., Juillard E. *Hum. Genet.*, **33**:213–222 (1976).

13.14 Talvik T., Mikelsaar A.V., Mikelsaar R., Kaosaar M., Tuur S. *Humangenetik*, **19**:215–226 (1973).

13.15 and 13.16 Hauksdottir N., Halldorssou S., Jensson O., Mikkelsen M., Mc Dermot A. *J. med. Genet.*, **9**:413–421 (1972).

13.17 Escobar J.I., Sanchez O., Yunis J.J. *Amer. J. Dis. Child.*, **128**:221–222 (1974).

13.18 Wilroy R.S., Summitt R.L., Martens P.R. New chromosomal and malformation syndromes, *Birth Defects Orig. Art. Ser.*, **XI-5**:217–222 (1975).

13.19 Moedjono S.J., Sparkes R.S. *Hum. Genet.*, **50**:241–246 (1979).

13.20 and 13.21 Allderdice P.W., Davis J.G., Miller O.J., Klinger H.P., Warburton D., Miller D.A., Allen F.H., Abrams C.A.L., McGilvray E.: *Amer. J. Hum. Genet.* **21**:499–512 (1969).

13.22, 13.23, and 13.24 Grace E., Drennan J., Colver D., Gordon R.R. *J. med. Genet.*, **8**:351–357 (1971).

13.25 Rethoré M.O., Praud E., Le Loch J., Joly C., Saraux H., Aussannaire M., Lejeune J. *Presse méd.*, **78**:955–958 (1970).

13.26 Niebuhr E., Ottosen J. *Ann. Génét.*, **16**:157–166 (1973).

13.27 Tolksdorf M., Wiedmann H.R., Goll U. *Arch. Kinderheilk.*, **181**:282–295 (1970).

13.28 M. Tolksdorf collection.

13.29 Fried K., Rosenblatt M., Mundel G., Krikler R. *Clin. Genet.*, **7**:203–208 (1975).

13.30 and 13.31 Personal collection.

14.1 Turleau C., Grouchy J. de, Bocquentin F., Roubin M., Chavin-Colin F. *Ann. Génét.*, **18**:41–44 (1975).

14.2 Raoul O., Rethoré M.O., Dutrillaux B., Michon L., Lejeune J. *Ann. Génét.*, **18**:35–39 (1975).

14.3 Reiss J.A., Wyandt H.R., Magenis R.E., Lovrien E.W., Hecht F. *J. med. Genet.*, **9**:280–286 (1972).

14.4 Short E.M., Solitare G.B., Breg W.R. *J. med. Genet.*, **9**:367–373 (1972).

14.5 Laurent C., Dutrillaux B., Biémont M.C., Genoud J., Bethenod M. *Ann. Génét.*, **16**:281–284 (1973).

14.6 Fryns J.P., Cassiman J.J., Berghe H. van den. *Humangenetik*, **24**:71–77 (1974).

14.7 Muldal S., Enoch B.A., Ahmed A., Harris R. *Clin. Genet.*, **4**:480–489 (1973).

14.8 to 14.11 Miller J.Q., Willson K., Wyandt H., Jaramillo M.A., McConnell T.S. *J. med. Genet.*, **16**:60–65 (1979).

14.12 Kovacs G., Mihai C. *Hum. Genet.*, **49**:175–178 (1979).

14.13 Young S.R., Donovan D.M., Greer H.A., Burch K., Potter D.C. *Hum. Genet.*, **33**:331–334 (1976).

14.14 Rethoré M.O., Couturier J., Carpentier S., Ferrand J., Lejeune J. *Ann. Génét.*, **18**:71–74 (1975).

14.15 Jenkins M.B., Kriel R., Boyd L. *J. med. Genet.*, **18**:68–71 (1981).

14.16 Turleau C., Grouchy J. de, Cornu A., Turquet M., Millet G. *Ann. Génét.*, **23**:238–240 (1980).

14.17 Johnson V.P., Aceto T., Likness C. *Amer. J. med. Genet.*, **3**:331–339 (1979).

14.18 Triolo O., Serra A., Bova R., Carlo Stella N., Caruso P. *Ann. Génét.*, **24**:236–238 (1981).

15.1 Hongell K., Livanainen M. *Clin. Genet.*, **14**:229–234 (1978).

15.2 Watson J., Gordon R.R. *J. med. Genet.*, **11**:400–402 (1974).

15.3 Castel Y., Riviere D., Boucly J.Y., Toudic L. *Ann. Génét.*, **19**:75–79 (1976).

15.4 Webb G.C., Garson O.M., Robson M.K., Pitt D.B. *J. med. Genet.*, **8**:522–527 (1971).

15.5 Magenis R.E., Overton K.M., Reiss J.A., MacFarlane J.P., Hecht F. *Lancet* **ii**:1365–1366 (1972).

15.6 and 15.8 Tzancheva M., Krachounova M., Damjanova Z. *Hum. Genet.*, **56**:275–277 (1981).

15.7 Zabel B., Baumann W. *Ann. Génét.*, **20**:285–289 (1977).

15.9 Turleau C., Grouchy J. de, Chavin-Colin F., Roubin M. *Ann. Génét.*, **20**:214–216 (1977).

15.10 Zabel B., Baumann W. *Ann. Génét.*, **20**:285–289 (1977). Same child as shown in Fig. 15.7.

15.11 Fujimoto A., Towner J.W., Ebbin A.J., Kahlstrom E.J., Wilson M.G. *J. med. Genet.*, **11**:287–291 (1974).

15.12 Kucerova M., Strakova M., Polivkova Z. *J. med. Genet.*, **16**:234–235 (1979).

15.13 Yunis E., Leibovici M., Quintero L. *Hum. Genet.*, **57**:207–209 (1981).

15.14 J.P. Fryns collection

15.15 Gardner R.J.M., Chewings W.E., Holdaway M.D. *N.Z. med. J.*, **91**:173–174 (1980).

15.16 Fryns J.P., Timmermans J., D'Hondt F., Francois B., Emmery L., Van den Berghe H. *Hum. Genet.*, **51**:43–48 (1974).

16.1 Leschot N.J., De Nef J.J., Geraedts J.P.M., Becker-Bloemko K.M.J., Talma A., Bijlsma J.B., Verjaal M. *Clin. Genet.*, **16**:205–214 (1979).

16.2 Dallapiccola B., Curatolo P., Balestrazzi P. *Hum. Genet.*, **49**:1–6 (1979).

16.3 Yunis E., Gonzalez J.T., Torres de Caballero O.M. *Hum. Genet.*, **38**:347–350 (1977).

16.4 Roberts S.H., Duckett D.P. *J. med. Genet.*, **15**:375–381 (1978).

16.5 Ridler, M.A.C., McKeown J.A. *J. med. Genet.*, **16**:317–320 (1979).

16.6 Rethoré M.A., Lafourcade J., Couturier J., Harpey J.P., Hamet M., Engler R., Alcindor L.G., Lejeune J. *Ann. Génét.*, **25**:36–42 (1982).

17.1 Berberich M.S., Carey J.C., Lawce H.J., Hall B.D. Same child at different ages. *Birth Defects, Orig. Art. Ser.*, **XIV-6C**:287–295 (1978).

17.2 Turleau C., Grouchy J. de, Bouveret J.P. *Clin. Genet.*, **16**:54–57 (1979).

17.3 Berberich M.S., Carey J.C., Lawce H.J., Hall B.D. *Birth Defects, Orig. Art. Ser.* **XIV-6C**:287–295 (1978).

18.1 to 18.3 J.P. Fryns collection.

18.4 to 18.9 Personal collection.

18.10 J.P. Fryns collection.

18.11 Fryns J.P., Lambrechts A., Jansseune H., Van den Berghe H. *Ann. Génét.*, **22**:30–32 (1979).

18.12 Turleau C., Grouchy J. de. *Clin. Genet.*, **12**:361–371 (1977).

18.13 Niazi M., Coleman D.V., Saldana-Garcia P. *J. med. Genet.*, **15**:148–151 (1978).

18.14 and 18.15 Castel Y., Rivière D., Nawrocki Th., Le Fur J.M., Toudic L. *Lyon méd.*, **233**:211–217 (1975).

18.16 and 18.17 Turleau C., Grouchy J. de. *Clin. Genet.*, **12**:361–371 (1977).

FIGURE CREDITS

18.18 Turleau C., Grouchy J. de. *Clin. Genet.*, **12**:361–371 (1977).

18.19 Kukolich M.K., Althaus B.W., Sears J.W., Mankinen C.B., Lewandowski, R.C. *Clin. Genet.*, **14**:98–104 (1978).

18.20 Vianna-Morgante A.M., Nozaki M.J., Ortega C.C., Coates V., Yamamura Y. *J. med. Genet.*, **13**:366–370 (1976).

18.21 Turleau C., Chavin-Colin F., Narbouton R., Asensi D., Grouchy J. de. *Clin. Genet.* **18**:20–26 (1980).

18.22 Stern L.M., Murch A.R. *J. med. Genet.*, **12**:305–307 (1975).

18.23 to 18.25 Turleau C., Chavin-Colin F., Narbouton R., Asensi D., Grouchy J. de. *Clin. Genet.*, **18**:20–26 (1980).

18.26 and 18.27 Personal collection.

18.28 and 18.32 Personal collection.

18.29 Giraud F., Hartung M., Mattei J.F., Passeron P., Coignet J. *Ann Génét.*, **14**:59–62 (1971).

18.30 Grouchy J. de, Lamy M., Thieffry S., Arthuis M., Salmon C. *C.R. Acad. Sci. (Paris)*, **256**:1028–1029 (1963).

18.31 Institut de Progénèse (J. Lejeune) collection.

18.33 and 18.35 Personal collection.

18.34 Institut de Progénèse (J. Lejeune) collection.

18.36 Grouchy J. de, Royer P., Salmon C., Lamy M. *Pathol. et Biol.*, **12**:579–582 (1964). Patient at 18 months and at 10 years.

18.37 Personal collection.

18.38 Institut de Progénèse (J. Lejeune) collection.

18.40 Personal collection.

18.41 Institut de Progénèse (J. Lejeune) collection. Mosaicism.

18.42 Fraccaro M., Hulten M., Ivemark B.I., Lindsten J., Tiepolo L., Zetterqvist P. *Ann. Génét.*, **14**:275–280 (1971).

18.43 Personal collection.

18.44 Coco R., Barreiro G.Z., Penchaszadeh V.B. *Ann. Génét.*, **18**:135–137 (1975).

18.45 Neu R.L., Watanabe N., Gardner L.I., Galvis A.G. *Ann. Génét.*, **14**:139–142 (1971).

18.46 G.B. Cote collection.

18.47 Deminatti M., Debeugny P., Croquette-Bulteel M.F., Delmas-Marsalet Y. *Ann. Génét.*, **13**:149–155 (1970).

19.1 and 19.2 Lange M., Alfi O.S. *Ann. Génét.*, **19**:17–21 (1976).

19.3 and 19.4 C. Pangalos collection.

20.1 Delicado A., Lopez-Pajares I., Vicente P., Gracia R. *Ann. Génét.*, **24**:54–56 (1981).

20.2 Schinzel A. Same child at 7 weeks and at 2 years 10 months. *Hum. Genet.*, **53**:169–172 (1980).

20.3 Schinzel A. *Hum. Genet.*, **53**:169–172 (1980). Same child as in Fig. 20.2.

20.4 Subrt I., Brychnac V. *Humangenetik*, **23**:219–222 (1974).

21.1 Personal collection.

21.2 Indian. Personal collection.

21.3 Black. Personal collection.

21.4 Arab. Personal collection.

21.5 Chinese. Personal collection.

21.6 Black. Personal collection.

21.7 21 years. Personal collection.

21.8 23 years. Institut de Progénèse (J. Lejeune) collection.

21.9 Institut de Progénèse (J. Lejeune) collection.

21.10 53 years. Institut de Progénèse (J. Lejeune) collection.

21.11 Lejeune J., Berger R., Rethoré M.O., Archambault L., Jérôme H., Thieffry S., Aicardi J., Broyer M., Lafourcade J., Cruveiller J., Turpin R. *C.R. Acad. Sci. (Paris)*, **259**:4187–4190 (1964).

21.12 Crandall B.F., Weber F., Muller H.M., Burwell J.K. *Clin. Genet.*, **3**:264–270 (1972) (Mental Retardation Research Center U.C.L.A.).

21.13 Richmond H.G., MacArthur P., Hunter D. *Acta paediat. Scand.*, **62**:216–220 (1973).

21.14 Challacombe D.N., Taylor A. *Arch. Dis. Childh.*, **44**:113–119 (1969).

21.15 Shibata K., Waldenmaier C., Hirsch W. *Humangenetik*, **18**:315–319 (1973).

21.16 Armendares S., Buentello L., Cantu-Garza J.M. *Ann. Génét.*, **14**:7–12 (1971).

21.17 Warren R.J., Rimoin D.L., Summitt R.L. *Amer. J. hum. Genet.*, **25**:77–81 (1973).

21.18 Holbek S., Friedrich U., Brostrom K., Petersen G.B. *Humangenetik*, **24**:191–195 (1974).

21.19 Institut de Progénèse (J. Lejeune) collection.

21.20 Dutrillaux B., Jonasson J., Lauren K., Lejeune J., Lindsten J., Petersen G.B., Saldana-Garcia P. *Ann. Génét.*, **16**:11–16 (1973).

21.21 Gripenberg U., Elfving J., Gripenberg L. *J. med. Genet.*, **9**:110–115 (1972).

21.22 Halloran K.H., Breg W.R., Mahoney M.J. *J. med. Genet.*, **11**:386–389 (1974).

21.23 Mikkelsen M., Vestermark S. *J. med. Genet.*, **11**:389–392 (1974). Mosaic.

21.24 Fryns collection.

22.1 Stoll C., Rohmer A., Sauvage P. *Ann. Génét.*, **16**:193–198 (1973).

22.2 Talvik T.A., Mikelsaar A.V.N. *Genetika*, **5**:129–133 (1969).

22.3 and 22.4 (Female twins) Lindenbaum R.H., Bobrow M., Barber L. *J. med. Genet.*, **10**:85–89 (1963).

22.5 to 22.8 Rethoré M.O., Noël B., Couturier J., Prieur M., Lafourcade J., Lejeune J. *Ann. Génét.*, **19**:111–117 (1976).

22.9 and 22.10 Crandall B.F., Weber F., Muller H.M., Burwell J.K. *Clin. Genet.*, **3**:264–270 (1972) Mental Retardation Research Center U.C.L.A.

22.11 Waleber R.G., Hecht F., Giblett E.R. *Amer. J. Dis. Child.*, **115**:489–494 (1968).

22.12 and 22.13 Dubowitz V., Cooke P., Colver D., Harris F. *J. med. Genet.*, **8**:195–201 (1971).

22.14 and 22.15 Warren R.J., Rimoin D.L., Summitt R.L. *Amer. J. hum. Genet.*, **25**:77–81 (1973).

23.1 Lin C.C., Gedeon M.M., Griffith P., Smink W.K., Newton D.R., Wilkie L., Sewell L.M. *Hum. Genet.*, **31**:315–328 (1976).

23.2 and 23.3 Personal collection.

23.4 and 23.5 Personal collection.

23.6 J.P. Fryns collection.

23.7 Institut de Progénèse (J. Lejeune) collection.

23.8 and 23.9 Personal collection.

23.10 Pena S.D.J., Ray M., Douglas G., Loadman E., Hamerton J.L. *J. med. Genet.*, **11**:211–215 (1974).

23.11 Telfer M.A., Richardson C.E., Helmken J., Smith G.F. *Amer. J. Hum. Genet.*, **22**:326–335 (1970).

23.12 Gardner R.J.M., Veale A.M.O., Sands V.E., Holdaway M.D. *Humangenetik*, **17**:323–330 (1973).

23.13 Telfer M.A., Richardson C.E., Helmken J., Smith G.F. *Amer. J. Hum. Genet.*, **22**:326–335 (1970).

23.14 and 23.15 Grouchy J. de, Brissaud H.E., Richardet J.M., Repéssé G., Sanger R., Race R.R., Salmon C., Salmon D. *Ann. Génét.*, **11**:120–124 (1968). Same child.

23.16 Grouchy J. de, Vialatte J., Chavin-Colin F., Roubin N., Turleau C. *Arch. fr. Pediat. (Suppl.)*, **35**:XLI-XLVII (1979).

23.17 Lejeune J., Abonyi D. *Ann. Génét.*, **11**:117–120 (1968).

23.18 Sergovich F., Vilenberg C., Pozsonyi J. The 49,XXXXX chromosome constitution: similarities to the 49,XXXXY condition. *J. Pediat.*, **78**:285–290 (1971).

23.19 Larget-Piet L., Rivron J., Baillif P., Dugay J., Emerit I., Larget-Piet A., Berthelot J. *Ann. Génét.*, **15**:115–119 (1972).

23.20 G.B. Côté collection.

23.21 to 23.23 Personal collection.

23.24 Institut de Progénèse (J. Lejeune) collection.

23.25 48,XXXY condition. Personal collection.

23.26 and 23.27 48,XXXY condition. Institut de Progénèse (J. Lejeune) collection.

23.28 and 23.29 XX male. Personal collection.

23.30 47,XYY condition. Personal collection.

23.31 Barr M.L., Carr D.H., Pozsonyi J., Willson R.A., Dunn G.A., Jacobson T.S., Miller J.R. *Canad. med. Ass. J.*, **87**:891–901 (1962).

23.32 Shapiro L.R., Briel C.B., Hsu L.Y.F., Calvin M.E., Hirschhorn K. *Amer. J. Dis. Child.*, **119**:79–81 (1970).

23.33 Kushnick T., Colondrillo M. *Clin. Genet.*, **7**:442–448 (1975).

23.34 A. Zaleski collection.

23.35 Zaleski W.A., Houston C.S., Pozsonyi J., Ying K.L. *Canad. Med. Ass. J.*, **94**:1143–1154 (1966).

23.36 Joseph M.C., Anders J.M., Taylor A.J. *J. med. Génét.*, **1**:95–101 (1964).

23.37 Institut de Progénèse (J. Lejeune) collection. XXXXY/XXXXXY mosaicism.

23.38 Joseph M.C., Anders J.M., Taylor A.I. *J. med. Genet.*, **1**:95–101 (1964).

23.39 Zaleski W.A., Houston C.S., Pozsonyi J., Ying K.L. *Canad. med. Ass. J.*, **94**:1143–1154 (1966).

23.40 Personal collection.

24.1 Wertelecki W., Graham J.M. Jr., Sergovich R.F. *Obstet. Gynecol.*, **47**:69–76 (1976).

24.2 Fryns J.P., Van de Kerckhove A., Goddeeris P., Van den Berghe H. *Hum. Genet.*, **38**:147–155 (1977).

24.3 Niebuhr E., Spawèvohn S., Henningsen K., Mikkelsen M. *Acta paediat. scand.*, **61**:203–208 (1972).

24.4 Fryns J.P., Vinken L., Geutjens J., Marien J., Deroover J., Van den Berghe H. *Ann. Génét.*, **23**:232–234 (1980).

24.5 Pitt D., Leversha M., Sinfield C., Campbell P., Anderson R., Bryan D., Rogers J. *J. med. Genet.*, **18**:309–311 (1981).